ENCYCLOPEDIA OF DRUG AND DRUG DELIVERY

ENCYCLOPEDIA OF DRUG AND DRUG DELIVERY

Vol. 2

DRUG CARRIER AND RESISTANCE

By

Dr. Kamlesh Dashora
Associate Professor
Institute of Pharmacy
Vikram University
Ujjain (M.P.)
(India)

&

Dr. Darshan Dubey
Assistant Professor
Institute of Pharmacy
Vikram University
Ujjain (M.P.)
(India)

D P H

DISCOVERY PUBLISHING HOUSE PVT. LTD.
NEW DELHI-110 002

Published by:
Tilak Wasan
DISCOVERY PUBLISHING HOUSE PVT. LTD.
4383/4A, Ansari Road, Darya Ganj
New Delhi-110 002 (India)
Phone: +91-11-23279245, 43596064-65
Fax : +91-11-23253475
E-mail: parul.wasan@gmail.com
discoverypublishinghouse@gmail.com
web : www.discoverypublishinggroup.com

***First Edition:* 2012**

ISBN: 978-93-5056-038-9 (Set)
978-93-5056-039-6 (Vol. 1)
978-93-5056-040-2 (Vol. 2)
978-93-5056-041-9 (Vol. 3)
978-93-5056-042-6 (Vol. 4)
978-93-5056-043-3 (Vol. 5)

Encyclopedia of Drug and Drug Delivery

Printed at:
Shree Balaji Art Press
Delhi

Preface

The Present title "**Encyclopedia of Drug and Drug Delivery**" is the amazing advancement of pharmaceutical studies. It provides the various fundamental aspects of drug designing to be understood adequately. It has been compiled for undergraduate, postgraduate students, research scholars, teachers, technicians, industrial microbiologists and scientists involved in pharmaceutical research and development. The basic concepts have been clearly explained and their functions are adequately highlighted. The presentation of text is simple and systematic. The selection of chapters and topics is according to the specified syllabi of several Indian Universities and are so structured as to enable the student to move easily from the fundamental to the complex. It describes the ideas and techniques that are creating or new domain of science and technology. It holds the promise of revolutionizing field ranging from medical science to pharmaceutical industries. It is our earnest hope that this title will be of great value to all our students.

Organizational and economic changes in the pharmaceutical industry have led to a marked reduction in the supply of new chemical entities. This combined with an increase in the expiration of patents and tighter regulatory control has lead many companies to rethink existing product design. The growing trend to improve the efficacy of existing drug products, to both extend parent life and allow competition with opportunistic generic companies, has allowed for a greater understanding of formulation design. This trend is set to continue for some time as economic and political pressure increases.

There have been many improvements in understanding the theoretical and practical aspects of drug discoveries. Numerous therapeutic compounds have been investigated to treat a variety of different illnesses. There are many pharmaceutical methods requiring a variety of formulation options. Advanced formulation design may be thought of as a way to optimize the mode of drug designing to a patient that will ultimately lead to a more effective therapy and approach an optimal therapeutic outcome for a particular drug. This book is designed to take the reader through various cutting-edge areas of advanced formulation design by describing therapeutic categories that may be beneficial. The reader will be introduced to some novel technologies and updated on the state of the art in areas designed to overcome poor efficacy issues.

To make the work more comprehensive and informative, the authors have consulted many authoritative books, research journals, abstracts, monographs etc., so there can be no claim to originality except in the manner of treatment.

The authors express their thanks to their friends and colleagues whose continue inspirations have initiated them to bring out this book.

The authors express their gratitude to Mr. Wasan and staff of M/s Discovery Publishing House Pvt. Ltd., for their whole hearted co-operation in the publications of this book.

The authors will remain sincerely responsible for any shortcomings of the book and be grateful to the readers for their suggestions and constructive criticism for the continuous betterment of the book. They have this opportunity to appeal to the readers to send their suggestions straightaway to this publisher.

–Authors

Contents

1

Introduction

While the experimental study of nanoparticle flow *in vivo* proves to be difficult, a variety of theoretical and practical techniques are becoming available to allow some understanding of the phenomena involved. These processes include (i) convective flow induced by the flow of blood, lymph or interstitial fluid, (ii) the influence of the interaction of nanoparticles with themselves or with biological components and the effect of this on their transport, and (iii) the effect of fluid flow and hence shear forces on particle access to, interaction with and removal from receptors. Diffusion and movement of particle suspensions in complex media such as interstitial tissue must also be considered. Much of the theoretical work which is relevant to this exploration of nanoparticle flow has not been directed towards biological endpoints, but this body of knowledge, and the analogous literature on the dynamic behavior of bacteria, erythrocytes and platelets provides the basis of a more rigorous analysis of the factors involved in drug carrier nanoparticle flow. The role of nanoparticles in drug, vaccine and gene delivery of great value, because their small dimension compared with microparticles allows them to interact more effectively with cells, be safely injected, and amongst other characteristics, diffuse further into tissues, and into and through individual cells.

The flow of nanoparticles in capillaries, lymphatics, tumor vessels, their extravasationl and movement in the cytoplasm of cells are all aspects of the topic covered in this overview, albeit from a phenomenological viewpoint. It is clear that particle diameter is a key parameter in the characterization and behavior of nanoparticle suspensions. In several of the analyses here, it becomes clear that another advantage of nanoparticles may be the relative lack of effect of shear stress, once particles adhere to surfaces as a prelude to uptake; this is contrasted with targeted microspheres whose residence on receptors and surfaces is size dependent, the larger particles being more vulnerable to detachment by shear forces. This chapter considers questions relating to the flow of nanoparticles *in vivo*, but which has often been simulated *in vitro* by chemical engineers and physicists interested in particle behavior in flow conditions.

Spherical particles are the norm, but not all nanosystems are spherical. The influence of asymmetry on the transport of nanoparticles *in vivo* is largely unknown, although the rheological characteristics of asymmetric particle suspensions have been understood for a long time. Flow behavior of nanoparticles in complex networks of narrow capillaries has relevance in the design and operation of microfluidic devices, as well as in drug delivery and targeting, and in toxicology; the extent to which it is relevant for delivery and targeting is explored here. In physical terms, the following situations could be considered:

1. Particle flow in rapidly flowing blood, including segregation and deposition of particles, and the behavior of particles at bifurcations in the capillary supply;
2. The effect of shear on adhesion of particles of different size and shape;
3. Particle flow in more static conditions, for example, in the tumor interstitium or the lymphatics;
4. Flow of particles in tissues, including the flow of particles into narrow pores; and
5. Flow or diffusion within the anisotropic interior of cells.

Added complications arise where bioadhesive or ligand-decorated particles are involved. In the latter case, binding and flow are linked. The enquiry can be divided into a discussion of the flow and movement of nanoparticles as a function of their size and route of administration, the influence of convection and blood or lymph flow, the influence of flow dynamics on the interaction of nanoparticles with target tissues and receptors, and the movement of nanoparticles once they have been absorbed and are making their way through individual cells and tissues towards a target. The routes of administration where flow is important potentially include all, even the oral route where particle flow and dynamics in the gut lumen and in the vicinity of both villi and microvilli is important. Particles below a critical size are taken up by the M-cells of Peyer's patches and by normal enterocytes, albeit in small quantities and find themselves in the lymph vessels, lymph nodes, blood, liver and spleen.

If the flow of nanoparticles away from their site of absorption is restricted due to the flow in lymph or blood being slow, this will reduce bioavailability and distribution. Rapid flow provides superior sink conditions and hence the use of everted gut sacs and cell monolayers *in vitro* can give unrealistic results for nanoparticle transit. The influence of flow dynamics on extravasation and perhaps on the enhanced permeability and retention (EPR) effect for nanocarriers has perhaps not been fully addressed. Both involve consideration of particle size, the diffusion and flow of nanoparticles through narrow channels, as well as navigation of tortuous environments. The availability of "extreme" nanoparticles in the form of dendrimers and quantum dots makes this topic a vital one in understanding the fate, toxicity' or accumulation of what is metrically a wide range of systems.

HISTORY

Our own interest in this field has resulted in part from studies on the size dependency of nanoparticle uptake after oral administration, where mesenteric lymphatic transport of 500 nm nanoparticles post absorption is determined by the flow of particles in a single file in the smallest mesenteric lymph vessels. In addition, studies on the flow of liposomes and niosomes, and the extrusion of flexible vesicles from glass capillaries under pressure, converting polyhedral vesicles into multilayer tubules, led us to consider the influence of stress forces on carrier integrity. Clearly, the elasticity of vesicles is important in their negotiation of capillaries when their diameter exceeds that of the capillary.

The flow of particles in fabricated capillaries which have a radius close to the particle radius is a challenge that has been tackled theoretically. We have suggested that multi-bilayer tubules can act as models for such flow experiments. Rheological examination of nanoparticle-blood mixtures and nanoparticle suspensions of mixed radius has also illustrated the potential complexity of particle movement in blood. In addition, erythrocyte blockage at bifurcations, or narrowing of vessels can lead to slowing down of blood flow and a change in rheology as the haematocrit increases. More recently, investigation of the transport of nanoparticles across cell monolayers and intracellular transport of dendrimers has assisted in defining some of the issues involved in targeting within cells.

FLOW OF NANOPARTICLES

The work of Fokin and colleagues on the transport of viral-sized colloids, following intravenous or intra-lymphatic injection, is relevant to drug delivery even if their objectives were different. 100-200 nm diameter sulphur colloid particles reach the lymph after IV injection in around 25 minutes; after intra-lymphatic injection particles appear in the venous blood only after 4 seconds. Following subcutaneous injection, similar particles reach the lymph after 2-9 min, although 95% of the particles remain at the injection site for at least 45 minutes. Here, the nanoparticles are being used as indicators of blood and lymph flow. What is also relevant to drug delivery is the influence of fluid flow on the movement and fate of nanoparticles. Ilium *et al.* observed uptake rates of 1.27 μm and 15.8 μm polystyrene particles in the lung and liver after IV injection. The sequestration in the lung was size dependent, but possibly affected because the smaller particles were taken up by the Kupffer cells of the liver, leaving the larger particles free to be taken up by the lung tissue. The rapidity of this suggests that, in effect, flow of the microparticles is solely determined by blood flow.

CONVECTIVE FLOW

Blood flow drives the convective flow of suspended particles. Diffusional transport occurs in static conditions or conditions of low fluid velocity. In a tube of flowing liquid, convective dynamics propel the particles in the direction of flow, but at the walls of the tube, there is the possibility of particle diffusion resulting in deposition. Blood velocity (mm s^{-1}) in arterioles and venules is a function of vessel diameter. In venules, the maximum velocity according to Jain is approximately 12 mm s^{-1}, while in arterioles, it can reach about 30 mm s^{-1}. Fluid velocity in tubes is not constant throughout the diameter of the tube, a feature that is important when the interaction of nanoparticles with epithelial cells or capillary walls is considered. The radial variation of shear is a factor that must be considered in polydisperse nanoparticulate systems and where nanoparticles adhere to erythrocytes, causing two distinct size distributions.

If nanoparticles adhere to erythrocytes or other blood elements, the translocation of the particles is controlled by the particular element to which it adheres. The rheology of suspensions of mixed particles is complex: viscosity reduces first with an increase in the fraction of larger particles in a suspension, and as the volume fraction increases, so does the viscosity. Ding *et al.* formulated a theoretical model examining particle migration in nanoparticle suspensions flowing through a pipe. "The model considers particle migration due to spatial gradients in viscosity and shear rate as well as Brownian motion. Particle migration due to these effects can result in significant non-uniformity in particle concentration over the cross section of the pipe" in particular for larger particles. Three mechanisms were proposed by the authors for migration in such non-uniform shear flow: (i) shear induced migration where particles move from regions of higher shear rate to regions of lower shear rate; (ii) viscosity gradient induced migration – particles move from regions of higher viscosity to regions of lower viscosity and (iii) self-diffusion due to Brownian motion. Diffusion inside microtubules has been studied to understand taxol binding to tubulin structures. The dimension of the tubulin lumen is of the order of 17 nm, approaching macromolecular dimensions, leading to friction between the inner walls and the moving macromolecules. This "*hindrance*" will also be an issue in the movement of nanoparticles in the smallest capillaries.

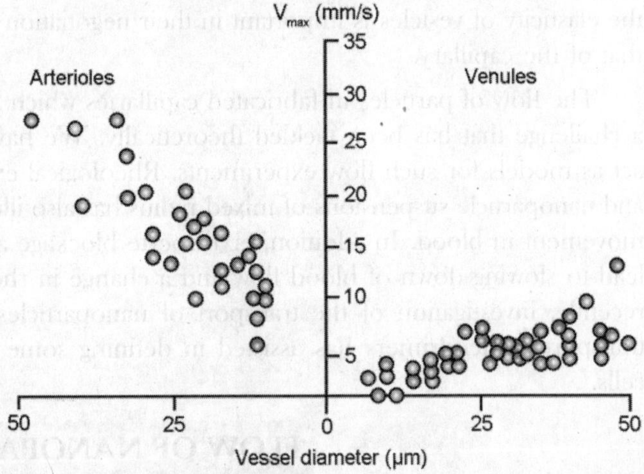

Fig. 1.1. Maximum blood velocity (mms^{-1}) in arterioles and venules as a function of vessel diameter.

With dendrimers whose diameters may be as small as 6 nm, the application of hindered theory to their movement could be relevant. No vessels are of this small radius, but the key parameter is the ratio of particle diameter to capillary diameter. The approach may well be important in cellular networks. It is not only capillary vessels that are the conduits of particle movement, but after extravasation, there is passage through cellular networks. The process could be considered to be akin to diffusion in porous networks. Binding of the moving particle (or macromolecule) to the luminal surface of the vessel will also hinder free flow or movement, a positive event in the case of specific ligand targeting of "decorated" systems. Polydisperse nano-systems can segregate during flow or migrate differentially leading to concentration differences. *Particle-image velocimetry* (PIV) has been used to track the flow characteristics of microparticles. The effects of flow on adhesion of monocytes to endothelial cells is relevant to the influence of flow and shear in particle interactions and uptake. The significance of flow can be demonstrated by the use of pharmacological

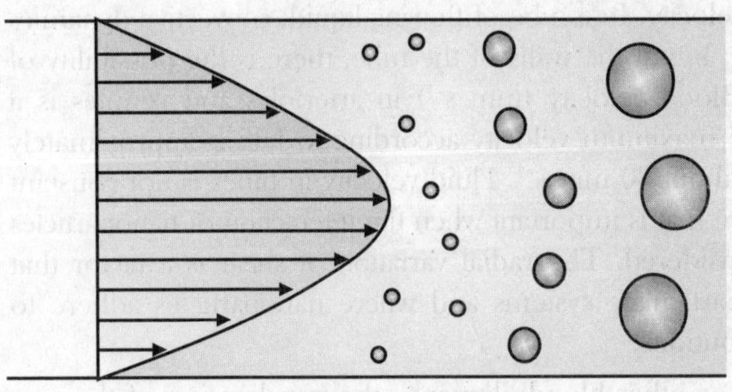

Fig. 1.2. Diagram showing the velocity pattern in a tube of flowing liquid. Particles of different size separate according to their diameter.

agents which change normal vessel patency, so that by the concomitant use of noradrenalin or angiotensin which constrict only normal vessels, the ratio of tumor to normal tissue blood flow can be optimized.

DIVERGENCE

Many theoretical studies of nanoparticle flow deal with linear tubes, whereas *in vivo* movement occurs through complex vessel architectures with bifurcations. Behavior at bifurcations in a vascular or capillary system is dependent not only on particle diameter, but also on the rigidity or flexibility of the particle concerned. Colloid transport in a bifurcating structure has been the subject of one recent paper. It is a process which depends on the orientation of the bifurcations, especially with particles whose density is greater than that of the medium, as well as on the different flow rates in the individual branches which are likely to be of different sizes. If nanoparticles are trapped or associate at bifurcations or indeed other obstacles in capillaries, then it is likely that they might associate more permanently, thereby changing their intrinsic theological behavior. Flexible particles do not of course suffer the same constraints in movement and progress, but their flexibility can lead to slow negotiation of movement around obstacles.

INTERACTIVE ROLE

Nanoparticles may interact with blood constituents: the adsorption of albumin, IgG and fibrinogen from blood onto hydrophobic particles is well known, but the effect of nanoparticles on blood has been less well studied. Kim's data indicate that the interaction of nanoparticles with erythrocytes changes the dynamics of flow of both erythrocytes and particles. Chambers and Mitragotri found that nanoparticles as large as 450 nm adhered to erythrocytes, and thus remained in the circulation for several weeks. The percentage of latex nanospheres in the circulation over a period of 6 hrs was highly dependent on particle size, retention times decreasing with increasing diameter from 220 nm to 1100 nm. These data are difficult to interpret on the basis of flow, as the erythrocytes with attached nanopar-ticles are eliminated somewhat faster than the native erythrocytes. Gorodetsky and colleagues explored interactions of carbo-platin nanoparticles (formed by CPt interaction with fibrinogen) with the fibrin mesh caused by the induction of clot formation.

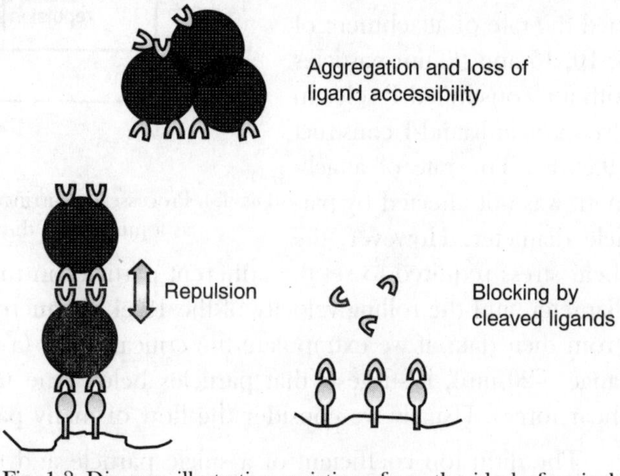

Fig. 1.3. Diagram illustrating variations from the ideal of a single ligand-decorated nanoparticle interacting with receptors spaced at an appropriate distance from the particles.

Rheological Studies

There appear to be no rheological studies comparing surface protein decorated nanoparticles with the unadorned forms. Certainly, it is possible that aggregation may be caused by the change in surface properties and that this will in turn change flow patterns and perhaps masking of ligands. Nanoparticles are of course sensitive to the medium in which they are placed even *in vitro* when cell media can cause significant increases in diameter because of particle flocculation. We have suggested that the interaction with surface receptors of nanoparticles decorated with ligands is more complex than intimated in discussions of targeting generally. The instability of plant lectins, frequently used as surface proteins on nanosystems, is discussed by Gabor *et al.*

NANOPARTICLES AND BODY FLUIDS

Nanoparticles *in vivo* flow in blood, lymph or tissue fluid at greater or lesser velocities, as discussed above. Deposition of particles which might occur in a static situation is itself a complex process, and will depend on the rugosity of the receiving surface. Particle deposition from flowing suspensions has been the subject of research which has considered not only diffusion, convection, geometrical interception and migration under gravity, but also the influence of tangential interactions. Patil *et al.* examined the rate of attachment of 5, 10, 15 and 20 μm particles with a reconstituted P-selectin glycoprotein ligand-1 construct 19.ek.Fc. The rate of attachment was not affected by particle diameter. However, the shear stress required to set the adherent particles in motion (S_e) decreased with increasing particle diameter, and the rolling velocity of the 19.ek.Fc microspheres increased with increasing diameter. From their data, if we extrapolate the critical shear (a plot of $1/S_e$ is linear with diameter over the range 5-20 μm), it suggests that particles below one micron in diameter will not be removed by shear forces. Usually we consider the flow of many particles in collective diffusion.

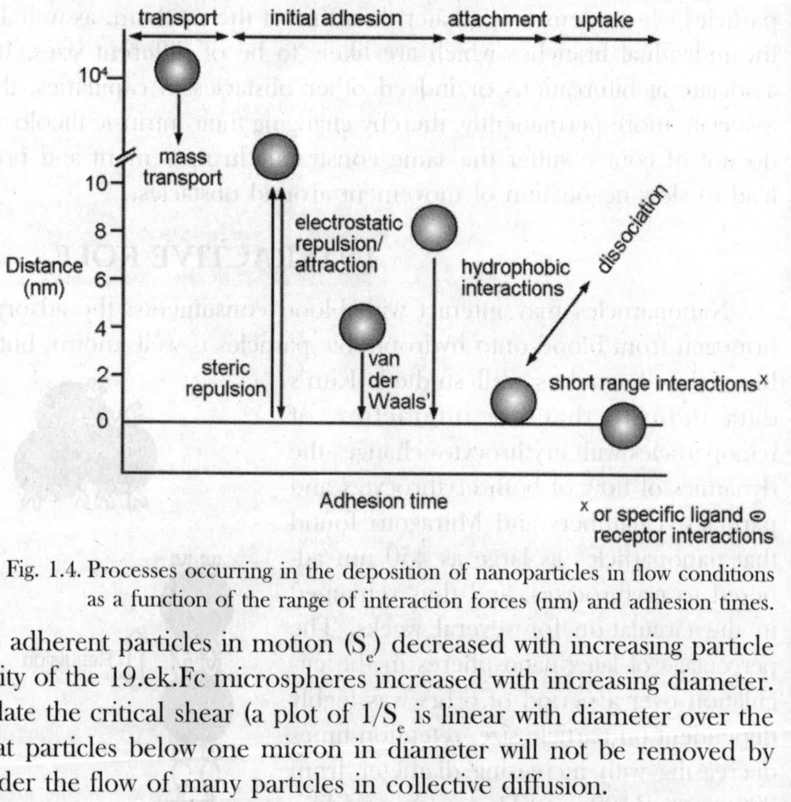

Fig. 1.4. Processes occurring in the deposition of nanoparticles in flow conditions as a function of the range of interaction forces (nm) and adhesion times.

The diffusion coefficient of a single particle and the collective diffusion coefficient coincides at infinite dilution, but can differ at higher concentrations. Cell adhesion mediated by not one but two receptors has been considered by Bhatia *et al.*; the analysis would also apply to decorated nanoparticles. In their study, the two receptors were selectin and integrin ICAM; "the state diagram" evolved shows the area of firm adhesion as opposed to rolling adhesion for leukocytes as a function of receptor densities and association rate constants. The fate of nanoparticles in flowing blood, their adhesion, extravasation and permeation into tumors, thus depends on a complex of factors such as diameter, surface ligand density and orientation, shape, capillary diameter and rugosity, bifurcations, viscosity and flow gradients.

NANOCRYSTALS

Nanosystems can be prepared in a variety of shapes. Nanocrystals are often irregular; there are asymmetric carbon nanotubes, and surfactant and lipid vesicles can be produced as discs, polyhedral structures, toroids and tubes. The vesicle constructs often have dimensions larger than 500 nm; it must be assumed that vesicles in the nanometer size range will be less affected. In these systems, shape is less important than membrane properties in controlling the release of encapsulated drug, but the flow properties of vesicular suspensions are clearly determined by shape and elasticity. As most particulate delivery vectors have been spherical, little attention has been paid to the influence of shape on fate; yet it is known that the shape of environmental particles and fibres, for example, influences their fate and toxicity. As discussed above, there are two different but related effects of particle flow: the effect of particle shape and size and characteristics on flow, as well as the effect of flow on flexible particles, as discussed by Bruinsma.

With elastic vesicles, we have argued that shape matters because it affects flow and potential fate *in vivo* through extravasation for instance; elasticity also allows vesicles to be transported in vessels which would be blocked by solid particles. The elasticity and visco-elasticity of such systems may be important in differentiating them from solid nanoparticles. Much of the debate on whether the shape of vesicles matters, is dependent on the knowledge of the nature of the capillary blood supply and the forces exerted on, and the damage done to, vesicles as they move in capillaries. In studies conducted in our laboratories with doxorubicin loaded niosomes, 60% of the drug remained in the vesicles 8 hrs after intravenous administration. The extent to which the drug loss was due to diffusion or to damage is not known, but vesicles subjected to deliberate stress can lose considerable amounts of their payload, simply by extrusion of the

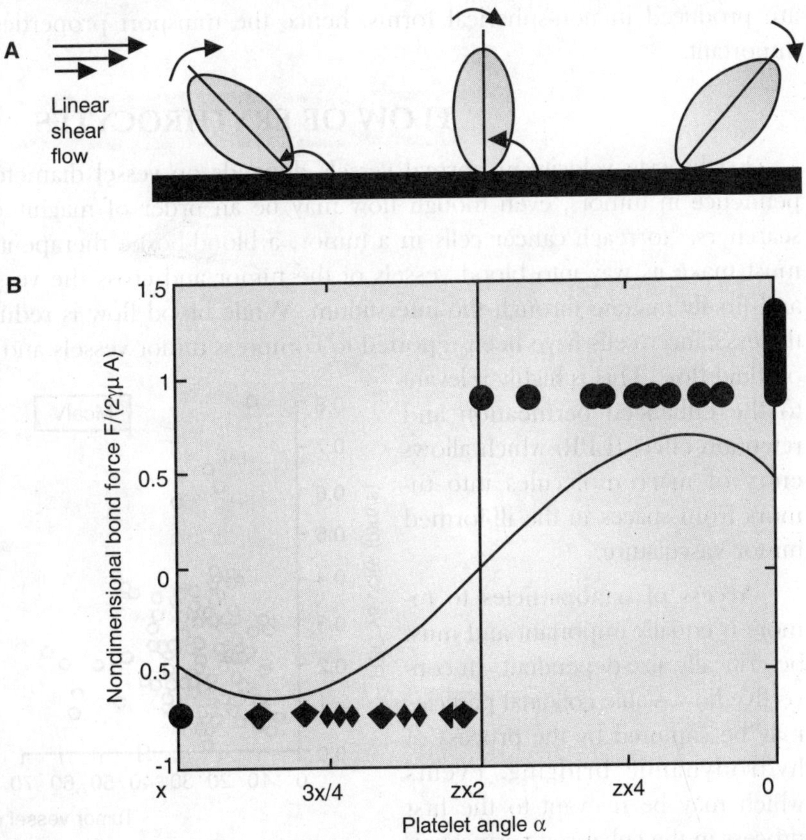

Fig. 1.5. The non-dimensional bond force as a function of the angle of an ellipsoidal platelet passing through zero when the platelet is 90° to the surface.

vesicles through capillaries of reducing diameter. Reduction in diameter of systems below 1 micron will clearly reduce such stresses and allow flexible systems to retain their loads intact. Vasanthi *et al.* treated the anisotropic diffusion of oblate spheroids, explaining that because non-spherical molecules rotate as they translate, their motion differs significantly from that of a sphere. For rods, theory predicts that the diffusion coefficient in the direction parallel to the major axis of the rod (D_{II}) is twice that in the perpendicular direction (D_{I}).

There are few studies which have considered the motion of ellipsoidal particles near a plane wall, although this is relevant to platelet flow and adhesion to the walls of vessels. Mody and colleagues have addressed the issue, observing the effects of shear stress on platelet adhesion. Platelets, unlike leukocytes, do not roll but display a flipping motion in the direction of flow, due to their flattened ellipsoidal structure. Flexible systems such as vesicles have been widely studied, while being forced under pressure in capillaries smaller than the vesicle diameter. The elasticity of the membranes can be estimated from the extent of deformation. Vesicle flow in linearly forced motion has been followed. Flexible vesicles adjust their shape to equilibrate the applied force; locally in some cases, two-dimensional flow of lipids in the vesicle membrane occurs, clearly influencing the position of the embedded surface ligands. There are many nanoparticulates which are produced in non-spherical forms, hence the transport properties of asymmetric particles is important.

FLOW OF ERYTHROCYTES

Erythrocyte velocity in normal vessels depends on vessel diameter, but there is no such dependence in tumors, even though flow may be an order of magnitude slower. According to researchers, "to reach cancer cells in a tumor, a blood-borne therapeutic molecule, particle or cell must make its way into blood vessels of the tumor and cross the vessel wall into the interstitium and finally migrate through the interstitium. While blood flow is reduced in tumor vessels, nonetheless cancer cells have been reported to compress tumor vessels and this will have consequences on fluid flow. This is highly relevant to the enhanced permeation and retention effect (EPR) which allows entry of macromolecules into tumors from spaces in the ill-formed tumor vasculature.

Access of nanoparticles to tumors is equally important and must be critically size-dependent. In convective flow, stable colloidal particles may be captured by the process of hydrodynamic bridging, events which may be relevant to the first process in the enhanced permeation and retention (EPR) effect. At high velocities but in the low Re regime,

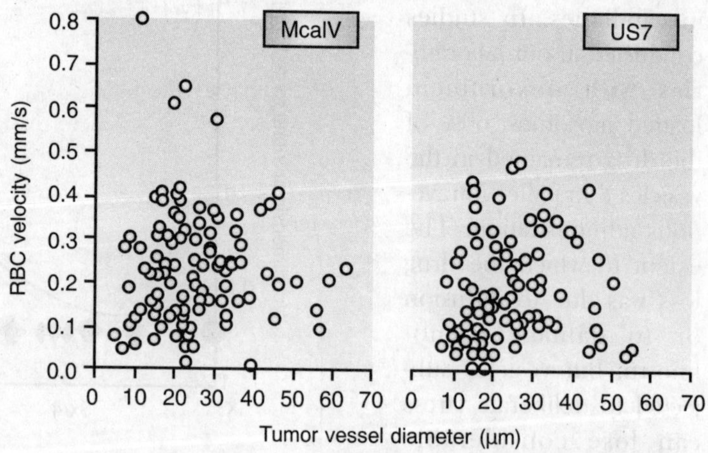

Fig. 1.6. Diagram from Jain showing the lack of a clear relationship between erythrocyte velocity and tumor vessel diameter in two tumor types, MCaIV and U87.

hydrodynamic forces acting on the particles at an entrance to a pore (or a defect in a tumor vessel) may overcome colloidal repulsive forces and result in flocculation of the particles and the plugging of the pore. The effects of velocity, particle concentration, and the ratio of pore size to particle size (the aspect ratio) on retention by hydrodynamic bridging have been studied. The effect of velocity on retention by bridging is opposite to that of retention by deposition. There is a critical flow velocity necessary for particle bridging to occur, a function of the net colloidal interparticle and particle–porous medium repulsion that must be overcome by the hydrodynamic forces for bridging to occur.

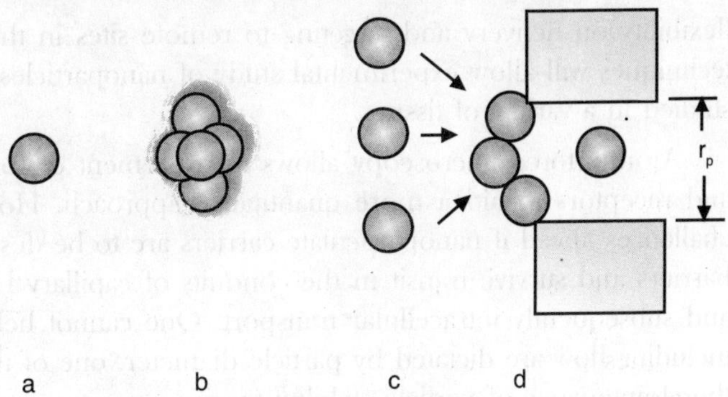

Fig. 1.7. Particle behavior prior to entry to a pore of radius, r_p: (a) a discrete nanoparticle, (b) aggregate, (c) individual particles converging on the pore opening demonstrating hydrodynamic bridging.

DRUG DELIVERY

Direct injection of delivery systems into tumors has both been a mode of experimental and clinical drug delivery. Solutions allow the drugs to diffuse or leach out of the tumor, especially through the needle track, whereas suspensions might allow some greater residence time. Viral vectors have been administered by intra-tumoral injection. To decrease the extent of viral dissemination into the systemic circulation, a viscous alginate solution was used as the viral vehicle. However, transgene expression was not increased perhaps because, as the authors speculate, the diffusion of the virus is reduced by the viscous medium once *in situ.*

The transport of particles of viral dimensions requires, according to Higuchi *et al.*, convective rather than diffusional transport. "The early transport of colloids into the vascular and lymphatic vessels relies largely on an extracellular pathway which depends on convective transport (i.e. *solvent drag*)". "Thus the particle uptake in the period immediately after injection is relatively insensitive to particle size; it is expected that viruses will be carried in the tissue towards lymphatics and microvessels with great efficacy leading to enhanced escape compared with the relatively low levels" for 1 and 0.4 μm particles. The question of how resistance to convective transport in the interstitial space (the interstitial fluid plus the extracellular matrix) has been considered at least for molecules. Clearly, the spacing between the cells or between fibres will be a significant factor in determining the size cut-off for transport.

CONCLUDING REMARKS

This phenomenological survey of possible factors affecting the flow and hence the mass transport of nanoparticles has explored a range of scenarios. It is by no means a comprehensive survey, but there is sufficient in the literature to stimulate further analyses to provide a better overall prediction of the influence of particle characteristics, particularly, diameter and surface nature, shape and flexibility on delivery and targeting to remote sites in the body. Confocal microscopy and other techniques will allow experimental study of nanoparticles so that their movement and fate can be studied in a variety of tissues.

Atomic force microscopy allows measurement of forces of interaction of particles with cells and receptors to aid a more quantitative approach. However, it is wrong to underestimate the challenges ahead if nanoparticulate carriers are to be designed to overcome the various biological barriers and survive transit in the conduits of capillary blood or lymph, extravasation and tissue, and subsequently intracellular transport. One cannot help but conclude that as many properties including flow are dictated by particle diameter, one of the most important strategies is to ensure the maintenance of particle stability *in vivo*.

2

Amphiphilic Molecules

The term microemulsion (originally introduced by Schulman *et al.* is used to denote a thermodynamically stable, fluid, transparent (or translucent) dispersion of oil, and water, stabilized by an interfacial film of amphiphilic molecules (i.e. molecules that possess within their structure a part that has an affinity for oil and a part that has affinity for water). The amphiphilic molecules used to stabilize the dispersion may be a pure surfactant, a polymeric surfactant, or a mixture of surfactants frequently in conjunction with a weakly amphiphilic molecule or cosurfactant, typically a medium chain alcohol. Although at first glance, a seemingly simple definition, that causes much confusion in the literature, and in particular, the pharmaceutical literature, as to what exactly constitutes a microemulsion such that every article on the subject always devotes some time to precisely defining what the authors mean by the term. This present article is no different.

INTERRELATIONSHIPS

Much of the confusion stems from the apparent relationship between a microemulsion and an emulsion. In fact, the two systems have little in common other than the fact that they are both dispersions of oil and water stabilized by an interfacial film of amphiphilic molecules. The most significant difference between the two systems, is that a microemulsion forms spontaneously, and unlike an emulsion, it requires no mechanical work for its formation. A microemulsion is therefore thermodynamically stable.

An emulsion, in contrast, generally requires considerable energy input for its preparation, and while it may possess (some) kinetic stability, it does not possess thermodynamic stability. This is an important distinction between the two systems, and one which has important consequences for their preparation and use. For example, complex equipment and considerable cost in terms of energy input are required for the formation of an emulsion, whereas a microemulsion can be prepared using simple mixing equipment. The requirement for energy input into a system for its formation provides a simple means of differentiating between a microemulsion and an emulsion. However, as it is virtually impossible to mix ingredients together without the input of some energy, the distinguishing feature is the energy level used, such that a microemulsion could be quite reasonably prepared using gentle heating or stirring to speed up its formation, whereas an emulsion would require the input of considerably more energy, such as that supplied by high pressure homogenization or ultrasonication for its preparation.

It is worth noting, however, that a number of papers incorrectly describe systems as being microemulsions, despite the fact that a considerable input of energy (e.g. homogenization) is required for the formation of droplets, which in some instances are even outside the generally accepted size range for a microemulsion. Other differences between a microemulsion and an emulsion include the size of the regions of disperse phase; in a microemulsion, the domains of disperse phase are typically less than 140 nm and relatively uniform in size, whereas an emulsion possess droplets of varying size in the general region of 1-10 µm. As a consequence of their small domain size, microemulsions are transparent (or translucent) while emulsions are usually cloudy. It is worth noting however that it is possible to obtain an optically clear emulsion if the refractive index of the two immiscible phases are of comparable values.

In addition, although it is frequently stated that one of the immiscible phases is water, it is possible to prepare either a microemulsion or an emulsion using either a non-aqueous polar phase such as ethylene glycol, gylcerol or ethanol, or two immiscible oils such as a fluorocarbon and a hydrocarbon or castor oil and silicone oil. One important outcome of the small domain size (and large interfacial area) of a microemulsion is the large amount of surfactant required to stabilize the system, in the order of 20 wt%, as opposed to a 1 to 2 wt% generally needed to stabilize an emulsion.

Nanoemulsion as a Drug Carrier

The confusion as to what differentiates a microemulsion and an emulsion has been further complicated by the emergence in the literature of another oil and water dispersion, known as a nanoemulsion, although the terms mini- or submicron emulsion have also been used. The distinction between a microemulsion and a nanoemulsion is even more blurred because the description of a nanoemulsion is very similar to that of a microemulsion in that they are both oil-in-water dispersions of small droplet diameter (for nanoemulsions a range of 20-200 nm is typically quoted) and of narrow droplet size distribution. Although the physical appearance of a nanoemulsion resembles that of a microemulsion, in that both systems are transparent (or translucent) and of low viscosity, there is an essential difference between the two systems, namely that a nanoemulsion (i.e. an emulsion) is, at best, kinetically stable, while a microemulsion is thermodynamically stable.

As a consequence, many of the nanoemulsions reported in the literature do not possess long-term stability. Some nanoemulsions *have* however exhibited sufficiently high levels of stability for them to be proposed as vehicles for drug delivery. It is worth commenting that, while the distinction between a nanoemulsion and an emulsion, in terms of their size, is rather arbitary, nanoemulsions because of their small droplet size, possess a higher stability against sedimentation or creaming than an emulsion. One supposed advantage of a nanoemulsion over a microemulsion is that it requires a lower surfactant concentration for its formation. For example, nanoemulsion droplets of radius 60-70 nm and containing 20 wt% isohexadecane, using only 4 wt% of a mixture of polyoxyethylene 4-dodecyl ether and polyoxyethylene 6-dodecyl ether surfactants. When comparing this surfactant concentration with the 20 wt% surfactant typically needed to prepare a microemulsion containing a comparable amount of oil, one should realize that the droplet size of a microemulsion thus produced would typically be ~10 nm. Consequently, in order to produce nanoemulsion droplets of the comparable size, the amount of surfactant required would increase (the surface area of the droplet varies with that of the square of the droplet radius) to a comparable value.

The pertinent question (in terms of drug delivery) is what is most beneficial or the optimal size of the droplets. Recent results suggest that small may not always be better, especially because of the need for large amounts of surfactant which, under certain circumstances, actually hinder drug absorption. Nanoemulsions, as a consequence of their relatively high kinetic stability, low viscosity, and transparency/translucency, are very attractive for a range of industrial applications, including the pharmaceutical field where they have been explored as drug delivery systems. It is worth noting however that the most stable nanoemulsions are generally, although not exclusively, prepared using expensive, high energy input methods such as microfluidization, ultrasonication, which makes their production expensive.

Surfactants

If a surfactant or a mixture of surfactants, and a cosurfactant are used to prepare the microemulsion, then it is usual to keep the ratio of the two amphiphiles constant and to plot them as a "single" component, thereby allowing the production of a (pseudo-)ternary phase diagram. It is possible to prepare microemulsion over a wide range of surfactant concentrations and oil-to-water ratios, although in most systems, a microemulsion exists only over a restricted range of concentrations. Indeed, a microemulsion is only one of a number of possible oil, water and surfactant association structures that can form, depending upon the chemical nature and concentration of each of the components, as well as the temperature and pressure (other structures include gels and various mesomorphic phases). Regardless of whether a microemulsion is formed or not, generally, at very low surfactant concentrations, a large multiphase region is seen, while at very high surfactant concentrations, liquid crystalline phases are frequently formed. It is clear that the microstructure of the microemulsion must vary over the range of possible microemulsion compositions.

At low oil or water volume fractions, microemulsions are generally considered to be a dispersion of either oil or water droplets stabilized by an interfacial film of surfactant and where appropriate, cosurfactant. These droplet structures are probably the most commonly encountered type of microemulsion microstructure. It is worth noting that both an emulsion and a nanoemulsion can only occur in the form of a droplet, either as an oil-in-water or water-in-oil droplet. At intermediate oil and water compositions, it is obviously not possible for the microstructure to be composed of droplets of one phase dispersed in the other. In these cases, it is thought that a bicontinuous structure exists, in which the water and oil domains are separated by a regular or topologically chaotic continuous amphiphile-rich interfacial layer.

A bicontinuous microemulsion is often the intermediate microstructure between an oil-in-water and a water-in-oil microemulsion, although a number of other microstructures such as cylinders and worm-like microemulsions have been reported to exist. In terms of its microstructure, a microemulsion is therefore a very complex system, and in instances where a microemulsion exists over a wide range of compositions, several different types of microstructure may be present. It is also important to remember that whatever the microstructure, a microemulsion is a dynamic system in which the interface is continuously and spontaneously fluctuating. For this reason, microemulsions stabilized by polymeric surfactants may be the most long lived.

Micellar System

There is much debate in the literature as to what exactly differentiates a microemulsion from a micelle at low volume fractions of disperse phase. Some investigators have perceived a difference between microemulsions and micellar systems containing solubilized oil or water, and have used the terms "*swollen*" micellar solutions or solubilized micellar solutions to describe such systems. These investigators argue that the term microemulsion should be restricted to systems in which the droplets are of large enough size such that the physical properties of the dispersed oil or water phase are indistinguishable from those of the corresponding oil or water phase, thereby theoretically making it possible to distinguish between oil-in-water (or water-in-oil) microemulsions and micellar solutions containing small amounts of solubilized oil (water).

However, in most cases, the transformation between micelles progressively swollen with oil (water) and a microemulsion containing an isotropic core of oil (water) appears to be gradual with no obvious transition point. As a consequence, there is no simple method available for determining the oil (water) content at which the core of the swollen micelle becomes identical to that of a bulk phase. Many researchers therefore use the term microemulsion to include swollen micelles, but not micelles containing no oil (or water). In biotechnological applications, water-in-oil microemulsions are frequently known as reverse micelles and or even as nonaqueous media.

System of Co-solvents

The above broad definition does not require a microemulsion to contain any microstructure. In other words, it includes systems that are co-solvents, i.e. systems in which the constituent components are molecularly dispersed. Most researchers in the field agree, however, that for a microemulsion to be formed, it is important that the system contains some definite microstructure. In other words, there is a definite boundary between the oil and water phases, and at which the amphiphilic molecules are located and that a co-solvent is not a type of microemulsion. The only way to distinguish a microemulsion from a co-solvent unambiguously is to perform either a scattering study (light, X-rays or neutrons) or PFG-NMR measurements. Regions of co-solvent formation generally appear at low concentrations of oil or water.

MICROEMULSIONS AS CARRIER

It is clear from its description that microemulsions possess a number of properties that make their use as drug delivery vehicles particularly attractive. Indeed, microemulsions were first studied with the view of using them as potential vehicles for poorly-water soluble drugs, in the mid 1970s by Elworthy and Attwood. However, it was not until the mid to late 1980s that they were widely investigated as drug delivery systems; this interest being largely the result of the arrival on the market of the cyclosporin A microemulsion preconcentrate, Neoral. Among the physical properties that make microemulsions attractive as drug delivery vehicle is their transparent nature, which means that the product is not only aesthetically pleasing, but allows easy visualization of any contamination.

The small size of the domains present means that a microemulsion can be sterilized by terminal filtration. Furthermore, depending on the composition of the microemulsion, it may be possible to heat sterilize the microemulsions.

Since oil-in-water microemulsions are able to incorporate lipophilic substances, they can be used to facilitate the administration of water-insoluble drugs. Significantly, the small droplet size provides a large interfacial area for rapid drug release, and so the drug should exhibit an enhanced bioavailability, enabling a reduction in dose, more consistent temporal profiles of drug absorption, and the protection of drug(s) from the hostile environment of the body. In addition to increasing the rate of drug release, microemulsions can also be used as a reservoir and actually slow the release of drug and prolong its effect, thereby avoiding high concentrations in the blood. Whether a drug is rapidly or slowly released from a microemulsion depends very much on the affinity of the drug for the microemulsion. Since microemulsions contain surfactants (cosurfactants) and other excipients, they may serve to increase the membrane penetration of drug.

Self-emulsifying Drug Delivery Systems (SEDDS)

Before discussing how microemulsions are being exploited in drug delivery, it is worth making one more distinction, namely the difference between a self-emulsifying drug delivery system (SEDDS) and a microemulsion. A SEDDS is a mixture of oil(s), and surfactant(s), ideally isotropic, sometimes containing cosolvent(s), which when introduced into aqueous phase under gentle agitation, spontaneously emulsifies to produce a fine oil-in-water dispersion. Typically, the size of the droplets produced by dilution of a SEDDS is in the range of 100 and 300 nm, while, upon dispersal in water, a SMEDDS formulation (a sub-group of the SEDDS) forms a transparent microemulsion with particle sizes <100 nm.

A SMEEDS is also known as a pre-microemulsion concentrate. It is worth noticing that this method of producing a fine oil-in-water emulsion using a S(M)EEDS is identical to the low energy emulsification method for producing oil-in water nanoemulsions. It is therefore likely that a diluted S(M)EDDS and nanoemulsion are identically the same. The technique of low-energy or self-emulsification has been commercially exploited for many years in the agrochemical industry, in the form of emulsifiable concentrates of lipophilic herbicides and pesticides. However, it has only recently been introduced in the pharmaceutical industry as a tool to improve the delivery of lipophilic drugs by incorporating the drug into a S(M)EDDS formulation which is then filled into capsules. Once the capsule has been swallowed and its contents come into contact with the GI fluid, the drug containing (micro)emulsion should be spontaneously formed. Once the drug containing (micro)emulsion is formed, there should be little difference between the fate of the drug thus administered and the same drug administered in a (pre-formulated) microemulsion, although the droplets formed from the S(M)EDDS tend to be of a larger size.

One advantage of administering a drug in a SMEEDS as opposed to a pre-formulated microemulsion, is its relatively small volume which can be incorporated into soft or hard gelatin capsules, convenient for oral delivery. To date, there has been a good amount of commercial success for the first self-microemulsifying drug delivery systems (SMEDDS) on the market, namely Neoral (cyclosporin A). In addition, the recent commercialization of two self-emulsifying formulations, namely Norvir (ritonavir) and Fortovase (saquinavir), has undoubtedly increased the interest in SEDDS and other emulsion-based delivery systems to improve the delivery of a range of drugs of varying physico-chemical properties. However, there are a number of reasons why S(M)EDDS are not in greater widespread use, but the main reason is probably the stability of the diluted SEDDS, which is in fact a thermodynamically unstable emulsion (although it may exhibit some limited kinetic or "meta" stability).

It should be noted however that as a SEDDS is either diluted just prior to administration or else in the body, the required droplet stability is less than 6 hrs (i.e. the transit time of materials down the small intestine). Although most studies of SEEDS have utilized isotropic liquids, the earliest reports of these self-emulsifying systems using pharmaceutical materials are in fact related to pastes based on waxy polyoxyethylene *n*-alkyl ethers. In the context of drug delivery via self-emulsifying systems, isotropic liquids are generally preferred to waxy pastes because if one or more excipient(s) crystallize(s) on cooling to form a waxy mixture, it is very difficult to determine the morphology of the materials. Despite this, there is currently a general move towards formulating semi-solid SEDDS. For example, attempts have been made to transform SEDDS into solid dosage forms by addition of large amounts of solidifying excipients such as adsorbents and polymers.

Unfortunately, as the ratio of SEDDS to solidifying excipients required for this approach is very high, this leads to problems in formulating drugs having limited solubility in the oil phase. Recent attempts have been made to reduce the amount of solidifying excipients by gelling the SEDDS with colloidal silicon dioxide. Khoo *et al.* have recently reported the preparation of a halofantrine-containing lipid-based solid self-emulsifying system using either Vitamin E TPGS or a blend of Gelucire 44:14:Vitamin E TPGS as the base. Upon dispersal, these systems produced dispersions that the authors described as microemulsions. Studies in fasted dogs showed that these solid dispersions exhibited a five- to seven-fold improvement in absolute oral bioavailability, when compared with the commercially available tablet formulation. In a different approach, Nazzal *et al.* have determined the potential of a reversibly induced re-crystallized semi-solid self-nanoemulsifying drug delivery system, based on a eutectic interaction between the drug and the carrying agent, as an alternative to a conventional SEDDS.

In these eutectic-based self-nanoemulsified systems, the melting point depression method allows the oil phase containing the drug itself to melt at body temperature from its semisolid consistency, and disperse to form emulsion droplets in the nanometer size range. Emulsion systems based on a eutectic mixture of lidocaine-prilocaine, and lidocaine-menthol have been used in the preparation of topical formulations. However, little is known of the use of eutectic mixtures for the preparation of self-(micro) emulsified formulations.

Putative Delivery Systems

There are a number of other *putative delivery* systems that are closely related to, or are prepared from, a microemulsion. These systems include a variety of gel formulations (including microemulsion-based gels, ringing gels, microemulsion gels) and double microemulsions.

Thickened microemulsion

Oil-in-water microemulsions can be readily gelled or thickened by the addition of a non-interacting, water-soluble polymer such as polyHEMA, Carbopol 940 or carrageenan to form clear "*microemulsion gels*". In these cases, it is the external aqueous phase that is gelled, while the microemulsion droplets are unperturbed. The structure of the resulting "*microemulsion gel*" is quite different, if it is prepared using an interacting polymer, such as stearate-polyethylene oxide-stearate. In this instance, the hydrophilic mid-block of the polymer is located in the continuous aqueous phase, while the hydrophobic end blocks are dissolved in the oil droplets, thereby connecting the various microemulsion droplets and resulting in the formation of a transient gel network. Clear, "*microemulsion gels*" are also sometimes obtained at surfactant and/or oil concentrations just outside the oil-in-water microemulsion region.

Sometimes, the resultant gel "*rings*" or vibrates when tapped. The ringing is due to the resonance of shear modes within the gel body. Neither of these "*microemulsion gels*", which are water continuous, are true microemulsions, which are fluid by definition. Clear gels can also be formed in oil continuous systems. For example, a gel can be formed when water is added to reverse micellar solutions of lecithin-in-oil. Here, the water causes the worm-like lecithin reverse micelles to intertwine and form a gel. In addition, gels, widely known as microemulsion-based gels, can be formed from water-in-oil microemulsions stabilized (predominately) by the di-chain surfactant sodium bis (2-ethylhexyl) sulfosuccinate (AOT), when gelatin, the natural amphiphilic polymer is added. Microemulsion-based gels have now been prepared in systems in which a large amount of the AOT has been replaced by nonionic surfactant; or more recently, using in place of AOT, the single chain surfactant, cetyltrimethylammonium bromide in combination with pentanol. In these gels, the gelatin is thought to form water-continuous channels between the microemulsion-droplets.

These microemulsion-based gels are very unusual in that, although they are oil continuous, they are electrically conducting. In addition, the continuous oil behaves as if it were still a fluid, even though placing a gel in a solution of the oil does not dissolve it. All of these "*microemulsion gels*" have potential, or are being explored for use as drug delivery systems. Of particular interest is the fact that the gels possess the properties of being transparent, infinitely stable and readily prepared using only the mildest of mixing. In addition the wide range of microemulsion gels available means that it is possible to select the gel of the required consistency for application to large areas of skin, the nasal membrane, vaginal and buccal membranes and for permeation enhancement. Microemulsion-based gels have been explored as vehicles for the iontophorectic delivery of drugs.

Potential drug delivery vehicle

Double (or multiple) *emulsions* have attracted much interest as potential drug delivery vehicles. For example, adding a water-soluble drug into the internal aqueous phase of a water-in-oil-in-water emulsion may allow the sustained release of the water-soluble drug. A double microemulsion should offer similar advantages over the rate of drug release of entrapped solutes. Double emulsions are notoriously difficult to formulate due to the requirement to have one surfactant (or mixture of surfactants) to stabilize the first (internal) emulsion and a second surfactant (or mixture of surfactants) of quite different physico-chemical properties to stabilize the second emulsion. Although a few papers have detailed the production of nanoparticles from systems they described as double microemulsions, the term "*double microemulsion*" in this context is very misleading, as it refers to the mixing of two water-in-oil microemulsions of comparable composition, but containing different solute in the aqueous phase.

There are however two papers which describe the preparation of double (oil-in-water-in-oil) microemulsions. In the first, Castro *et al.*, report spectroscopic studies of nifedipine in Brij 96 based oil/water/oil multiple microemulsions. In the second, Carli *et al.,* detail the preparation of an oil-in-water-in-oil microemulsion from an oily phase of either polyglycolized glycerides or a mixture of mono-, di- and tri-glycerides, which is microemulsified using a mixture of water and surfactant (soy lecithin and Tween 80). The resultant o/w microemulsion is subsequently re-dispersed in an oily phase to produce the double (o/w/o) microemulsion.

Dry emulsions

In practical terms, a solid dosage form is preferable to a liquid dosage form in respect of convenience, ease of handling and accurate dosing. Consequently, a number of researchers have attempted to develop powdered, re-dispersible emulsion-derived formulations, known as solid state or dry emulsions. Such solid-state emulsions can be used to modulate the release rates of emulsified compound. Dry emulsions have been variously prepared by removing water from an oil-in-water emulsion, using water-soluble or -insoluble solid carriers or indeed a mixture of both water-soluble and water-insoluble carriers, by rotary evaporation, lyophilization, or spray drying. Attempts have also been recently made to prepare dry microemulsions using similar methodology. For example, Moreno *et al.* have lyophilized an amphotericin B-containing lecithin-based oil-in-water microemulsion in the presence of 5 wt% mannitol. The lyophilized product was an oily cake from which the microemulsion could be easily reconstituted over several months.

The rationale for developing the water-free formulation was to avoid the hydrolysis of lecithin, which occurs upon its dispersal in water, thereby preventing any deterioration of the formulation upon storage. Overall, the lyophilized lecithin based oil-in-water microemulsions appear to be valuable systems for the delivery of amphotericin B, with regard to ease and low-cost of manufacturing and their stability and safety, compared with other formulations already in the market. In a recent paper, Carli *et al.* reported an alternative approach to prepare a "dry" formulation known as a nanoemulsified composite of the coenzyme Q10, ubidecarenone. This composite is prepared by incorporating the ubidecarenone into the inner phase of the double microemulsion, which is then deposited onto a solid microporous carrier such as cross-linked polyvinylpyrrolidone. Among the advantages offered by this approach are good processing and storage properties, easy re-dispersibility in water, high bioavailability and maintenance of the submicron size of the released droplets. Kim *et al.,* have prepared entric-coated solid state premicroemulsion concentrates by first preparing a pre-microemulsion concentrate containing 10 wt% of the drug cyclosporin A, 18.5 wt% of a medium chain triglyceride, 51 wt% of surfactant and 20 wt% of cosurfactant. The pre-microemulsion concentrates were then enteric-coated as films using polymers, such as sodium alginate, Eudragit L100 and cellulose acetate phthalate, and the resulting films were pulverized to produce powdered, dry, enteric coated premicroemulsion concentrates. Using this approach, the authors successfully prepared a once-a-day oral dose form of cyclosporin A.

MICROEMULSION FORMULATION

Microemulsions are far more difficult to formulate than emulsions because their formation is a highly specific process involving spontaneous interactions among their constituent molecules. In addition, in a number of cases, effects due to the order of the mixing of the component molecules have been observed. Since no adequate theory currently exists to predict from which molecules microemulsions can be formed, mainly because of the requirement to determine a number of unknown parameters, microemulsion formulations are generally developed empirically, although some useful practical guidance as to the choice of the constituent components can be found in the literature. A recognized and classical approach to microemulsion formulation is to undertake a systematic study of the phase behavior of the systems understudying utilizing of phase diagrams. A major drawback of this approach is the considerable time it takes to develop the phase diagram, especially considering the combination of possible oil, surfactant and cosurfactant, and the fact that time may be necessary for a system to equilibrate.

Heat and sonication are therefore often used, particularly with systems containing nonionic surfactants, to speed up the formation process. While there are now commercially available automated systems to prepare phase diagrams, the chief drawback of these systems is their cost. A number of attempts have been recently made to use modeling to predict microemulsion formation, thereby aiding in the formulation of microemulsions. A range of modeling techniques have been used including artificial neural networks, genetic algorithms and a combination of data mining, computer-aided molecular modeling, descriptor calculation and multiple linear regression techniques. Unfortunately, however, all of these techniques require a considerable amount of work prior to prediction, thereby restricting their potential usefulness. Furthermore, the amount of work required for the predictions increases as the number of components of the microemulsion increase; microemulsions formulated from five components (i.e. oil, water, surfactant, cosurfactant, electrolyte and drug) are not uncommon in pharmaceutical use. To the authors' knowledge, to date, no work has been performed predicting how much drug can be incorporated into a microemulsion and whether the presence of drug has any effect on microemulsion phase behavior. This is an important ommision as microemulsions cannot be considered to be inert, since the presence of drug in some instances (greatly) influences phase behavior.

Preparation of Microemulsions

The selection of components for the preparation of microemulsions suitable for pharmaceutical use involves a consideration of their toxicity, and if the systems are to be used topically, their irritancy and sensitizing properties as well. There are a number of surfactant and cosurfactants that are considered acceptable for use as excipients in pharmaceutical formulation. Strickley has recently reviewed those surfactants and cosurfactants currently used in commercially available oral and intravenous formulations. In the general scientific literature, by far the most widely used surfactant to prepare a microemulsion is the double chain, ionic surfactant, sodium bis (2-ethylhexyl) sulfosuccinate (AOT), although a large number of studies have used the single chain, nonionic surfactants of the type C_iE_j, where i is the number of carbons in the alkyl chain, C, and j is the number of ethylene oxide units in the polyoxyethylene chain, E.

Both AOT and the C_iE_j surfactants possess the important advantage of being able to form microemulsions in the absence of a cosurfactant, unlike most other types of surfactant such as the widely studied single chain, ionic surfactant, sodium dodecyl sulphate (SDS), which will only form microemulsions in the presence of an alcohol cosurfactant. Neither AOT nor SDS would be considered to be apprpropriate for the preparation of pharmaceutically acceptable microemulsions. As a general rule, nonionic and zwitterionic surfactants tend to be less toxic than ionic surfactants and are therefore more widely used as pharmaceutical excipients. Assuming that the surfactants do not degrade into toxic materials, surfactants that posses biodegradable/chemically unstable linkers tend to exhibit less chronic toxicity than those that are chemically stable. For example, as a group, the polyoxyethylene *n*-acyl surfactants exhibit ten times less chronic toxicity than their *n*-alkyl counterparts, mainly due to their quicker degradation time of days as opposed to weeks. When it comes to comparing acute toxicity, the two groups of surfactants exhibit comparable toxicity. Perhaps the most widely used nonionic surfactants in pharmaceutical formulations are the polyoxyethylene sorbitan *n*-acyl esters, i.e. the Tweens and in particular, Tween 20 and Tween 80, both of which are used parenterally and orally. In addition, polyoxyethylene derivatives of the triglyceride, castor oil have acceptability for intravenous administration.

Other pharmaceutically acceptable surfactants are the polyoxyethylene *n*-alkyl ethers and *n*-acyl esters, although both these groups of surfactant tend to be restricted for topical use. Other nonionic surfactants that are currently attracting much pharmaceutical interest, although they do not yet have acceptability, are the polyglycerol *n*-acyl esters, the *n*-alkyl amine N-oxides and the n-alkyl polyglycosides (or sugar surfactants). The *n*-alkyl polyglycosides have attracted much pharmaceutical interest, not because of their excellent biodegradability, but because they can be manufactured from renewable resources. All of the aforementioned surfactants have been used to prepare microemulsions, generally as sole surfactant, the only exception being the *n*-alkyl polyglycosides, which tend to require the presence of a cosurfactant. Pluronics (or poloxamers) of the type poly(ethylene oxide)-poly(propylene oxide)-poly(ethylene oxide) (PEO-PPO-PEO) are another class of pharmaceutically attractive surfactant. Interestingly, most reports detailing the use of polymeric surfactants to stabilize a microemulsion describe the preparation of water-in-oil, generally in conjunction with a second surfactant.

Siebenbrodt and Keipere have reported the formation of a triacetin-in-water microemulsion using Pluronic L 64 as sole surfactant. Lettow *et al.*, have used Pluronic PI23 as sole surfactant to prepare oil-in-water microemulsions, incorporating a 1:1 oil:P123 weight ratio of either 1,3,5-trimethylbenzene or 1,2-dichlorobenzene. Finally, the pharmaceutically acceptable zwitterionic lecithin has been extensively used as a surfactant, however, with very few exceptions, it is not possible to prepare a microemulsion using lecithin as sole surfactant. Generally, lecithin is combined with another surfactant such as Tween 80, or a cosurfactant such as ethanol, when formulating microemulsions. Although ethanol is considered to be pharmaceutically acceptable, typical cosurfactants such as propanol and butanol are not. In addition to toxicity issues, the use of such cosurfactants, which may possess partial oil and water-solubility, can lead to problems with the dilutability of the microemulsion. This is a particular issue if the microemulsion is to be administered orally or parenterally. Consequently, a number of researchers have explored the use of a second surfactant as cosurfactant when formulating a microemulsion.

Microemulsions thus prepared tend to be very stable against dilution, as the "*cosurfactant*" generally has little solubility in either the oil or aqueous phase. Alternately (pharmaceutically acceptable), short chain mono- and di-glycerides have been used in place of a short chain alcohol to successfully prepare microemulsions. In a number of instances, short chain fatty acids such as sodium caprylate have been used as cosurfactants, primarily for the formation of microemulsions for oral delivery; sodium caprylate is known to enhance absorption of drugs across the gastrointestinal tract.

A number of researchers have also used cosolvents such as the polyhydric alcohols, sorbitol, glycerol and propylene glycol to aid microemulsion formation. In a number of instances, these materials have been described as "*cosurfactants*", which quite clearly do not sit in the interfacial surfactant monolayer. Rather, they tend to exert their effect by altering the solvent properties of the polar phase.

Aromatic Oils

Most reports in the chemical literature detail the preparation of microemulsions using aromatic oils such as benzene and short chain alkanes such as hexane. "*Pharmaceutical*" oils, unlike those used in the chemical and agrochemical industries, tend to be large in terms of molecular weight and therefore volume, and are relatively polar. Both of these properties tend to work against the oil when it comes to formulating it in a microemulsion, as it is well established that smaller molecular volume oils are easier to solubilize and are solubilized to a greater extent than larger oils. Although there are reports that in some systems, particularly those containing surfactants with long, unsaturated hydrophobes such as polyoxyethylene oleyl ether, the largest molecular volume oil is solubilized to a greater extent than some of the smaller molecular volume oils.

The most commonly used "*pharmaceutical*" oils are medium and long chain triglycerides, and esters of fatty acids such as ethyl oleate, isopropyl myristate are popular. It has become common practice for researchers to screen the solubility of drug in the various components of the microemulsion, in order to predict the optimal composition of the final formulation. However, extreme care has to be exercised when using this approach, as very often, the solubility in the final microemulsion formulation does not correlate well with that seen in the various components.

Important Features

It is noticeable that in contrast to their ease of preparation, it is very difficult to establish the microstructure of a microemulsion. Yet, such information is important as it may influence the drug behavior of the microemulsion in use. For example, it is known that the microstructure of the microemulsion may alter the release rate of any incorporated solute. Currently, a range of physico-chemical techniques are used to characterize microemulsions. These techniques are often used in tandem to obtain a better picture of the system, as it is unlikely that any one technique alone will give sufficient information.

Scattering techniques (light, neutron and X-ray) and pulsed field gradient NMR are generally used to determine the microstructure of the microemulsion. One serious limitation with characterizing microemulsions is that most techniques rely on the concentration of disperse phase being low enough to avoid particle-particle interactions, as an estimated volume fraction of 1 vol% is suitable. The requirement is a particular problem with microemulsions that contain cosurfactants that partition between the oil and water phases, because these systems frequently undergo a change upon dilution.

ROUTES OF DRUG DELIVERY

Although most of the original work exploring microemulsions as drug delivery vehicles examined their potential for oral drug delivery, microemulsions have now been explored as vehicles for most routes of administration. Currently, they are probably most widely studied for their potential as transdermal delivery vehicles.

Through Mouth

Microemulsions (and SMEDDS) have been widely studied as oral drug delivery vehicles. Indeed, the first commercially available "*microemulsion*" formulation was a premicroemulsion concentrate of the lipophilic peptide, cyclosporin A. This formulation, known commercially as Neoral, was introduced onto the market in the late 1980s and immediately attracted much attention, mainly because of the high and reproducible bioavailability it produced, but also because developments in biotechnology at that time meant that it had never been easier to produce on a large scale therapeutically-relevant protein and peptides. Unfortunately, because of their physico-chemical properties, in particular their large size and poor stability, proteins and peptides are very difficult to formulate. Microemulsions offered an attractive solution to this problem, and consequently, most of the original exploratory studies on microemulsions as drug delivery vehicles were spent developing oral protein/peptide microemulsion formulations.

Water soluble molecules

As the majority of therapeutic proteins and peptides are hydrophilic and water- soluble, most studies utilizing microemulsions as vehicles for such molecules have exploited water-in-oil microemulsions. After cyclosporin A, which is unusually highly lipophilic, for a therapeutic peptide, the most widely studied peptide is insulin, with much of the early work in this area being performed by Ritschel. For example, Kraeling and Ritschel compared the peroral microemulsion formulation of insulin and capsule forms and determined that the microemulsion formulation increased the bioavailability of the insulin. Recently, more complex microemulsion-based systems have been developed in an attempt to improve the extent of insulin absorption. For example, a recent study performed by Natnasirichaiku *et al.* showed a significant improvement in the oral bioavailability of insulin (in diabetic rats) when administered in nanocapsules dispersed in a water-in-oil microemulsion. Santiago *et al.* have developed a new, enteric oral dosage form of insulin, in which an association of insulin and cyclodextrin contained within a microemulsion is processed into granules. In the most recent study aimed at developing an oral formulation of insulin, Iek *et al.* used a conventional lecithin-based water-in-oil microemulsion formulation prepared from 21.6 wt% water, 37.6 wt% Labrafil M 1944 CS as oil and stabilized by 40.8 wt% of a 1:1 weight ratio of lecithin (Phospholipon 90G) and ethanol.

In addition to insulin (21.6 IU/g water), some of the microemulsions contained the enzyme inhibitor aprotinin (2500 KIU/g water). Although it is the first time that a microemulsion formulation has contained both a protein/peptide and an enzyme inhibitor, the concept of adding an enzyme inhibitor, to a formulation containing a peptide in an attempt to reduce its degradation is not new. The plasma glucose and insulin levels of the rats after intragastric administration of the formulations to both diabetic and non-diabetic rats were significantly different from those obtained after oral administration of an aqueous insulin solution. Although the addition of aprotinin to the microemulsion containing insulin increased bioavailability when compared with those not containing it, the difference was insignificant. Other peptides formulated as water-in-oil microemulsions in an attempt to improve their oral absorption include RGB peptides, and more recently, *N*-acetylglucosaminyl-*N*-acetylmuramyl dipeptide (GMDP).

The poor bioavailability of GMDP has been attributed to both its poor stability in the lumen of the gastrointestinal tract and its poor intestinal permeability. When GMDP was administered intraduodenally in a water-in-medium-chain trigylceride microemulsion, a ten-fold increase in bioavailability was observed, i.e. a bioavailability of 80.2% was achieved as opposed to 8.4%, seen after administration of an aqueous solution of GMDP. This increase in bioavailability is consistent with the work of Constantinides *et al.* who utilized a similar medium chain triglyceride based microemulsion to increase the oral bioavailability of the water-soluble peptide SK&F 106760, after intraduodenal administration to rats. Ke *et al.* have recently reported an attempt to develop water-in-oil microemulsions suitable for the incorporation of therapeutic proteins and peptides using a medium chain triglyceride, water and tocopheryl polyethylene glycol 1000 succinate (TPGS) as the primary surfactant. However, as TPGS could not form microemulsions when used as sole surfactant, it was mixed with a second surfactant, either Tween 20, 40, 60 or 80, at a weight ratio in the range of 4:1 to 1:4.

A range of glycols and polyols were examined as cosurfactants. Although stable, transparent microemulsion and gel regions were identified, the extent of these regions was influenced by the precise nature and the amount of the secondary surfactant and cosurfactant. For example, Tween 80, which is an ester of the *unsaturated* C18 fatty acid, oleic acid, was more effective in forming a microemulsion than Tween 60, which is an ester of the *saturated* C18 fatty acid, stearic acid. In this study, although the microemulsions were ultimately intended for use as delivery vehicles for protein or peptide drugs, they were not examined for this purpose.

Water soluble molecules other than proteins

Other water-soluble therapeutic molecules that have been administered in microemulsions include the aminoglycoside antibiotic, gentamicin and the biologically active polysaccharide, heparin. In common with all aminoglycosides, gentamicin is highly polar and is therefore considered unlikely to be absorbed from the gastrointestinal tract via simple diffusion. In order to facilitate the trans-mucosal delivery of the drug, Hu *et al.* prepared a SMEDDS formulation of gentamicin using a range of surfactants. When Labrasol was used as surfactant, a 54.2% bioavailability of gentamicin was obtained, compared with only 8.4 and 3.4% when Tween 80 and Transcutol P were respectively used. Labrasol was also found to inhibit intestinal secretory transport from the intestinal enterocytes, providing the formulation with the additional benefit of inhibiting the efflux of gentamicin out of the enterocytes into the GI lumen. Due to its low bioavailability, heparin is generally administered by injection. In an attempt to formulate an orally active version of heparin, Kim *et al.* synthesized a low molecular weight heparin (LMWH)-deoxycholic acid (DOCA) conjugate (termed LMWH-DOCA) and formulated it in a water-in-oil microemulsion using as oil, the medium chain trigylceride, tricaprylin, a mixture of Tween 80 and Span 20 surfactants, LMWH-DOCA and water (volume ratios of 5:3:1:1 respectively).

Oral administration of LMWH-DOCA in the water-in-tricaprylin microemulsion to mice resulted in a bioavailability of 1.5%. Toxicity studies suggested that the enhancement in bioavailability, observed with the DOCA-conjugated LMWH, was administered in a microemulsion not due any local toxicity such as disruption or damaging of the intestinal membrane.

Water insoluble drugs

A number of poorly water-soluble, low molecular weight, lipophilic drugs have also been formulated in microemulsions (or SMEEDS) for oral delivery including nitrendipine, danzol halofantrine and biphenyl dimethyl dicarboxylate. These studies serve as an illustration of how important it is to understand the influence on microemulsion formation of the various formulation components. It is worth commenting that the main use of SMEEDS formulations is for the oral administration of lipophilic drugs. Formulating nitrendipine in a SMEEDS formulation, composed of a 1:1 (w/w) mixture of glycerol monocaprylic ester (MCG) and propyleneglycol dicaprylic ester (DCPG) and nonionic surfactant (various), was observed to significantly enhance its absorption when compared with a suspension or an oil solution, and served to reduce the effect of the presence of food on its absorption. However, the absorption profile of nitrendipine was seen to vary with the type of surfactant used; absorption was rapid from the Tween 80-stabilized formulation, while the HCO-60-based formulation gave a prolonged plasma concentration profile.

No absorption of nitrendipine was observed from the formulation containing BL-9EX (polyoxyethylene alkyl ether, $C_{12}E_9$). Damage to the gastrointestinal mucosa also differed with the type of surfactant employed. HCO-60 and Tween 80-based formulations were mild to the organs, while BL-9EX-based formulation caused serious damage. The study of Porter *et al.* appropriately demonstrates the effect of changing the nature of the trigylceride involved in the formulation on drug absorption. These workers studied three lipid-based danazol formulations; namely a long-chain triglyceride solution (LCT-solution), a SMEDDS based on long (C18) chain lipids (LC-SMEDDS) and a SMEEDS formulation containing medium (C8-C10) chain lipids (MC-SMEDDS). These formulations were administered to fasted beagle dogs and their absorption, compared with that obtained with a micronized danazol formulation administered postprandially and in the fasted state. Although both the LCT-solution and LC-SMEDDS formulations were found to significantly enhance the oral bioavailability of danazol, when compared with fasted administration of the micronized formulation, the MC-SMEDDS produced little improvement in danazol bioavailability. This result was partly attributed to the fact that upon digestion of the medium-chain formulation, significant drug precipitation was observed. Khoo *et al.* also considered the effect of formulating halofantrine as a pre-microemulsion concentrate in a formulation based on either a medium- or long-chain triglyceride.

Both formulations, which were administered as soft-gelatin capsules, contained the same amount of medium or long chain trigylceride and were stabilized by the same surfactant/cosurfactant mixture, consisting of Cremophor EL and ethanol. Although the plasma levels of the drug were not significantly different between the two formulations, the amount of drug absorbed lymphatically varied in that 28.3% of the dose administered in the long-chain trigylceride formulation was transported lymphatically, as opposed to only 5.0% of the dose administered in the medium-chain formulation. Kim *et al.* attempted to improve the solubility and bioavailability of biphenyl dimethyl dicarboxylate, a drug used in treating liver diseases, by formulating it as a premicroemulsion concentrate. In order to optimize drug loading in the formulation, these workers screened drug solubility in a range of surfactants and oils, and on the basis of these results selected: Tween 80 and Neobee M-5. However, care must be taken when using this approach to optimize the formulation with respect to drug loading, as it has shown that solubility of drug in the bulk components is not a reliable indicator of solubility, in the final microemulsion formulation.

The danger of predicting drug solubility in the final formulation, on the basis of bulk solubility, can be seen in the study of Kim *et al.* where the solubility of the drug in a formulation consisting of a 2:1 weight ratio of Tween 80 to Neobee M-5 was 7 times that of the formulation containing a Tween 80:Neobee M-5 weight ratio of 1:4, despite the solubility of the drug in Neobee M-5 being 10 times that seen in Tween 80. The final formulation, which consisted of 35 wt% triacetin and 65 wt% Tween 80 and Neobee M-5 at a weight ratio of 2:1, greatly enhanced the oral bioavailability of BDD, possibly due to the increased solubility of the drug and its immediate dispersion in the gastrointestinal tract. Itoh *et al.* optimized the formulation of the poorly water-soluble drug N-4472, N-[2-(3,5-di-tert-butyl-4-hydroxyphenethyl)-4,6-difluorophenyl]-N-[4-(Nbenzylpiperidyl)] urea, by complexing it with L-ascorbic acid and incorporating the complex into a SMEEDS comprising Gelucire 44/14, HCO-60 and sodium dodecyl sulfate. Upon dilution with water, the SMEEDS formulation produce a fine dispersion of 18 nm droplets which were stable over the pH range of 2.0 to 7.0. The oral bioavailability of the drug was between 2-4 times that which was obtained with an aqueous solution of the complex.

Buccal Delivery

To date, very little work has been performed on investigating the use of microemulsions as vehicles for buccal delivery. In 1988, Ceschel *et al.* showed that the penetration of the essential oil, Salvia sclarea L. through porcine buccal mucosa *in vitro* was increased when formulated as a microemulsion, as opposed to the pure essential oil. Scherlund *et al.* investigated the potential of lidocaine and prilocaine thermosetting microemulsions and mixed micellar solutions as drug delivery systems for anesthesia of the periodonlal pocket. The formulations contained between 2-10 wt% of a eutectic mixture of lidocaine or prilocaine (melting point 18°C), while the block copolymer surfactants, Pluronic F127 and F68, were present at between 13 and 17 wt% for F127, and between 2 and 6 wt% for F68. F127 was chosen, as it is known to gel at body temperature and it is important that the formulation is easy to apply, remain at the application site, have a fast onset time, be non-irritant, and stable under normal storage conditions. The pH of the formulations was varied between 5 and 10. Most of the combinations were found to result in clear solutions, presumably oil-in-water microemulsions or mixed micellar solutions, depending on the pH of the system. At low pH, lidocaine and prilocaine are positively charged, and they could be expected to behave largely as water-soluble cationic surfactants, hence possibly forming mixed micelles. On the other hand, at high pH, the drug substances are poorly soluble and could be expected to act largely as hydrophobic solutes and form the core of the microemulsion droplets.

Intravenous Delivery

In recent years, considerable emphasis has been given to the development of injectable microemulsions (o/w) for the *intravenous delivery* of drug, in order to increase the solubility of the drug to reduce drug toxicity, to reduce hypersensitivity, and to improve drug solubility and reduce pain upon injection. A very recent development is the formulation of microemulsions as long circulating vehicles, and more recently, as drug tageting agents. In addition, water-in-oil microemulsions have been investigated as depot vehicles for the intramuscular delivery of drugs. The first published study which established the potential of microemulsions for use in intravenous delivery was probably that of von Corswant *et al.* These researchers prepared a pharmaceutically acceptable, bicontinuous microemulsion from a medium-chain triglyceride oil, poly(ethylene glycol) 400 and ethanol cosolvents and stabilized by soybean phosphatidylcholine and poly(ethylene glycol)(660)-12-hydroxystearate.

Prior to administration, the microemulsion required dilution with a suitable aqueous phase. Upon dilution, the microemulsion formed an oil-in-water microemulsion with droplets of size between 60 and 200 nm, smaller than the size of the droplets in a commercial intravenous emulsion, namely Intralipid. From their animal studies, the authors concluded that the microemulsion they developed was suitable for administion by intravenous infusion to conscious rats. Unfortunately, although the researchers did determine drug solubility in the bicontinuous microemulsions, they did not report this. Park and Kim also investigated the formulation of poorly water-soluble flurbiprofen at ~8 times its aqueous solubility into an oil-in-water microemulsion suitable for intravenous administration. The microemulsions were prepared using varying weight ratios of oil (ethyl oleate) to surfactant (Tween 20), and contained a range of isotonic solutions as the polar (aqueous) phase. Unfortunately, insufficient information was supplied regarding the precise compositions of the microemulsions, in particular, how much oil and surfactant were present, so as to draw conclusions about the formulation; (perhaps surprisingly) the ratio of oil to surfactant used did not seem to have any effect on the amount of drug solubilized and that the presence of too much drug had a destabilizing effect on the microemulsion.

Disappointingly, the pharmacokinetic parameters of flurbiprofen, after intravenous administration of flurbiprofen-loaded microemulsion to rats, were also not significantly different from those of flurbiprofen in phosphate buffered saline solution. In a later publication, Park *et al.* overcame the problem of stability seen in their earlier study by replacing the surfactant Tween 20 with lecithin and distearoylphosphatidyl-ethanolamine-*N*-poly(ethyleneglycol) 2000 (DSPE-PEG) and using ethanol as a cosolvent. Due to the presence of the long chain polyoxyethylene groups on the exterior surface of the microemulsion droplets, it was perhaps unsurprising that the biodistribution of flurbiprofen administered in this microemulsion was quite different. In particular, reticuloendothelial uptake of flurbiprofen decreased, suggesting that it may ultimately be possible to target drugs incorporated in this microemulsion to different sites of the body. As part of a series of papers, Brime *et al.* and Moreno *et al.* prepared a novel amphotericin B lecithin-based oil-in-water microemulsion, in an attempt to produce a formulation with less toxic effects than the currently available commercial formulation, Fungizone.

The microemulsion which contained as oil isopropyl mystriate and a mixture of either Tween 80 or Brij 96 with lecithin as surfactant. In some instances, formulation was lyophilized in an attempt to increase its stability. The overall results of the toxicity studies were encouraging as the amphotericin B-containing microemulsions exhibited a low toxicity, suggesting a potential therapeutic application. Zhang *et al.* prepared a lecithin-based SMEDDS formulation of the drug norcantharidin. Upon dilution, the release rate of norcantharidin contained in the SMEEDS formulation was found to be dependent on the size of the disperse phase and the type of lecithin used. Interestingly, although norcantharidin was poorly soluble in the ethyl oleate and only slightly soluble in water, microemulsions containing ethyl oleate oil exhibited a significant increase in solubilization over the corresponding aqueous solution.

Clonixic acid is currently marketed in salt form because of its poor water-solubility. However, the commercial dosage form causes severe pain after intramuscular or intravenous injection. To improve the apparent aqueous solubility of clonixic acid and to reduce the pain it causes on injection, Lee *et al.* (2000) incorporated 3 mg/mL clonixic acid into oil-in-water microemulsions (size 120 nm) prepared from pre-microemulsion concentrate of castor oil, and a mixture of Tween 20 and Tween 85 surfactants (present in a weight ratio of 5:12:18). Although the microemulsion formulation significantly reduced the number of rats licking their paws as well as the total licking time, suggesting less pain induction by the microemulsion formulation; the pharmacokinetic parameters of clonixic acid after intravenous administration were not significantly different from those of the commercial formulation, lysine clonixinate. The results of the study suggested that a microemulsion formulation is an alternative vehicle for clonixic acid. Paclilaxel (Taxol) injection is known to cause hypersensitivity reactions. Consequently, He *et al.* explored whether it was possible to prepare a non-sensitizing paclitaxel microemulsion using egg phosphatidylcholine, Piyronic F68 and Cremophor EL as surfactants, and ethanol as cosurfactant. Note that there was no mention of the presence of a specific oil.

The study showed that for an equivalent dose, the paclitaxel microemulsion did not cause any hypersensitivity reaction, whereas Taxol did. In addition, the bioavailability of the paclitaxel in the new microemulsion was significantly higher and the elimination rate slower than that achieved with Taxol. The authors suggested that the drug molecules, trapped in the oil droplets, diffused into the systemic circulation slowly. Furthermore, the small particle size of the droplets (10-50 nm) meant that the microemulsion droplets could escape from uptake and phagocytosis of RES. Infact it was previously suggested that it should be possible to modify the surface of the microemulsion droplets, with polyoxyethylene chains, to significantly improve circulation time. Kanga *et al.* have recently explored the possibility of optimizing the release of paclitaxel from a SEEDS formulation using the polymer, PLGA. The SEEDS formulation, which was a mixture of drug, tetraglycol, Cremophor ELP, and Labrafil 1944 also contained PLGA of varying molecular weight. The droplet size of the microemulsions was in the range of 45-270 nm, with the systems without PLGA exhibiting the smaller size. The release rate of paclitaxel decreased in the order of PLGA, PLGA 8K, PLGA 33K, and PLGA 90 Kg/mol, suggesting that the molecular weight of PLGA in microemulsion could control the release rate of paclitaxel from microemulsion.

Alternative formulation

Long circulating microemulsions have been suggested as an alternative formulation to long circulating vesicles on the basis of their small size, thus avoiding uptake by the RES, their stability and their ability to solubilize lipophilic compounds more effectively than vesicles, and their ease of preparation. Wang *et al.*, and Junping *et al.*, have determined the potential of intravenous delivery systems of emulsion/ microemulsion systems based on vitamin E, cholesterol and PEG_{2000}-lipid. In their first study, Wang *et al.,* prepared emulsions containing 1 part drug, 3 parts vitamin E, 3 parts cholesterol and 3 parts PEG_{2000}-DSPE with the final formulation containing 5 mg of drug in 10 mL of saline solution. Although the emulsion was reported to form spontaneously on the addition of the required amount of saline, the formulation was homogenized to produce a more uniform particle size distribution of 123.0 ± 1.2 nm; no information was given as to the size of the droplets prior to homogenization.

The zeta potential and drug loading efficiency of the sub-micron emulsion were –12.67 + 1.35 my and 96.3 + 0.3. Although the size and loading efficiency of the formulation remained uncharged when stored at 7 to 8°C for a year, ~6.5% decomposition of the drug was observed. The plasma area under the curve (AUC) of the drug in the sub-micron emulsion was significantly greater than that of free drug. Overall, the drug in the emulsion had a lower acute toxicity and greater potential antitumor effects than the free drug, suggesting that the formulation is a useful tumor-targeting sub-micron emulsion drug delivery system. In a follow-up study, Junping *et al.* prepared microemulsions of vincristine suitable for injection using vitamin E, PEG_{2000}-DSPE and cholesterol, adding oleic acid to it. The weight ratio of components used was I part drug, 5 parts *oleic* acid, 5 parts vitamin E, 5 parts cholesterol and 5 parts PEG_{2000}-DSPE. No homogenization was used in the preparation of the microemulsion which yielded microemulsion droplets of 138.1 ± 1.2 nm, when prepared using saline at pH 7.4. Note that 10 mL of microemulsion solution contained 1 mg of drug.

The adjustment the pH of the aqueous phase pH and the presence of oleic acid was essential for a high drug loading (94.3 ± 0.3%), while the vitamin E was required for long-term storage of the formulation at 7 to 8°C. The formulation was stable, with respect to particle size, when stored at 78°C in the dark for 1 year, while the loading efficiency of drug decreased by approximately 3%, and 7.4% decomposition of the drug was observed. The plasma AUC of the vincristine in the microemulsion was significantly greater than that of free drug. As with the previous formulation, the drug in the microemulsion exhibited a low acute toxicity and a high potential antitumor effect.

Delivery of lipophilic

Shiokawa *et al.* recently reported the development of a novel, tumor targeted microemulsion formulation suitable for delivery of the lipophilic antitumor antibiotic, aclacinomycin A. Tumor targeting was achieved via folate linked to the exterior surface of long circulating (pegylated) microemulsions. Folate was selected because the folate receptor is abundantly expressed in a large percentage of human tumors, but it is only minimally distributed in normal tissues. The basic composition of the microemulsion was PEG_{2000}-DSPE/cholesterol/vitarnin E/drug (present at a 3:3:3:1 weight ratio or 7:48.3:43,3:1.5 molar ratio). In one microemulsion, 0.24 mol% of folate linked PEG_{2000}-DSPE was present, another contained 0.24 mol% of folate linked PEG_{5000}-DSPE. In a third, the folate was linked directly to the DSPC and in the final one, no folate was present. The association of the folate-PEG_{5000}-linked microemulsion and folate-PEG_{2000}-linked microemulsion with the target cells was 200-and 4-fold higher, whereas their cytotoxicity was 90- and 3.5-fold higher than those of nonfolate microemulsion respectively. The folate-PEG_{5000}-linked microemulsions showed 2.6-fold higher accumulation in solid tumors 24 hrs after i.v. injection and greater tumor growth inhibition than free drug. These findings suggest that a folate-linked microemulsion is a feasible means for tumor-targeted delivery of lipophilic drug. This study shows that folate modification with a sufficiently long PEG chain on the exterior of a microemulsions is an effective way of targeting the carrier to tumor cells.

Delivery through Skin

Integumentary administration

The dermal and transdermal routes of administration offer several advantages compared with other routes of administration. However, the poor permeability of the stratum corneum often limits the possibilities for choosing the topical administration route. Therefore, novel innovative formulations such as microemulsions that have the potential to facilitate skin permeation are of great interest. The investigation of microemulsions as vehicles for cutaneous drug delivery is increasingly common as their potential is realized. Indeed, the cutaneous route is currently the most popular route of adminstration for a microemulsion.

Microemulsions offer significant potentials as transdermal delivery vehicles, since they are robust, frequently stable to the addition of significant amounts of soluble enhancers, excipients and depending on their molecular architecture. Kreilgaard has reviewed the use of microemulsions as cutaneous drug delivery vehicles in 2002. In the present review, work prior to 2002 will not be dealt with in any detail. In addition, due to the large amount of research in the area, the review is not exhaustive.

Proteins and peptides

Recently, the transdermal route has received attention as a promising means to enhance the delivery of drug molecules, particularly peptides, across the skin, using harsh physical enhancement techniques such as iontophoresis and sonophoresis. Very little research has been performed, investigating microemulsions as vehicles for peptide delivery. Getie *et al.* examined the skin penetration profiles of 0.75 wt% desmopressin acetate released from a water-in-oil microemulsion comprising 5 wt% water, 20 wt% Tagot 02:Span 80 3:2 and 74.25 wt% isopropyl myristate. However, the profile was comparable to that obtained using a standard amphiphilic cream. Although the amount of drug that penetrated the upper layers of the skin was significantly higher from the cream than from the microemulsion at all time intervals, within 6 hrs 6% of the applied dose reached the acceptor compartment from the microemulsion instead of 2% from the cream within 300 min, suggesting that the water-in-oil microemulsion has potential for the systemic administration of the drug.

Water soluble drugs

Water-in-oil microemulsions have been used to enhance the penetration of water-soluble drugs. For example, Alvarez-Figueroa and Blanco-Mendez reported the *in vitro* delivery of water-soluble methotrexate from hydrogels using iontophoresis, and passively from oil-in-water and water-in-oil microemulsions prepared using either a 3:1 v:v Labrasol: Plurol Isostearique mixture or a 3:1:1.2 v:v:v Tween 80:Span 80:1,2-octanediol mixture as surfactant/cosurfactant, and either ethyl oleate or isopropyl myristate as oil. All microemulsion formulations studied were more effective than passive delivery from aqueous solution of the hydrophilic drug, although for the microemulsions, delivery was greater from the oil-in-water systems. However, delivery from the microemulsions was less than that using iontophoresis, probably because of the lower solubility of drug in microemulsions than in simple aqueous solution. Escribano *et al.* attempted to improve the transdermal permeation of sodium diclofenac. Four formulations were studied. One was an oil-in-water microemulsion based on transcutol (19 wt%), plurol oleique (19.5 wt%), water (30.6 wt%), isostearyl isostearate (10.9 wt%) and Labrasol (19 wt%).

The other three formulations were *"co-solvent"* systems prepared from various of the ingredients used for the microemulsion formulation. In this study, the microemulsion performed less well than the various co-solvent formulations and in a similar manner to an aqueous solution of the drug. This observation is perhaps not surprising as various enhancers were involved in the microemulsion droplets and were not available to improve drug penetration. Also, as it is likely that the drug was predominately in the continuous phase of the microemulsion, it is not surprising that the formulation behaved in a similar manner to an aqueous solution. The *in vitro* transdermal permeation of the antineoplastic, 5-fluorouracil, incorporated at 1.25 mg/mL in water-in-oil microemulsions prepared using AOT/water/ isopropylmyristate has been studied by Gupta *et al.* These researchers found that as the water content increased from 0.9, 1.8, 2.7 and 3.6% w/w, microemulsions prepared with a surfactant to oil ratio of 5:95 showed 1.68, 2.36, 3.58 and 3.77-fold increases respectively in the skin flux of 5-fluorouracil, compared with an aqueous solution of drug. Increasing the surfactant: oil weight ratio from 5:95 through 9:91 to 13:87, at fixed water:surfactant content of 15, gave 3.58-, 5.04- and 6.3-fold enhancements of drug flux. In their study used attenuated total reflectance-Fourier transform infrared spectroscopy to determine that the microemulsions exerted their enhancement by interacting and perturbing the architecture of the statun corneum. The extent of this perturbation was dependent upon the concentrations of water and AOT in the microemulsion. Preliminary toxicity studies suggested that the microemulsions were a suitable vehicle for transdermal delivery.

Delivery of amphiphilic drugs

Jurkovic *et al.* have investigated the formulation of the amphiphilic antioxidant ascorbyl palmitate in a microemulsion, with a view to using the formulation as a protectant against free radical formation due to UV irradiation. Both oil-in-water and water-in-oil microemulsions were prepared using a medium chain triglyceride as oil, and PEG-8 caprylic/capric glycerides (Labrasol) and polyglyceryl-6-dioleate, (Plurol oleique) as surfactant and cosurfactant. The ascorbyl palmitate was incorporated into the microemulsions at various concentrations between 0.5-5.0 wt%. The microemulsions were gelled using either xanthan gum (water-in-oil) or Aerosil 200 (water-in-oil). The effectiveness of the ascorbyl palmitate in the microemulsions depended on both the concentration and type of microemulsion. Regardless of the type of microemulsion, efficacy was significantly higher at the higher ascorbyl palmitate concentrations. Overall, the oil-in-water microemulsions were more effective at protecting against UV irradiation, although they delivered ascorbyl palmitate to the skin at a slower rate than the water-in-oil microemulsions. The effect of formulation composition on the *in vitro* release rate of the amphiphilic drug, diclofenac diethylamine, from a range of microemulsion vehicles containing PEG-8 caprylic/capric glycerides (surfactant), polyglyceryl-6 dioleate (cosurfactant), isopropyl myristate and water was determined by Djordjevic. The phase behavior of the microemulsions was determined in the absence of drug.

In the microemulsions selected for further study, the level of water present ranged from 10 to 60 wt% while the amount of oil varied from 8 to 46.6 wt%. The physico-chemical characterization studies indicated the microstructure to be either bicontinuous or non-spherical, and despite its amphiphilic nature, the drug was partitioned mainly in the water phase. The non-linearity of the drug release profile from the bicontinuous microemulsions was thought to be due to a complex distribution of drug within the microemulsion. The flux of the drug increased by >4 times, from a water-in-oil to an oil-in-water microemulsion, the release of drug from the bicontinuous microemulsion, suggesting that the microstructure hampers the release of the drug.

Delivery of water insoluble drugs

Dalmora and Oliveria and Dalmora *et al.,* investigated the release of piroxicam encapsulated in β-CD in cationic oil-in-water microemulsions, in an attempt to optimize the drug's delivery. The results demonstrated the potential of the reservoir *in vivo* system following the use of a microemulsion. The high degree of retention of the active substance can provide a means for modulating the anti-inflammatory effect, by greatly extending the release period relative to those formulations where the piroxicam is only dissolved or dispersed in a homogeneous aqueous medium. In conclusion, both microemulsions and β-CD-containing microemulsions can offer many promising features for their use as topical vehicles for piroxicam delivery. Some of the microemulsions gelled using carbopol 940. Paolino *et al.* examined the potential of oil-in-water microemulsions as topical drug vehicles for the percutaneous delivery of ketoprofen. Microemulsions were prepared using triglycerides as oil, and were stabilized by a mixture of lecithin and *n*-butanol as a surfactant/co-surfactant system. The percutaneous enhancer, oleic acid, was added to some of the microemulsions.

Physicochemical characterization of the microemulsions yielded a mean droplet size of 35 nm and a negative zeta potential of –19.7 mV in the absence of oleic acid and –39.5 mV in its presence. The ketoprofen-loaded microemulsions showed an enhanced permeation through excised human skin with respect to conventional formulations, although no significant percutaneous enhancer effect was observed in the presence of oleic acid. Microemulsions showed a good human skin tolerability on volunteers. Shukla *et al.* have investigated the potential of oil-in-water (o/w) microemulsions as vehicles for the dermal delivery of a eutectic mixture of lidocaine (lignocaine) and prilocaine, which acted as the oil phase. The microemulsion was stabilized by a blend of a 2:3 ratio Tween 80 and Poloxamer 331, a mixture of water and propylene glycol were used as the hydrophilic phase. These microemulsions were able to solubilize up to 20 wt% of the eutectic mixture. In an attempt to enhance the transdermal delivery of the poorly water soluble drug, triptolide, and to reduce the toxicity problems associated with its usage, a water-in-oil microemulsion was compared with that of solid lipid nanoparticles.

The microemulsion which was formulated using 40 wt% isopropyl myristate, 50 wt% Tween-80:1,2-propylene glycol (5:1, v/v) and water and contained 0.025 wt% of triptolide, gave a steady-state flux (for over 12 hours) and a permeability coefficient of triptolide of 6.4 ± 0.7 mg/cm^2 per h and 0.0256 ± 0.002 cm/h; a value which was approximately double that of the solid liquid nanoparticles and 7 times higher than that of triptolide solution of the same concentration. In another study, Chen *et al.* also studied the incorporation of the drug, into a similar microemulsion using oleic acid as oil.

Oleic acid was added because it is a known penetration enhancer, although there was no evidence of it acting as such in the present formulation. The addition, however, of 1 wt% menthol to the formulation slightly increased penetration from 1.58 ± 0.04 to 2.08 ± 0.06 μg/ cm^2 per h ($p < 0.05$). Encouragingly, no obvious skin irritation was observed for the formulation studied, suggesting that microemulsions are promising vehicles for the transdermal delivery of triptolide. Ross *et al.* examined the transdermal penetration, across full thickness hairless mouse skin, of the insect repellant, *N,N*-diethyl-m-toluamide (DEET), contained in either a 1:1 v /v ethanol:water solution (containing 20 wt% DEET) or one of two commercially available microemulsion formulations (3M Ultrathon Insect Repellant (containing 31.6 wt% DEET; 3M, St. Paul, MN), and Sawyer Controlled Release DEET Formula (19.0%; Sawyer Products, Safety Harbor, FL). Both formulations were of interest because they were marketed as retarding the absorption of DEET due to being microemulsions.

All of the DEET preparations exhibited considerable penetration, e.g., the ethanolic DEET formulation had a time to detection of approximately 30 min with steady stale at 85 min. The penetration obtained with the Sawyer was no different from that obtained from the ethanolic solution. The other microemulsion formulation (3M) demonstrated a different profile; despite being a higher concentration of DEET (30 wt% versus 20 wt%) and a comparable time to detection (40 min), the time to reach steady state was delayed, although there was still substantial absorption at steady state. Sintov and Shapiro prepared a high surfactant lidocaine microemulsion, containing as surfactant a mixture of glyceryl oleate and either PEG-40 stearate or PEG-40 hydrogenated castor oil, isopropyl myristate as oil, tetraglycol as cosurfactant, water, and up to 10 wt% of drug, although 2.5 wt% was generally used. The microstructure of the microemulsion went from oil-in-water, through bicontinuous to water-in-oil.

The penetration of the drug from the various formulations showed that the surfactant mixture containing PEG-40 stearate was best, while the water and surfactant/co-surfactant concentration was also important. Significantly, the lag time for penetration was reduced, suggesting that these microemulsions loaded with drug would provide rapid local analgesia. Priano *et al.* investigated the delivery from a water-in-oil microemulsion, of apomorphine present as ion-pair complex with octanoate to increase its lipophilicity and to diminish its dissociation. As the drug was present at a high concentration, the dispersed phase acted as a reservoir, making it possible to maintain an almost constant concentration in the continuous phase and therefore achieving pseudo-zero-order release kinetics.

The composition of the microemulsion was complex, containing 18.2 wt% water, 42.1 wt% of oily phase of isopropyl miristate-decanol 1:1.5 v/v, 3.9 wt% R-apomorphine hydrochloride, 7.3 wt% Epikuron 200, 7.1 wt% benzyl alcohol, 4.6 wt% octanoic acid 3.5 wt% sodium octanoate, 5.7 wt% sodium taurocholate, 7.6 wt% 1,2-propanediol. The microemulsion was thickened by the addition of 5.9 wt% Aerosil 2000. The microemulsion was able to provide *in vitro*, through hairless mouse skin, a flux of 88 g/h per cm^2 for 24 hrs, with a kinetic release of pseudo-zero-order, and was chosen for *in vivo* study; all the components were biocompatible and safe. The flux gave a first approximation of the feasibility of the transdermal administration in man. The pain and discomfort caused by the injection of local anesthetics has stimulated research into developing topical anesthetics. However, the currently available formulations, such as Ametop @gel, (4 wt% amethocaine base preparation) have a number of disadvantages, in particular a long delay of typically 40-60 min between application and anesthetic effect and the requirement for a plastic occlusive dressing. Arevalo *et al.* have recently developed a decane-in-water microemulsion stabilized by lauromacrogol 300 and containing 4 wt% of amethocaine in an attempt to achieve faster drug permeation, thus reducing the time to reach optimum anesthetic effect. The amethocaine microemulsion proved to be a promising fast-acting analgesic in experimental preclinical studies.

Transport of hydrophilic and hydrophobic drugs

Although microemulsions have long been suggested as suitable formulations for the co-adminstration of drugs of very varying physico-chemical properties, it is only very recently that anyone has reported doing so. Lee *et al.* have developed a novel micro-emulsion enhancer formulation for the co-administration of hydrophilic (lidocaine HCl, diltiazem HCl) and lipophilic (lidocaine free base, estradiol) drugs. The microemulsions composed of isopropyl myristate and water, and were stabilized by the nonionic surfactant, Tween 80. Transdermal enhancers such as *n*-methyl pyrrolidone (NMP) and oleyl alcohol were incorporated into all systems without apparent disruption of the system. Unfortunately, the authors did not give the precise, composition of the microemulsions tested; it was only mentioned that they contained a 1:1 v:v mixture of water and ethanol, isopropylmyristate as oil and Tween 80 as surfactant, and were either oil-in-water or water-in-oil. Interestingly, regardless of the physico-chemical nature of the drug, the oil-in-water microemulsions provided significantly better flux for all drugs studied ($p < 0.025$). Enhancement of drug permeability from the oil-in-water systems was 17-fold for lidocaine base, 30-fold for lidocaine HCl, 58-fold for estradiol, and 520-fold for diltiazem HCl. Significantly, the simultaneous delivery of estradiol with diltiazem hydrochloride did not affect the transport of either drug ($p > 0.5$).

Topical immunization

Traditionally, vaccines have been administered by injection using needles, although the concept of topical immunization through *intact* skin has attracted much attention. Cui *et al.* recently hypothesized that a fluorocarbon-based microemulsion system could be one possible way to deliver plasmid DNA across the skin. Cui *et al.* screened a range of fluorosurfactants for their ability to form ethanol-in-perfluorooctyl bromide microemulsions. Note that the authors provided no evidence of a microemulsion being formed. The stability of plasmid DNA in the formulations was also examined. From the surfactant screen, the commercially available Zonyl® FSN-100, an ethoxylated nonionic fluorosurfactant, was selected for further study. Significant enhancements in luciferase expression and antibody and T-helper type-1 based immune responses, relative to those of "naked" pDNA in saline or ethanol, were observed after topical application of plasmid DNA in ethanol-in-perfluorooctyl bromide microemulsion system.

From these studies, it can be concluded that fluorocarbon-based microemulsions are suitable for DNA vaccine delivery, although the mechanism(s) of the immune response induction is not known. It is possible that the transport of the molecules across the skin is via the hair-follicles, because DNA is too large and highly charged to cross intact stratum corneum.

Topical Administration of Ophthalmic Drugs

Conventional ophthalmic dosage forms tend to be either simple solutions of water- soluble drugs or suspension or ointment formulations of water-insoluble drugs. Unfortunately, as these delivery vehicles generally result in poor levels of drug absorption across the cornea, most of the applied drug does not reach its intended site of action. However, because of the relative safety and convenience of topical application in ophthalmology, as well as the relatively low risk (compared with other routes of administration) of systemic side-effects, topical administration of ophthalmic agents is the preferred route of delivery. Microemulsions and sub-microemulsions should offer a possible solution to the problem of poor delivery to the cornea, by sustaining the release of the drug, as well as by providing a higher penetration of drug into the deeper layers of the eye. In addition, they offer the potential of increasing the solubility of the drug in the ophthalmic delivery vehicle. Gallarate *et al.* were probably the first to examine the potential of microemulsions as vehicles for ophthalmic delivery. Since then, a number of groups have successfully demonstrated the ability of microemulsions (sub-microemulsions) to prolong the ocular delivery of drug. In their study, Gallarte *et al.* were able to further prolong the release of timolol by forming an ion pair with octanoic acid.

Garty and Lusky demonstrated that the delivery of pilocarpine from an oil-in-water microemulsion was delayed to such an extent that the instillations of the microemulsion formulation twice daily were equivalent to four times daily the applications of conventional eye drops. A similar result was reported by Muchtar *et al.* who determined *in vitro* that the corneal penetration of indomethacin formulated in a sub-microemulsion was more than three times that obtained using commercially available drops. A number of researcher have investigated the potential of positively charged microemulsions to retain the delivery vehicle in the eye, thereby sustaining delivery. To date, a range of drugs have been formulated in a microemulsion in an attempt to sustain release including adaprolol maleate, timolol, levobunolol, chloramphenicol tepoxalin, piroxicam, delta-8-tetrahydrocannabinol, pilocarpine, indomethacin, antibodies and dietary iso-flavonoids and flavonoids. In general, these studies showed that it was possible to delay the effect of drug incorporated in a microemulsion, thereby improving bioavailability. The proposed mechanism of the delayed action is that microemulsion droplets are not eliminated by the lachrymal drainage, thereby acting as drug reservoirs.

The first studies conducted on man with microemulsions containing adaprolol maleate and pilocarpine, confirmed the results of the earlier studies performed mainly using rabbits. Vandamme has recently reviewed the use of microemulsions as ocular delivery system, and thus only studies since then will be considered in the present review. Fialho and da Silva-Cunha recently investigated the long term application of a microemulsion system in rabbits intended for the topical ocular administration of dexamethasone. The formulation contained 5 wt% isopropyl myristate as oil, 15 wt% Cremophor EL as surfactant, and a polar phase of water and 15 wt% propylene glycol, with dexamethasone present at a concentration of 0.1 wt%. Significantly, ocular irritation tests in rabbits suggested that the microemulsion did not provide significant alteration to eyelids, conjunctiva, cornea and iris over a Fe3-month period. In addition, the formulation exhibited greater penetration of dexamethasone in the anterior segment of the eye and longer release of the drug when compared with a conventional preparation.

The area under the curve obtained for the microemulsion system was more than two-fold that of the conventional preparation ($p < 0.05$). Gulsen and Chauhan have recently developed a disposable soft contact lens of a drug-containing microemulsion dispersed in a poly 2-hydroxyethyl methacrylate (HEMA) hydrogel, suitable for ophthalmic delivery, in an attempt to reduce drug loss and side-effects. Upon insertion into the eye, the lens will slowly release the drug into the pre lens (the film between the air and the lens) and the post-lens (the film between the cornea and the lens) tear films, thus providing a sustained delivery of drug. Assuming the size and drug loading of the microemulsions is low, the lenses should be transparent. It was found using these microemulsion-containing lenses, with and without a stabilizing silica shell, that drug could be released for a period of >8 days. By altering droplet size and loading, it is possible to tailor release.

Intravaginal Delivery

In their 2001 review, D'Cruz and Uckun proposed that microemulsion gel formulations had great potential as intravaginal/rectal drug delivery vehicles for lipophilic drugs, such as microbicides, steroids, and hormones, because of their high drug solubilization capacity, increased absorption, and improved clinical potency, as long as a non toxic formulation could be prepared. In their review, D'Cruz and Uckun reported the formulation of two microemulsion-based gels using commonally available pharmaceutical excipients. Repeated intravaginal applications of formulations to rabbits and mice were found to be safe and did not cause local, systemic, or reproductive toxicity. D'Cruz and Uckun investigated the potential of the microemulsion-based gels as delivery vehicles of two lipophilic drugs, WHI-05 and WHI-07, which exhibit potent anti-HIV and contraceptive activity.

As AIDS is spread largely through sexual intercourse, the development of a dual action vaginal spermicidal microbicide to curb mucosal viral transmission, as well as to provide fertility control would have a tremendous impact world wide. D'Cruz and Uckun investigated the formulation of 2 wt% of the lipophilic drugs in a microemulsion-based gel, composed of Phospholipon 90G and Captex 300 as the oil phase, with Pluronic F68 and Cremophor EL as surfactants, and seaspan carragennan and Xantral as gelling agents. The microemulsions were gelled to obtain the necessary viscosity for the gel-microemulsion formulation. Under the conditions of their intended use, intravaginal application of the gel-microemulsions containing 2 wt% of drug in a rabbit model resulted in marked contraceptive activity, as well as exhibiting a lack of toxicity. Therefore, as a result of its dual anti-HIV and spermicidal activities, the drug-containing gels shows unique clinical potential as a vaginal prophylactic contraceptive for women who are at a high risk of acquiring HIV by heterosexual transmission.

Delivery via Nasal Route

Nasal route has been demonstrated as being a possible alternative to the intravenous route for the systemic delivery of drugs. In addition to rapid absorption and avoidance of hepatic first-pass metabolism, the nasal route allows the preferential delivery of drug to the brain via the olfactory region, and is therefore a promising approach for the rapid-onset delivery of CMS medications. The solution-like feature of microemulsions could provide advantages over emulsions in terms of the sprayability, dose uniformity and formulation physical stability.

Li *et al.* developed a diazepam-containing ethyl laurate-in-water microemulsion, stabilized by Tween 80 and containing propylene glycol and ethanol as cosolvents for the rapid-onset intranasal delivery of diazepam. A single isotropic region, which was considered to be a bicontinuous microemulsion, was seen at high surfactant concentrations but at various Tween 80:propylene glycol: ethanol ratios. Increasing Tween 80 concentration increased the microemulsion area, microemulsion viscosity, and the amount of water and oil solubilized. In contrast, increasing ethanol concentration produced the opposite effect.

A microemulsion consisting of 15 wt% ethyl laurate, 15 wt% water and 70 wt% Tween 80:propylene glycol:ethanol at a 1:1:1 weight ratio contained 41 mg/mL of the poorly-water soluble diazepam. The nasal absorption of diazepam from the formulation was fairly rapid with a maximum drug plasma concentration being obtained within 2 to 3 min, while bioavailability at 2 hrs post-administration was ~50% of that obtained with intravenous injection. Zhang *et al.* attempted to prepare an oil-in-water microemulsion, containing a high concentration of nimodipine, suitable for brain uptake via the intranasal route of delivery. Three microemulsion systems stabilized by either Cremophor RH40 or Labrasol, and containing a variety of oils, namely isopropyl myristate, Labrafil M 1944CS and Maisine 35-1, were developed and characterized. The nasal absorption of the drug from the three microemulsions was studied in rats. The formulation composed of 8 wt% Labrafil M 1944CS, 30 wt% Cremophor RH 40/ethanol (3:1 weight ratio) and water solubilized up to 6.4 mg/mL of drug and exhibited no ciliotoxicity.

After intranasal administration, the peak plasma concentration was obtained of 1 hr, while the absolute bioavailability was ~32%. Significantly, uptake of the drug in the olfactory bulb after nasal administration was three times that which was obtained from intravenous injection. In addition, the ratios of the AUC in brain tissues and cerebrospinal fluid to that in plasma obtained after nasal administration were significantly higher than those seen after administration. In conclusion, the microemulsion system appears to be a promising approach for the intranasal delivery of nimodipine. Richter and Keipert investigated the *in vitro* permeability of the highly lipophilic material, androstenedione, across excised bovine nasal mucosa, porcine cornea and an artificial cellulose membrane.

In order to control release, the two microemulsion formulations studied contained either hydroxypropyl-β-cyclodextrin or propylene glycol. Both microemulsions were prepared from 5 wt% isopropyl myristate, 20 wt% Cremophor EL and water. The permeation of the drug through the three tissues was influenced by the microemulsion. For example, the apparent permeability coefficients (P_{app}) of the drug from the microemulsions across nasal mucosa did not differ from the P_{app} of the drug contained in solution. In the case of the other two membranes, release from both of the microemulsion formulations exhibited extended time lags, so no P_{app} could be calculated. It seems that the composition of the microemulsion had a greater impact on the P_{app} of cornea than on the P_{app} of the other tissues. The structure of the different membranes is probably responsible for the observed differences in permeation.

Pulmonary Delivery

Emulsions and (to a far lesser extent) microemulsions have been investigated as vehicles for pulmonary delivery. By far, the most widely studied systems are those containing fluorocarbon oil and are stabilized by a (predominately) fluorinated surfactant. Fluorocarbon oils are of pharmaceutical interest because of their biological inertness and their high (and unique) ability to dissolve gas, which means they can support the exchange of the respiratory gases in the lungs. In addition, a fluorocarbon oil, namely perfluorooctylbromide, is in Phase II:III clinical trials in the United States, for the treatment of acute respiratory distress by liquid ventilation. It should be noted that en-large hydrocarbon surfactants are ineffective solubilizers in fluorocarbon-based systems. Instead, fluorocarbon surfactants are required. To date, fluorocarbon-based (micro)emulsions have been investigated for use as oil-in-water systems for *in vivo* oxygen delivery (blood substitutes), targeted systems for diagnosis and therapy, and water-in-fluorocarbon systems for pulmonary drug delivery.

Water-in-perfluorooctyl-bromide microemulsions have been shown to deliver homogeneous and reproducible doses of a tracer (caffeine) using metered-dose inhalers (pMDI) pressurized with hydrofluoroalkanes (HFAs). Lecithin-based reverse microemulsions have also been investigated as a means of pulmonary drug delivery. In these studies, dimethylethyleneglycol (DMEG) and hexane were used as models for the propellants, dimethyl ether (DME) and propane respectively. A combination of equilibrium analysis and component diffusion rate determination (by pulsed-field gradient [PFG]-NMR) and iodine solubilization experiments were used to confirm the formation of a microemulsion. Water soluble solutes, including selected peptides and fluorescently labeled poly-α,β-[N-(2-hydroxyethyl) D,L-aspartamide] were dissolved in the microemulsions in a lecithin- and water-dependent manner. Experiments with DME /lecithin demonstrated microemulsion characteristics similar to those in the model propellant and produced a droplet size and a fine particle fraction suitable for pulmonary drug delivery.

Patel *et al.* have prepared water-in-hydrofluorocarbon (specifically 134a) micro-emulsions using a combination of fluorinated polyoxyethylene ether surfactants and a short chain hydrocarbon alcohol such as ethanol. In the absence of a hydrocarbon alcohol, only cosolvent systems, but not microemulsions, were formed. Due to the high molecular weight of the fluorocarbon surfactant, large concentrations of fluorocarbon surfactant are required to solubilize relatively small amounts of water compared with comparable hydrocarbon-based surfactants. This has obvious implications for the pharmaceutical application of such systems. To date, very little on the potential of oil-in-water microemulsions for pulmonary drug delivery has been investigated, yet they are attractive vehicles because of their ability to solubilize high amounts of drug.

Antibiofilm activity

Al-Adham *et al.* demonstrated that microemulsion formulations have a significant antimicrobial action against planktonic populations of both *Pseudomonas aeruginosa* and *Staphylococcus aureus* (i.e. greater than a 6 log cycle loss in viability over a period as short as 60 s). Transmission electron microscopy studies indicated that this activity may in part be due to significant losses in outer membrane structural integrity. Nevertheless, these results have implications for the potential use of microemulsions as antimicrobial agents against this normally intransigent microorganism. More recently, the same group have determined the antibiofilm activity of an oil-in-water microemulsion, prepared from 15 wt% Tween 80, 6 wt% pentanol and 3 wt% ethyl oleate, by incubating the microemulsion with an established biofilm culture of *Ps. aeruginosa* PA01 for a period of 4 hrs. The planktonic MIC of sodium pyrithione and the planktonic and biofilm MICs of cetrimide were used as positive controls and a biofilm was exposed to a volume of normal sterile saline as a treatment (negative) control. The results showed that exposure to the microemulsion resulted in a three log-cycle reduction in biofilm viability, as compared to a one long-cycle reduction in viability observed with the positive control treatments, suggesting that microemulsions are highly effective antibiofilm agents.

3

Ionic Carriers

The "*classical*" organic anion (OA) or organic acid secretory system is thought to be involved fundamentally in the excretion of common drugs, toxins, and endogenous metabolites into the urine, thereby maintaining the equilibrium of anionic metabolites in the body. Organic anion transporters (OATs), the proteins that underlie the OA secretory system, typify secondary (and/or tertiary) active membrane transporters, which translocate substrates across the cell membrane by utilizing a transmembrane electrochemical gradient of the substrate itself or of another solute. To date, the OATs identified include OAT1 (originally, NKT), OAT2 (originally, NLT), OAT3 (originally, Roct), OAT4, OAT5, mouse OAT6, and URAT1 (originally, RST) as well as a number of orphan family members. OATs belong to the largest family of secondary active membrane transporters: the major facilitator superfamily (MFS), which is conserved from bacteria to mammals.

MFS transporters function as transmembrane uniporters, symporters, and antiporters, and transport a wide range of hydrophilic and amphiphilic substrates, including inorganic ions (Na^+, Cl^-, HCO_3^-, etc.), endogenous metabolites (amino acids, sugars, neuro-transmitters, etc.) and xenobiotics (drugs and toxins). However, OATs specialize in transporting organic anions, and most of the accumulated knowledge concerning OAT function has been found by studying prototypical substrates such as *p*-aminohippurate and estrone 3-sulfate. Within this superfamily, the OATs are members of the solute carrier family SLC22a, which includes the structurally and functionally related organic cation transporters (OCTs), organic carnitine transporters (OCTNs), unknown soluble transporters (USTs), and flylike putative transporters (Flipts). OATs are heavily expressed in excretory organs such as the kidney, suggesting a critical role in maintaining endogenous homeostasis.

Of note, the OATs are implicated in several clinical disorders and are important modulators of drug efficacy and toxicity, due to their ability to transport a large number of the most commonly prescribed pharmaceuticals, such as angiotensin-converting enzyme (ACE) inhibitors, loop and thiazide diuretics, β-lactam antibiotics, methotrexate, and nonsteroidal anti- inflammatory drugs. In recent years there has been considerable progress in understanding the molecular basis of the renal secretory process, owing to the continued identification of organic anion transporters that mediate basolateral and apical organic anion uptake. Furthermore, these transporters have also been found to be present in numerous epithelial tissues other than the kidney, including the olfactory mucosa, liver, choroid plexus, retina, and placenta, which potentially implicates OATs more broadly in secretory processes. Moreover, the sequencing of the human genome has uncovered the chromosomal organization of OAT genes, enabling a deeper picture of the OATs at the molecular, evolutionary, and regulatory levels.

GENOMICS OF OATs

Aside from their obvious structural and functional conformity, the OATs also manifest similar tissue expression patterns. OATs are expressed predominantly in the kidney; however, functionally similar genes such as OAT1 and OAT3 are also both found in choroid plexus, and hOAT5 and UST3 are both found in the liver. Therefore, there may be regulatory elements that function to link expression of these genes in certain tissues. A relatively unique feature of the SLC22a family genes is that many members tend to appear as pairs or clusters in the genome. Six of the eight known OATs are found in three tightly linked pairs (i.e., as adjoining neighbors without other genes interposed between them): OAT1-OAT3, OAT4-RST/URAT1, and UST3-hOAT5.

The dendrogram of the OAT family reveals that these physical pairs are also closely related phylogenetic pairs. Both *OAT1* and *OAT3* are found on human chromosome 11q12.3 in a tandem repeat, separated by a mere 8 kilobase pairs; *OAT4* and *URAT1* are found on 11q13.1 in a tandem repeat, separated by 20 kb. These observations suggest that the pairing of OAT genes (which are expressed primarily in the kidney) might exist to facilitate the coordinated transcription (coregulation) of pair

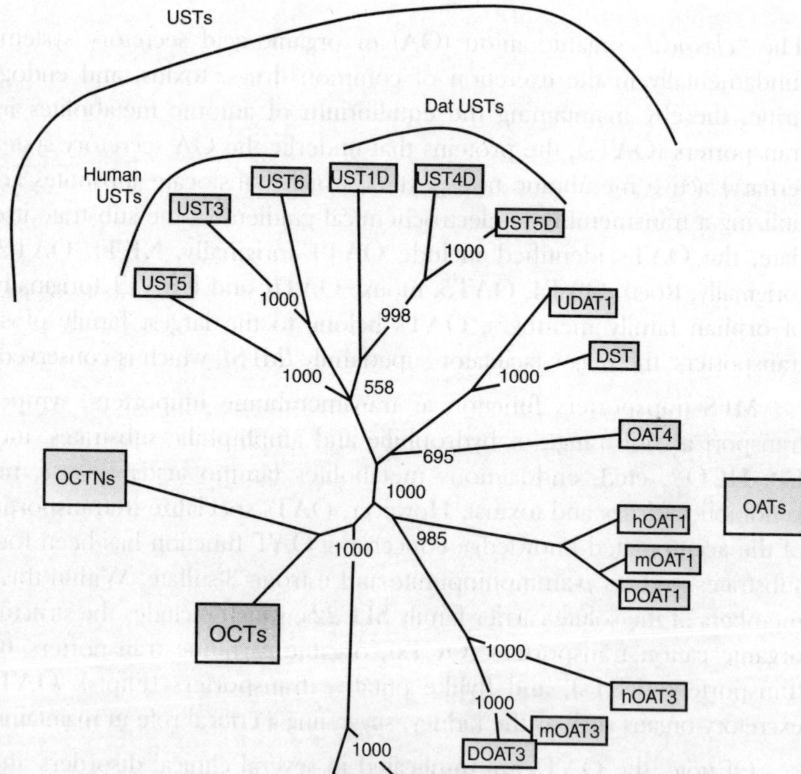

Fig. 3.1. Schematic illustrating the SLC22 family generated by aligning the sequences of rat, mouse, and human OATs, and representative OCTs and OCTNs, and using the alignment output to generate a dendrogram.

members. In the post-genomic era, tools for identifying potential regulatory elements and transcription binding sites have been utilized to explore potential control regions and determine consensus sequence binding sites for transcription factors integral to biological processes during kidney development (e.g., PAX1, PBX, WT1, TCF, and HNF1). For example, some putative transcription factor binding sites present in OAT noncoding regions may play a role in proximal tubule differentiation. Furthermore, experiments have shown that liver-enriched hepatocyte nuclear factor HNF-4α (nuclear receptor 2A1) can bind to and transactivate the *SLC22A7/OAT2* promoter. Although we are only at the early stages of understanding regulation of the expression of these transporters, earlier physiological studies have provided some insight into the regulation of OATs.

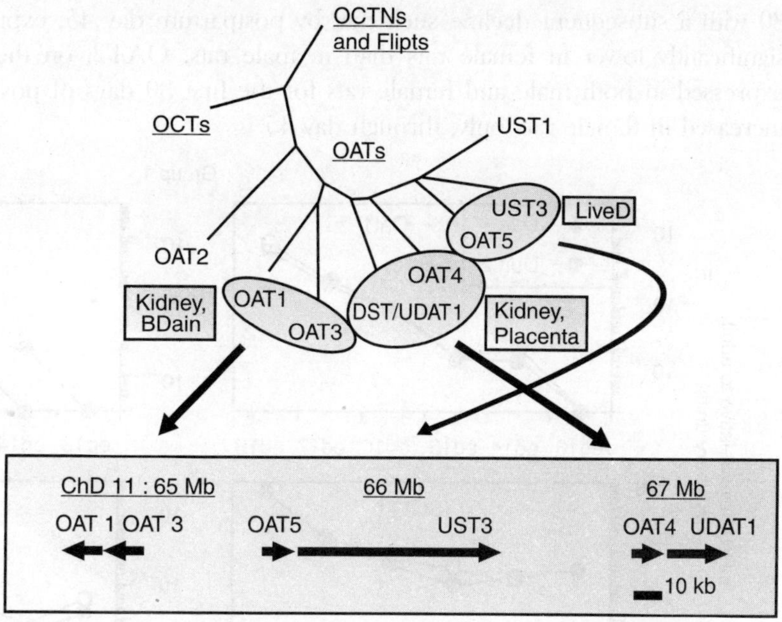

Fig. 3.2. Pairing, phylogeny, and tissue distribution of the OATs.

Postnatal maturation of OA transport has been found to be influenced by a number of factors, including substrate availability, age, and gender. For example, increased OA uptake into renal cortical slices can be induced by prior exposure to organic anions. Since the increase could be prevented by the protein synthesis inhibitor, cycloheximide, substrate exposure is believed to induce the synthesis of either the OATs themselves or of other proteins required for OA uptake. OA transport has also been shown to be under endocrine regulation. Transport of the prototypical OAT substrate, *p*-aminohippurate (PAH), can be stimulated by treatment with dexamethasone or thyroid hormones, particularly in young rats. Such regulation may be one potential mechanism by which organic anion transport is activated during a critical period for OAT synthesis: neonatal development.

IDENTIFICATION OF ANIONS

Identification of the specific proteins mediating organic ion secretion at the molecular level has allowed for analysis of the pre- and postnatal ontogeny of OATs, which may have clinical implications for the dosing of drugs to premature infants and full-term newborns. Analysis of the postnatal expression of these transporters showed that OAT1 expression in the postnatal kidney correlated with the changes in OA transport observed previously in physiological studies and that the expression levels for the OATs increased progressively postnatally in the rat. However, significant differences were observed between male and female rats. For example, while mRNA levels of OAT1 increased progressively in male rats, its expression peaked in female rats at postpartum day

30 with a subsequent decline such that by postpartum day 45, expression levels of OAT1 were significantly lower in female rats than in male rats. OAT2, on the other hand, was minimally expressed in both male and female rats for the first 30 days of postnatal development, but then increased in female rats only, through day 45.

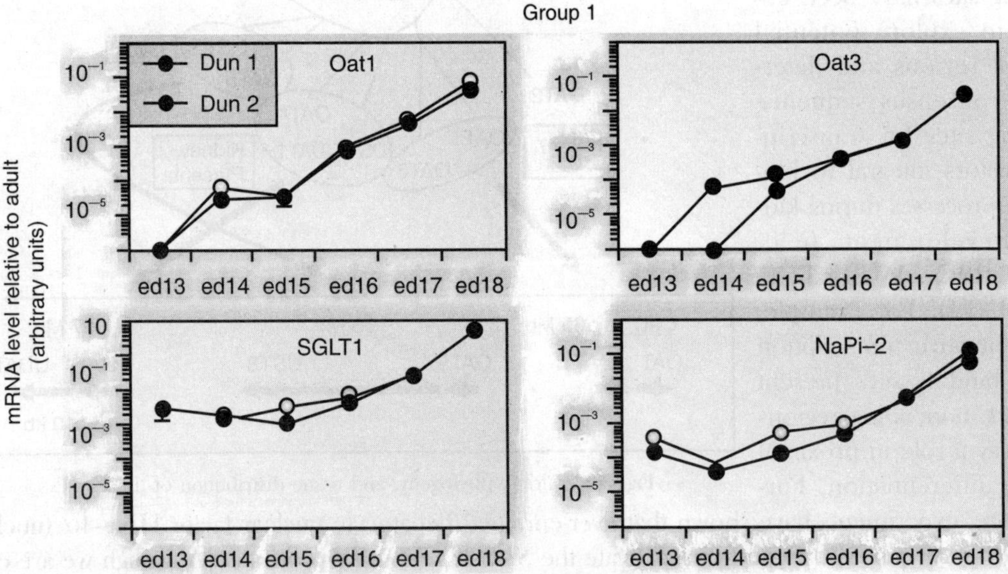

Fig. 3.3. Relative gene expression levels of OAT genes in embryonic kidney as determined by QPCR, illustrating the up-regulation of SLC family genes as rat kidney maturation progresses.

Finally, OAT3 was found to increase progressively over the observed course of postnatal development in both male and female rats. Taken together with the results of the physiological studies described above, these data (despite any gender differences in expression) support the notion that the postnatal maturation of renal OA transport observed may depend on increased expression of the transporters. Despite numerous physiological studies, few data exist on the embryonic ontogeny of these critically important transporters. When OAT1 was first identified by our group (as NKT), its expression was determined in the embryonic kidney. Subsequently, using a combination of Northern blot analysis and in situ hybridization, the spatiotemporal expression patterns of the organic anion transporters (OAT1, OAT2, and OAT3) during murine development were examined. Interestingly, the expression patterns for all of these genes in the kidney were remarkably similar.

The mRNA encoding these proteins was detectable in the kidney at days 14 to 16 of embryogenesis, and the expression levels continued to rise through embryogenesis, with the highest level detectable (by in situ hybridization and Northern blot analysis) in adult kidney. Similar patterns of expression were observed in a separate study of the ontogeny of OAT1 in the embryonic rat kidney. The roughly cotemporal expression of these transporters in the developing proximal tubule of the kidney suggests the existence of a common transcriptional regulatory pathway for the expression of OATs. Recently, detailed QPCR examination of the prenatal renal ontogeny of the OATs in the rat was reported, in which OAT1 and OAT3, as well as markers of proximal tubular differentiation, Na⁺/glucose cotransporter (SGLT1) and Na⁺/phosphate co-transporter (NaP2), were found to increase progressively over the embryonic period examined (ed13-ed18). Transporter expression and function was also examined in two commonly used models of nephrogenesis: (i) culture of the whole embryonic kidney and (ii) coculture of the uninduced metanephric mesenchyme (an embryonic progenitor tissue of kidney which gives rise to the nephrons of the kidney) with a heterologous inducing tissue (i.e., embryonic spinal cord). Both of these culture systems are invaluable tools for the study of kidney development since much of early nephrogenesis is reproduced in vitro with remarkable fidelity.

The expression patterns of OAT1, OAT3, SGLT1, and NaPi2 were found to closely resemble those observed in vivo. Moreover, using fluorescein uptake as an in vitro functional assay of OAT expression, inhibition of fluorescein accumulation by probenecid was observed within the tubular structures of the cultured organs. Together with the expression studies, these findings indicate not only that the OATs are expressed functionally during kidney development and might be markers of proximal tubule differentiation, but also raise the possibility that these organ culture systems may represent convenient and reliable in vitro models for study of the developmental induction of the OATs. Given the increasingly lower gestational ages at which infants are being delivered, these types of studies may eventually shed light on the ability of kidneys in premature babies to handle drugs and toxins.

SIGNIFICANCE OF ANIONS

Although OATs are thought to be similar to other MFS transporters, having 12 α-helical transmembrane domains (TMDs) as well as functional similarity, the crystal structure of OATs is not known, and hence their transport mechanism remains unclear. However, close examination of the primary structure reveals integral clues; weak sequence homology between the two six-helix halves suggests that MFS proteins originated from the duplication of a six-TMD ancestor. There are two large inter-connecting loops in OAT1, one between TMD 1/2, and another between TMD 6/7. The first large loop is extracellular and contains multiple consensus N-glycosylation sites. The role of these sites in the regulation of transport function has not been determined, although glycosylation of OAT1 was shown to be required for protein trafficking to the membrane. The second large loop is intracellular and contains several canonical protein kinase C (PKC) phosphorylation sites.

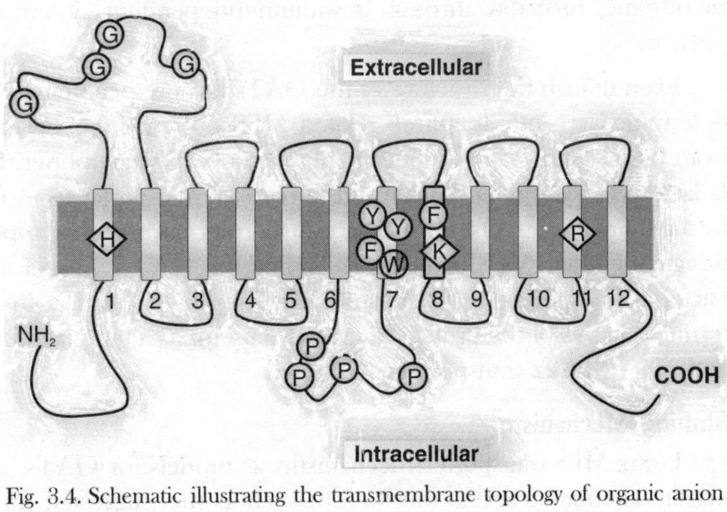

Fig. 3.4. Schematic illustrating the transmembrane topology of organic anion transporters.

Despite the fact that PKC down-regulates OAT function,

the phosphorylation sites of the second loop are probably not involved directly in this down-regulation. OATs are multispecific transporters that bind a vast array of substrates, with binding being based on general substrate physicochemical properties (charge, hydrophobicity, hydrogen-bonding ability) rather than distinctive molecular characteristics. However, there are no direct data on the molecular interactions of substrate ions with the amino acid residues lining the channel of the OATs which might indicate the exact properties that determine substrate binding and transport. Mutagenesis studies have been conducted targeting highly conserved amino acid residues believed to be essential for transport function.

These studies have identified conserved aromatic and cationic residues, which appear to interact, respectively, with hydrophobic and anionic moieties of substrates. For example, in rat OAT3, five conserved aromatic residues (located in TMDs 7 and 8) and three conserved basic residues (in TMDs 1, 8, and 11) are required for transport activity, as identified by expression studies in *Xenopus* oocytes. Importantly, in the OCTs (cation-transporting homologs of the OATs), acidic or neutral residues are located in the positions corresponding to the critical basic residues in the OATs. Therefore, these conserved basic residues are believed to determine the substrate charge specificity of the OATs. Remarkably, a rat OAT3 double mutant with the Lys370 and Arg454 residues substituted by one neutral and one acidic residue (K370A/R454D) has been reported to change its substrate orientation from anions to cations. Although the exact mechanism for substrate transport is not known, considerable data indicate that it is based on ion exchange across the membranes of the tubular cell.

Basolateral transport is driven by the concentration gradient of intracellular dicarboxylates, mostly α-ketoglutarate (αKG), which are exchanged for extracellular anions in an adenosine triphosphate (ATP)-independent mechanism. As a requisite, the high concentration gradient of dicarboxylates is maintained by the sodium-dicarboxylate cotransporter, which in turn is driven by the inwardly directed sodium gradient across the basolateral membrane generated by Na^+/K^+-ATPase. Basolateral entry of anions is followed by apical efflux through the brush border membrane, probably through a sodium-independent system whose exact mechanism remains unclear.

Even though the structure of the OATs has not been elucidated, the crystal structure has been determined recently for three related MFS proteins: the glycerol 3-phosphate (G3P) transporter from the *Escherichia coli* inner membrane (a G3P/Pi antiporter, GlpT), the *E. coli* lactose permease (a lactose/H^+ symporter, LacY), and the oxalate transporter from *Oxalobacter formigenes* (an oxalate-formate antiporter, OxlT). These transporters have similar topology, suggesting similar structural design for other MFS proteins, including the OATs. Therefore, the resolved structure of these bacterial transporters, with substrate-binding sites located at the interface between the N- and C-terminal halves of the protein, may prove useful as a template for structural modeling of the OATs and other SLC22 transporters.

Shifting Mechanism

Using MFS transporter mechanisms as models for OATs, one might postulate a single-binding-site alternating access mechanism such as that suggested for the MFS transporters and elaborated in more detail in thermodynamic and kinetic studies. In this mechanism the transporter switches between two alternating conformations, inward facing (C_i) and outward facing (C_o), thus allowing translocation of substrate across the membrane. Recently, a more detailed mechanism of the conformational change has been proposed based on the crystal structure of a bacterial antiporter, GlpT. According to this mechanism, substrate binding within the transporter binding site reduces the energy barrier between the inward- and outward-facing conformations (C_i and C_o) and thus facilitates a transition between them, involving a swiveling movement of the two six-TMD halves of the transporter relative to each other ("*rocker-switch*" mechanism).

Analysis of the density map of another bacterial transporter, OxlT, suggests the existence of a third, "closed" conformation of the transporter (C_c) with the substrate-binding site isolated from both cytoplasm and periplasm, as an intermediate state in the transition between the C_i and C_o conformations. Importantly, in anion-transporting MFS proteins, including OATs, the binding site supposedly comprises two positively charged amino acid residues. These two key residues are, for example, Arg45 (located in the α-helix H1) and Arg269 (H7) in GlpT, Arg46 (H1) and Arg275 (H7) in UhpT (sugar phosphate transporter of *E. coli*), Arg272 (H8) and Lys355 (H11) in OxlT, and Lys370 (H8) and Arg454 (H11) in rOAT3. Based on the structure of GlpT, the two key positively charged residues in the substrate-binding site were suggested to form hydrogen bonds simultaneously with a negatively charged substrate ion. The formation of such a "bridge" complex (e.g., Arg45... P_i... Arg269 in GlpT) upon binding a substrate ion to a transporter in the conformation open to one side of the membrane was suggested as a mechanism to pull the N- and C-terminal domains closer to each other and make further tilting easier.

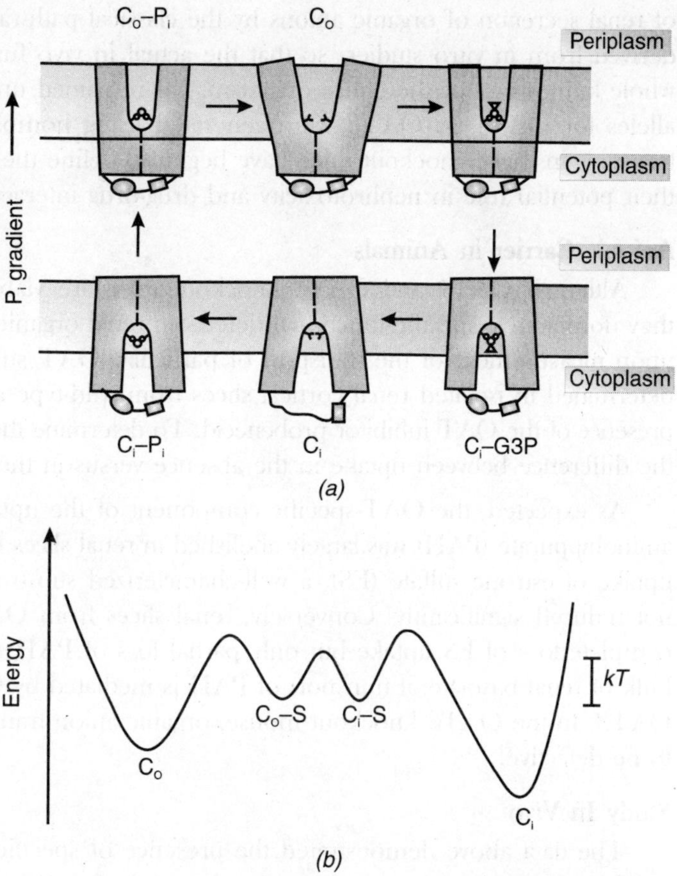

Fig. 3.5. Proposed single binding site, alternating access mechanism with a rocker-switch type of movement. (a) Reaction cycle of substrate translocation. (b) Schematic drawing to illustrate the free-energy levels of various conformations of G1pT in the translocation cycle.

EVIDENCES

As noted earlier, the evidence accumulated indicates that OAT1 and OAT3 manifest functional properties and anatomical localization consistent with a critical role in the basolateral uptake step of renal secretion of organic anions by the classical pathway. However, most of this evidence has derived from in vitro studies, so that the actual in vivo function of OATs, in the context of the whole kidney and/or the entire organism, has remained uncertain. Recently, mice containing null alleles for OAT1 or OAT3 were generated using homologous recombination. Studies of OA transport in these knockout mice have begun to define the role of OAT1 and OAT3 in vivo and their potential role in nephrotoxicity and drug-drug interactions.

Anionic Carrier in Animals

Although OAT1 and OAT3 knockout mice are viable, fertile, and appear grossly normal, they do manifest specific functional defects in renal organic anion handling which were detectable upon measurement of the transport of particular OAT substrates. Uptake of organic anions was determined in isolated renal cortical slices from wild-type and knockout mice in the absence and presence of the OAT inhibitor probenecid. To determine the OAT-specific component of transport, the difference between uptake in the absence versus in the presence of inhibitor was estimated.

As expected, the OAT-specific component of the uptake of the prototypic organic anion *p*-aminohippurate (PAH) was largely abolished in renal slices from OAT1 knockout animals, whereas uptake of estrone sulfate (ES), a well-characterized substrate of OAT3 (but not of OAT 1), was not reduced significantly. Conversely, renal slices from OAT3 knockout mice manifested nearly complete loss of ES uptake but only partial loss of PAH uptake. These findings suggest that the bulk of renal basolateral transport of PAH is mediated by OAT1, while that of ES is mediated by OAT3. In the OAT3 knockout mouse, organic anion transport in choroid plexus was also found to be defective.

Study In Vivo

The data above demonstrated the presence of specific deficits in the capacity for basolateral uptake of organic anions into the renal cortex of OAT knockout mice. To determine whether these transport deficits translated to diminished renal secretion, the urinary excretion of organic anions in OAT1 knockout and wild-type mice was compared by performing clearance experiments under anesthesia. Although there was no significant decrement in ES clearance in the OAT1 knockout mouse (as expected given the ex vivo results above), the renal clearance of PAH was severely reduced. Since there were no differences in glomerular filtration between the genotypes (as determined by measurement of inulin clearance), all of this reduction was attributable to loss of net secretion.

A long body of physiological work supports the notion that PAH is the prototype of the highly diverse set of organic anions that are handled in common by the classical pathway. Therefore, our finding that OAT1 is responsible for the bulk of renal secretion of PAH in vivo suggests an integral role in the functioning of this pathway, and thus in the excretion of the numerous clinically important compounds counted among its substrates. The naturetic response to the diuretic furosemide was also blunted in OAT1 knockout mice. Perhaps most interesting was the accumulation of a number of endogenous metabolites in the serum of these animals. Several of these compounds were found to inhibit PAH transport by OAT1 in*Xenopus* oocytes, suggesting that they may be "true" OAT substrates in vivo.

PHARMACEUTICAL APPLICATIONS

Because the OATs are at crucial cellular interfaces for various excretory organs (such as the liver and kidneys), variation or polymorphisms in these genes may account for clinically important differences in drug efficacy and renal drug handling of commonly prescribed pharmaceuticals [e.g., ACE inhibitors, methotrexate, and nonsteroidal anti- inflammatory drugs (NSAIDs)] as well as elimination of key metabolites. Therefore, polymorphisms in the OAT genes may account partially for differing disease prevalence and risk of adverse drug reactions (ADRs) among different human subpopulations. ADRs are of particular interest because there are over 2 million incidents and approximately 100,000 deaths related to ADRs in the United States each year. Differences in the expression or function of OATs may alter the elimination of many pharmaceutical agents significantly, thereby increasing the risk of a significant ADR. In particular, there has been considerable interest in single-nucleotide polymorphisms (SNPs) in this family of genes and their potential role in drug handling by the kidney, thereby affecting drug concentrations and half-lives in the serum.

These considerations also apply to transport across all the barrier epithelia in which OATs are expressed: renal clearance (OAT1, OAT2, OAT3, OAT4, and URAT1), transport across the blood-brain barrier via the choroid plexus (OAT 1 and OAT3), transport across the placenta (OAT4), and transport across the olfactory mucosa (mouse OAT6). It has been shown that certain coding region polymorphisms affect drug transport by OATs. For example, certain OAT1 and OAT3 SNPs are known to affect function in *Xenopus* oocyte expression systems. The effects of these coding region SNPs on in vivo drug handling remain unclear. However, coding region polymorphisms seem relatively uncommon, raising the possibility that noncoding region SNPs, which could affect OAT transcriptional activity, may be clinically important. Such polymorphisms have recently been identified.

The pairing of many OATs in the genome also raises the possibility that a polymorphism in the transcriptional region of one OAT could also affect the expression of the paired OAT. Moreover, given the redundancy of OATs as demonstrated by both transport studies and in knockout mice, it is probably most useful to consider multiple OAT SNPs together. Furthermore, since net transport of organic anions across an epithelial barrier (e.g., the renal proximal tubule) depends on both apical and basolateral transport mechanisms, it is probably important to consider the entire set of apical and basolateral transporters in order to obtain a more accurate perspective on how various polymorphisms affect such transport. Thus, the combinatorial effects of certain SNPs in different OATs on organic anion transport in vivo could be quite complex, but perhaps a more clinically relevant approximation. However, genomic variation (either at the level of expression or function) may have clinical implications (change in drug efficacy or toxicity) only with certain disease states (an environment–gene interaction). For example, expression of OAT3 is thought to be directly related to the clearance of cefazolin, a commonly prescribed antibiotic.

This relationship, however, was only observed in patients with mesangial proliferative glomerulonephritis, in whom decreased expression of OAT3 mRNA was strongly associated with decreased renal clearance of cefozolin. This implies that certain patients are at a higher risk of developing cefazolin-related drug toxicity (e.g., hepatitis), depending on individual differences in expression of the OAT3 gene. Therefore, genomic variation may be especially important in stressed or disease-specific environments. There is also a growing body of research that is delineating the importance of OATs in organ systems other than the kidney, and these are mentioned here only briefly.

Efflux of xenobiotics and organic anion substrates across the blood-brain barrier, for example, is thought to occur in a three-step process in the epithelial cells of the choroid plexus, similar to the transport of substrates across the renal proximal tubule cell: apical uptake, transport across the cell, and basolateral transport into the blood. Consistent with this model, OAT3 is located on the apical surface of epithelial cells of the choroid plexus, and knockout mice demonstrate decreased organic anion transport, suggesting an important role for OATs in protecting the central nervous system from toxic endogenous and exogenous substrates. Similarly, the placenta protects the fetus from exposure to toxic substrates and allows the delivery of pharmaceutical drugs (such as zidovudine prophylaxis against HIV) by similar mechanisms. OAT4, for example, has been localized to the placenta and is known to transport many drugs, such as zidovudine, antibiotics, and antihypertensives. The discovery of the olfactory OAT, OAT6, also raises the possibility that SNPs in this gene could be important in determining the efficacy of nasally administered drugs as well as in olfaction.

Additional Clinical Applications

OAT1 and OAT3, expressed primarily on the basolateral surface of renal proximal tubular cells, are responsible for the uptake from the blood into the cell, whereas OAT4 and RST/URAT1, located on the apical surface, are considered candidate genes involved in efflux from the cell into the proximal tubule and/or substrate reabsorption into the proximal tubular cell. One disease thought to be a result of altered organic anion clearance is gout. Gout can be associated with uric acid crystal deposition in the kidney (resulting in nephropathy) or within joints (resulting in an acutely painful inflammatory arthritis). Uric acid is a product of purine metabolism and undergoes net reabsorption in the renal proximal tubule. URAT1, the human homolog of the gene first identified as RST, is thought to transport uric acid across the apical proximal tubular cell. Accordingly, case-control and cohort studies have suggested that loss of function polymorphisms on URAT1 are associated with *hypouricemia*.

One study found decreased renal uric acid associated with a human polymorphism close to the N-terminus (C426T), and uncovered another human polymorphism, 313A (a deletion from 313D to 333P), that had no significant uric acid transport in *Xenopus* ooyctes. Nevertheless, the molecular basis of renal urate handling in vivo remains poorly understood. A detailed understanding of molecular mechanisms underlying uric acid clearance could result in novel pharmaceutical treatments for gout that target uric acid transporters. Competition for OAT-binding sites may explain certain types of toxic reactions. Most of the drugs in plasma bind competitively to OATs at the level of the substrate- binding site and influence the pharmacokinetics of other drugs. Since this transporter system has the capacity to recognize and bind a variety of endogenous substrates, OATs are likely to be involved in certain toxic or nephrotoxic drug reactions. For example, ochratoxin A, a mycotoxin that causes Balkan nephropathy, has been shown experimentally to be accumulated via OAT1 in the renal proximal tubular cells. Another example is the nephrotoxicity due to the antiviral drug cidofovir (used to treat ophthalmic cytomegalovirus infection and thought to be active against orthopox viruses such as smallpox), which is known to be mediated by active secretion of cidofovir in the renal tubules. This effect has shown to be counteracted by co-administration of probenecid, a potent inhibitor of organic anion secretion in the proximal renal tubular cells.

Methotrexate clearance has been reported to be reduced with NSAIDs acting as competitive substrates for the binding sites of renal organic anion transporters. The net effect of this competition is an increase in plasma concentration of methotrexate and the manifestation of severe toxicity, in the form of bone marrow suppression and damage to intestinal epithelium. Alternatively, the multispecificity of OATs can be used to prolong the action of drugs by competitive inhibition of secretion at the OAT binding site. Such inhibition decreases renal excretion and enhances retention of drugs, an example being the common practice of coadministering probenecid to prolong the action of β-lactam antibiotics. In summary, it may be appropriate to consider the classic secretory transporter system as a target for the treatment of various drug-related side effects and that modification at this cellular level could open up a new era of therapeutic strategies for clinically significant diseases.

Since the identification of NKT (later called OAT1), OATs have been shown to play key roles in mediating the renal absorption and excretion of drugs, xenobiotics, and endogenous metabolites. Thus, it has become readily apparent that targeting of this family of transporters, and an understanding of their transport mechanisms, is critical for elucidating mechanisms of drug handling and nephrotoxicity. Given that most of the OAT family members have overlapping substrate specificities and tissue distribution, and both primary sequence and structural homology, it would not be that surprising if they exhibited complex, coupled regulation, as if a clustered unit. Widespread genetic profiling of humans and a more comprehensive list of substrate affinities may ultimately lead to individualized patient care and usher in an era where renal drug, xenobiotic handling, and certain types of toxicity are seen in light of OAT transport capacity.

Further work will focus on the structural basis of substrate transport as well as the creation of a list of transported endogenous and exogenous molecules that could potentially uncover the true physiological functions of the OATs and other SLC22 family members. Finally, the complex regulatory mechanisms likely to underlie OAT family transcription and cell surface regulation must be understood in vivo, and to do this, animal models, including single and multiple OAT knockouts, must be created. These approaches should lead the field closer to a more integrated understanding of renal drug handling and toxicity.

CATIONIC CARRIERS

Renal elimination, including both passive glomerular filtration and active tubular secretion, is a major route of clearance for many drugs and drug metabolites. Active secretion of drugs by the kidney takes place primarily in the renal proximal tubule, where drug transporters facilitate the flux of drug molecules from the blood to the tubular lumen for excretion. As early as 1947, it was recognized that transport systems exist that aid in the elimination of a variety of structurally diverse organic cations, including drugs and toxic xenobiotics, as well as their metabolites. Prior to the molecular cloning of organic cation transporters, it had been shown that uptake of low-molecular-weight organic cations into the renal proximal tubule (the first step in active secretion of organic cations) is facilitated by a polyspecific membrane potential-sensitive transport system. Efflux of organic cations into the tubular lumen, the second step in active tubular secretion, was shown to occur via an H^+- or cation-exchange mechanism. Additionally, uptake of organic cations into other tissues, such as hepatic uptake, or transport across the blood-brain barrier, was known to display properties of protein-mediated transport across the plasma membrane (e.g., saturability, and sensitivity to temperature, membrane potential, and small-molecule inhibitors). The first member of the organic cation transporter (OCT) family [rat OCT1 (rOCT1)] was identified in rat by expression cloning from rat kidney. This was followed quickly by homology cloning of other OCT isoforms in rodents and human.

Of the membrane transporters cloned to date, three have been identified as members of the basolateral (membrane potential-driven) organic cation transporter system: OCT1 (*SLC22A1*), OCT2 (*SLC22A2*), and OCT3 (*SLC22A3*). All are members of the *SLC22A* family of solute carrier (SLC) transporters, which includes the novel organic cation transporters OCTN1 and OCTN2 (*SLC22A4* and *SLC22A5*) and the organic anion transporters OAT1-4 (*SLC22A6, SLC22A7, SLC22A8,* and *SLC22A11*), as well as several transporters that are less well characterized. The molecular identification of the organic cation transporters (OCTs) has led to more detailed studies of the transport mechanism, substrate specificity, and tissue distribution of these transporters, and to a greater understanding of their roles in organic cation disposition and elimination.

PHYSIOCHEMICAL CHARACTERISTICS

In terms of structural and functional characteristics, all three of the OCTs share several common features: (i) a similar transmembrane topology, (ii) a shared group of preferred substrates, and (iii) a common transport mechanism. The OCTs also share a common genomic structure of 11 coding exons. The genes encoding the three OCTs are located in close proximity on human chromosome 6q26, and presumably arose via gene duplication from a single ancestral OCT gene. The predicted secondary structure of the OCTs, based on sequence or hydropathy analysis, consists of 12 transmembrane domains (TMDs) with cytoplasmic amino and carboxy termini. The OCTs share a large extracellular loop between TMDs 1 and 2 which contains potential N-glycosylation sites, as well as a large intracellular loop between TMDs 6 and 7 which includes predicted phosphorylation sites. The potential importance of N-glycosylation and phosphorylation for OCT activity is discussed below.

The transmembrane domains are believed to be important for substrate recognition by the OCTs; specifically, recent evidence suggests that the fourth and tenth trans-membrane domains are critically involved in substrate recognition by the OCTs, and differences between isoforms in terms of substrate specificity may be related to differences in these critical regions. However, it is important to note that given the broad substrate selectivity of the OCTs, the key domains or residues involved in substrate recognition may be different from one substrate to another, even within the same protein. For example, in mutational analysis studies of rat OCT1, mutation of two residues in the fourth TMD (Trp218Tyr and Tyr222Leu) resulted in increased affinity for both

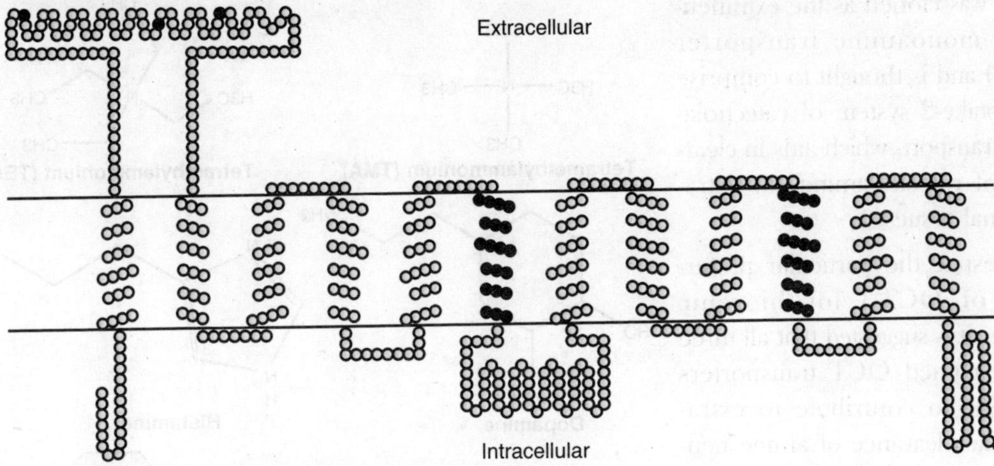

Fig. 3.6. Secondary structure predicted for OCT1.

tetraethylammonium (TEA) and 1-methyl-4-phenylpyridinium (MPP^+), whereas a third mutant (Thr226Ala) had increased affinity for MPP^+ but no change in affinity for TEA. This suggests that OCT1, and probably all of the OCTs, contain multiple overlapping but nonidentical recognition sites for the various structurally diverse substrates. Transport of organic cations by the OCTs occurs by facilitated diffusion and is driven by the inside-negative membrane potential.

Positively charged cations are taken up into cells according to the electrochemical gradient, and this process is membrane potential sensitive [i.e., artificially reducing the membrane potential (as through replacement of extracellular Na^+ with K^+, or treatment with ionophores such as valinomycin) reduces the rate of transport by OCTs]. Electrophysiological studies using *Xenopus laevis* oocytes microinjected with OCT1, OCT2, or OCT3 cRNA show directly that transport of organic cations by OCTs is electrogenic (i.e., addition of substrate to the extracellular media results in inward currents). OCT activity is insensitive to changes in the Na^+ gradient. Transport of weak bases by OCTs has also been shown to be sensitive to extracellular pH; however, this is explained by the decrease in fractional ionization of the substrate at high pH and does not indicate that transport of organic cations is driven by proton exchange. OCT transport can also occur by cation-cation exchange. Preloading OCT1-injected *X. laevis* oocytes with unlabeled TEA enhances uptake of [^{3}H]MPP^+, and similar results have been found for OCT2.

SELECTION OF SUBSTRATES

Common substrates of all OCTs include low-molecular-weight relatively hydrophilic organic cations such as the prototypical cation TEA, the neurotoxin MPP^+, and the endogenous compound *N*-methylnicotinamide (NMN). Several clinically important drugs have been shown to interact with all of the OCTs, including the antidiabetic drug metformin, and the peptic ulcer drug famotidine, demonstrating the broad potential for influence of OCTs on drug disposition and drug action. Included among OCT substrates are endogenous compounds such as the biogenic amine neurotransmitters. Dopamine, epinephrine, norepinephrine, histamine, and serotonin have all been shown to interact with one or more OCT isoforms. Of particular interest in this regard is OCT3, which was cloned as the extraneuronal monoamine transporter (EMT) and is thought to comprise the uptake-2 system of catecholamine transport, which aids in clearance of catecholamines in extraneuronal tissues.

Despite the particular preference of OCT3 for biogenic amines, it is suggested that all three of the cloned OCT transporters combine to contribute to extraneuronal clearance of amine neurotransmitters. Although the OCT family shows broad overlap in substrate specificity, there are examples of relatively isoform-specific substrates and inhibitors. Examples include corticosterone, which has approximately 40-fold greater affinity than Oct1 for rat Oct2, and conversely, mepiperphenidol and *O*-methylisoprenaline, which show a 70-fold greater affinity for rOct1 than for rOct2. OCT3 appears to have the most unique substrate selectivity among the OCT family, with a preference for endogenous monoamines such as dopamine and norepinephrine.

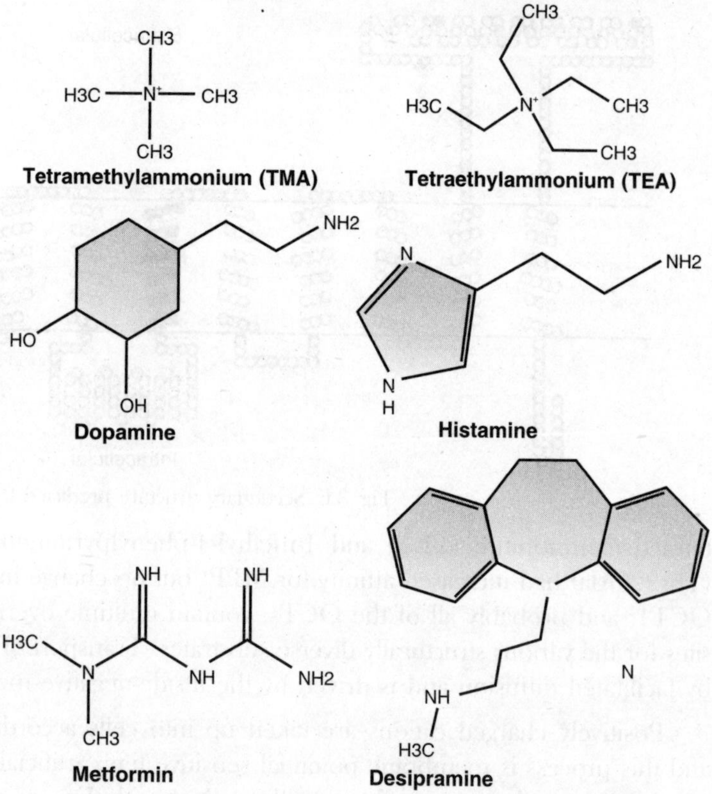

Fig. 3.7. Representative substrates of organic cation transporters.

Notably, although most known substrates of the OCTs are cations, some OCT substrates are anionic or neutral compounds at physiological pH. For example, the anions prostaglandin E_2 and prostaglandin $F_{2\alpha}$ have been shown to be substrates for both OCT1 and OCT2, and the neutral steroid β-estradiol interacts with OCT3 with high affinity. Thus, a net positive charge does not appear to be an absolute requirement for interaction with the OCTs. Importantly, OCT orthologs from different species may exhibit differences in substrate specificity. For example, OCT1 orthologs from human, rabbit, mouse, and rat all transport the low-molecular-weight tetraalkylammonium compounds tetramethyl-ammonium (TMA) and TEA. In contrast to human and rabbit, rat and mouse Oct1 do not transport the larger tetrapropylammonium (TPA) or tetrabutylammonium (TBA), suggesting that molecular mass or hydrophobicity may effect differences in recognition of OCT substrates across species.

CATIONIC DIVERSITY

Despite the similarities in structure and function, there are important differences in the tissue distribution of mRNA transcripts of the OCTs in human tissues. OCT1 is expressed predominantly in human liver, with lower levels of expression in other tissues. OCT1 is thus characterized as a *liver-specific* OCT. Analogously, OCT2, which is expressed almost exclusively in kidney, is described as the *kidney-specific* OCT. OCT3, also referred to as the *extraneuronal monoamine transporter* (EMT), has a broad tissue distribution, with moderate levels of mRNA found in the aorta, skeletal muscle, prostate, adrenal gland, salivary gland, liver, term placenta, and fetal lung. Both OCT1 and OCT2 have been localized to the basolateral membrane in renal proximal tubular epithelial cells in rat as well as in cell culture models of kidney (GFP-rOCT2-transfected MDCK cells).

In human proximal tubules, OCT2 expression is restricted to the basolateral membrane; however, OCT1 is expressed at very low levels in human kidney. Despite early controversy regarding OCT2 localization, the emerging consensus is that OCT2 is indeed a basolateral organic cation uptake transporter in kidney. The pharmacological significance of OCT2 thus appears primarily to be its role in active secretion of organic cations in the kidney, facilitating the first step (uptake of organic cations across the basolateral membrane) of active secretion. Efflux of organic cations across the apical membrane is thought to occur by a separate transporter-mediated process, either by the primary active efflux pump, MDR1 (which generally transports bulky hydrophobic cationic compounds) or by an organic cation-H^+ exchange mechanism for small, hydrophilic organic cations.

The molecular identity of the organic cation-H^+ exchanger is currently unclear; the novel organic cation transporters (OCTN1 and OCTN2) and the more recently discovered multidrug and toxin exclusion transporters (MATE1 and MATE2) are likely candidates, as all of these are highly expressed in kidney and have been shown to transport the typical cation TEA in a manner consistent with H^+ exchange. It is possible that all of these transporters participate in apical organic cation efflux, with differential importance depending on the substrate under question. Among the apical organic cation transporters, only OCTN2 has been shown to influence renal organic cation (TEA) secretion in vivo; however, appropriate animal models to study the in vivo role of OCTN1, MATE1, and MATE2 in renal organic cation secretion do not yet exist. In liver, the primary site of OCT1 expression, OCT1 was found to be expressed at the sinusoidal (basolateral) membrane by immunohistochemistry and Western blot. This suggests that OCT1 acts as a sinusoidal uptake transporter in liver, aiding in the transfer of organic cations from blood into the hepatocyte for elimination by metabolism or biliary excretion.

In addition, for drugs that target the liver for their pharmacological activity, OCT1 may be a limiting step in drug access to hepatocytes and may therefore influence drug action. It is important to note that species differences exist in the tissue-specific expression of OCTs. Most notably, whereas in human kidney OCT2 is by far the most abundant OCT in terms of mRNA expression, rodent kidney expresses nearly similar levels of OCT1 and OCT2. As with substrate-specificity differences, differences between species in terms of tissue distribution of OCTs are important to consider when interpreting the results of studies performed in animal models.

CONTROL OF CATIONIC ACTIVITY

As mentioned previously, the large intracellular loop between transmembrane domains 6 and 7 contains several predicted phosphorylation sites that are conserved among the several OCT isoforms. These include target sequences for protein kinase A (PKA), protein kinase C (PKC), protein kinase G (PKG), and tyrosine kinase. There exists ample evidence for the modulation of activity of the OCTs by activation of these kinases, presumably through phosphorylation of the OCTs at the protein level.

Several N-glycosylation sites in the large extracellular loop between transmembrane domains 1 and 2 are also potential sites of protein-level regulation. Additionally, hormonal regulation, which is thought to reflect effects on transcription, has been described for OCT2. Regulation of OCT expression is also modulated during development and in certain pathophysiological states. Current knowledge of the regulation of the OCTs will be discussed individually for each isoform.

OCT1

Early studies of rat OCT1 showed that stimulation of either PKA or PKC results in a significant increase in organic cation transport, as measured by uptake of the fluorescent compound 4-[4-(dimethylamino)-styryl]-*N*-methylpyridinium (ASP^+) in HEK-293 cells stably transfected with rOCT1 PKC stimulation also results in increased affinity for TEA, TPA, and quinine. The effect of PKC stimulation on rOCT1 activity is thought to occur by phosphorylation of the rOCT1 protein, resulting in a conformational change that enhances substrate binding. rOCT1 activity is significantly reduced by treatment with aminogenistein, a specific inhibitor of $p56^{lck}$ tyrosine kinase, suggesting that this kinase is an endogenous activator of OCT1; however, inhibitors of other tyrosine kinases do not affect OCT1 function. In contrast to rOCT1, stimulation of PKA results in decreased activity of human OCT1 in two different expression systems.

In the same studies, PKC activation has no appreciable effect on activity of human OCT1. However, as with rOCT1, human OCT1 appears to be positively regulated by the $p56^{lck}$ tyrosine kinase, as evidenced by reduced hOCT1 activity after treatment with aminogenistein. For both rOCT1 and hOCT1, PKG signaling does not appear to influence OCT1 activity. Human OCT1 has further been shown to be regulated by the Ca^{2+}-calmodulin complex. Specifically, inhibition of Ca^{2+}- and calmodulin-dependent protein kinase II (CaMKII) results in reduced ASP^+ transport by hOCT1, suggesting that this kinase consitutively stimulates hOCT1 activity. In both PKA stimulation and CaMKII inhibition, the reduction in transport activity by hOCT1 is the result of decreased affinity for the substrates tested. These results suggest, again, that the observed effects of protein kinases are due to phosphorylation of the OCT1 protein, causing a conformational change that results in either decreased or increased substrate-binding affinity.

OCT2

Studies on regulation of basolateral organic cation transport have been performed in isolated human proximal tubules, in which OCT2 is known to be the major OCT isoform based on mRNA expression and relative affinity for various substrates. These studies have shown that activation of PKA, PKC, or PKG leads to reduced uptake of the fluorescence organic cation 4-(4-(dimethylamino)styryl)-*N*-methylpyridinium (ASP^+) into renal proximal tubules, suggesting an inhibitory effect of these kinases on OCT2 activity. Inhibition of PKC by calphostin C reverses the inhibition of OCT2 activity, demonstrating the specificity of the PKC-mediated pathway. However, when heterologously expressed in HEK293 cells, hOCT2 was shown to be inhibited only by PKA activation, not by PKC. These conflicting results may be explained by differences between these experimental systems in terms of the expression of genes involved in the PKC pathway. Further studies using heterologously expressed OCT2 have identified a role for G protein-coupled receptor (GPCR) signaling in regulation of OCT2.

The muscarinic receptor agonist carbachol causes a dose-dependent decrease in OCT2 activity (i.e., uptake of ASP^+) that is not explained by direct inhibition of the OCT2 protein by carbachol. Similarly, activation of the phospholipase C (PLC) pathway through purinergic receptor agonism by ATP leads to dose-dependent decreases in OCT2 activity. As both of these pathways result in increased intracellular Ca^{2+} concentrations, it has been hypothesized that the mechanism for OCT2 regulation by GPCRs involves the Ca^{2+}-calmodulin complex. Reduction of intracellular Ca^{2+} concentration or treatment with a Ca^{2+} chelator results similarly in decreased OCT2 activity, and inhibition of calmodulin and its downstream targets CaMKII and myosin light-chain kinase (MLCK) produces a similar effect. Although the effect of muscarinic receptor stimulation on OCT2 activity is partially explained by its effects on intracellular Ca^{2+}, additional studies have shown that the majority of this effect is due to activation of phospatidylinositol 3-kinase (PI3K), as inhibition of PI3K almost completely prevents the inhibitory effect of carbachol on OCT2-mediated transport.

These studies suggest that PKA, PI3K, and the Ca^{2+}-calmodulin complex are endogenous regulators of OCT2; however, in contrast to the effect of protein kinases on OCT1, the inhibition of transport activity is not explained by effects on substrate-binding affinity. It is therefore thought that these proteins regulate other aspects of OCT2, such as maturation and trafficking of OCT2 to the plasma membrane, or stimulation of endocytosis. Glycosylation of OCT2 may be an additional mechanism for regulation of organic cation transport. The rabbit OCT2 isoform was shown to contain three N-glycosylation sites (Asn71, Asn96, and Asn112), each of which was found to be necessary for optimal transport activity by mutational analysis. The Asn112Gln mutant resulted in a significant decrease in membrane expression (to approximately 20% of wild-type) in stably transfected Chinese hamster ovary (CHO) cells. Asn71Gln and Asn96Gln did not alter OCT2 surface expression, but the unglycosylated triple mutant (Asn71Gln/Asn96Gln/Asn112Gln) was retained intracellularly.

Affinity for TEA was increased approximately twofold by mutation of each of the three glycosylation sites. Maximal transport rate was decreased fivefold by the Asn112Gln mutation (explained by the reduced surface expression of Asn112Gln) and threefold by Asn96Gln (presumably, due to a lower transporter turnover number). Thus, N-glycosylation may regulate maturation and trafficking of OCT2 as well as substrate recognition and transporter turnover. There is also evidence for hormonal regulation of OCT2. Functional studies of rat renal cortical slices showed higher basolateral transport of TEA in slices from male rats versus female rats, and rOCT2 mRNA and protein expression was also shown to be higher in male rats than in female rats. Further, treatment of rats with testosterone increased rOCT2 and protein expression in the kidney of both male and female rats. In Madin-Darby canine kidney (MDCK) cells, treatment with dexamethasone, hydrocortisone, or testosterone increased both OCT2 mRNA and protein levels as well as uptake of TEA across the basolateral membrane. Conversely, treatment with estradiol led to decreased OCT2 expression and activity in male rats and in MDCK cells. The physiological significance of hormonal regulation of OCT2 is unclear but may be related to gender-specific differences in clearance of endogenous metabolites.

OCT3

Although OCT3 contains several conserved target sequences for PKA, PKC, and PKG, studies of heterologously expressed OCT3 have shown that activation of these pathways has no apparent effect on OCT3 activity. However, inhibition of the p44/42 mitogen-activated protein kinase (MAPK) has been shown to decrease MPP^+ transport by OCT3, suggesting that MAPK is a positive regulator of OCT3. Similarly, inhibition of calmodulin as well as of the downstream effector CaMKII results in decreased OCT3 activity. Selective inhibition of the calmodulin-dependent phosphodiesterase PDE1 also results in decreased OCT3 activity, suggesting PDE1 stimulation as an additional mechanism for Ca^{2+}- and calmodulin-dependent regulation of OCT3.

ANIMAL MODELS

Targeted genetic disruption of the organic cation transporters in mice (i.e., OCT knockout mice) provided the first direct evidence of the significance of organic cation transporters to drug disposition in vivo.

In Animals (Oct1 Knockout)

Oct1-/- knockout mice were shown to be viable and fertile, showing no obvious physiological abnormalities compared with their wild-type littermates, suggesting that Oct1 is dispensible for normal physiology. However, significant differences were observed in the disposition of organic cations in these mice. For example, when administered the typical organic cation, TEA, *Oct1-/-* mice showed significantly (approximately fivefold) reduced uptake of TEA into the liver, the site of highest OCT1 expression. Biliary excretion of TEA was 2.5-fold lower in *Oct1-/-* mice, which was explained by the reduced hepatic uptake of TEA. Additionally, direct intestinal excretion of TEA was reduced by approximately 50%.

Renal excretion of TEA was paradoxically increased (~50%) in *Oct1-/-* mice; however, this was explained by the lack of OCT1 in the liver of *Oct1-/-* mice; that is, decreased hepatic uptake of TEA in these mice resulted in increased availability to the kidney for renal excretion. Further, the coexpression of Oct1 and Oct2 in rodent kidney may have led to a confounding effect of Oct2 activity in *Oct1-/-* mouse kidney. In addition to the pharmacokinetic differences observed for TEA, *Oct1-/-* mice showed similar decreases in hepatic uptake of other OCT1 substrates, including the neurotoxin MPP^+ (~60% reduced) and *meta*-iodobenzylguanidine (MIBG) (~75% reduced). Cimetidine, an OCT1 inhibitor (but not a substrate), did not show significant differences in hepatic uptake in *Oct1-/-* mice compared with wild-type mice.

Similar results were found after intravenous injection of [^{14}C]choline, which is an Oct1 substrate; however, it was suggested that rapid metabolism of the radiolabeled compound to species that are not Oct1 substrates, as well as redundancy (i.e., other transporters capable of transporting choline), may have masked any possible effect of Oct1 on the pharmaco-kinetics of this compound. Further studies of *Oct1-/-* mice have focused on the antidiabetic drug metformin, which exerts its pharmacological effects in the liver. *Oct1-/-* mice showed a greater than 30-fold decrease in metformin uptake into liver compared with wild-type littermates. Further studies investigated the role of Oct1 in the development of metformin-induced lactic acidosis, a leading toxicity from this drug. A significant increase in serum lactic acid concentration was observed after administration of metformin to wild-type mice, but only slight elevations in serum lactate were seen in *Oct1-/-* mice, despite similar pharmacokinetic profiles between genotype groups. Taken together, these results suggest that Oct1-mediated metformin transport is a limiting step in metformin uptake into liver, and that the lactic acidosis induced by metformin is related to the availability of the drug to this target organ.

Oct2 Knockout

Similar to *Oct1-/-* mice, *Oct2-/-* mice were viable and fertile, showing no apparent physiological defects. However, unlike the *Oct1-/-* mice, pharmacokinetics of intravenous [^{14}C]TEA were quite similar in *Oct2-/-* mice and wild-type mice, with the exception of a slight decrease in distribution of TEA into brain in Oct2-deficient mice. Renal clearance of TEA in *Oct2-/-* mice was not different from that in wild-type mice, with active secretion being the primary mechanism of clearance. The explanation for these findings is that in mouse kidney, Oct1 expression is sufficient to maintain the capacity for active secretion of organic cations even in the absence of Oct2. Thus, to develop a mouse model of renal disposition of organic cations, it was necessary to generate an *Oct1/2-/-* double-knockout mouse.

Oct1/Oct2 Knockout Mice

Generation of *Oct1/2-/-* double-knockout mice revealed that the absence of both genes, as with each of the single knockouts, was compatible with normal physiology (i.e., normal viability, fertility, and life span were observed, with no apparent physiological abnormalities). However, unlike the single knockouts, *Oct1/2 -/-* mice show significant impairment in the active tubular secretion of organic cations in the kidney. Specifically, renal tubular secretion of TEA was effectively abolished in *Oct1/2-/-* mice, with renal clearance approximating glomerular filtration. This resulted in significantly elevated plasma levels of TEA in these mice compared with wild-type or *Oct1-/-* single-knockout mice. After steady-state infusion of TEA, plasma levels were elevated approximately sixfold in the *Oct1/2-/-* double-knockout mice compared to *Oct1-/-*, *Oct2-/-*, or wild-type mice.

Oct3 Knockout

As mentioned earlier, OCT3 is believed to be primarily responsible for the transport of endogenous monoamines in extraneuronal tissues, also known as the *uptake-2 transport system.* Zwart et al. generated an Oct3-deficient mouse in order to determine the importance of this transporter in the peripheral disposition of monoamine neurotransmitters. Although the *Oct3-/-* mice were viable and fertile and showed no deficiency in dopamine or norepinephrine metabolism, they did show impaired uptake-2 activity as evidenced by a >70% reduced uptake of the synthetic monoamine MPP$^+$ into heart in *Oct3-/-* mice compared to wild-type mice. Additionally, it was shown that in *Oct3+/-* heterozygous females impregnated in a heterozygous cross, intravenous injection of MPP$^+$ resulted in a three-fold reduction of MPP$^+$ uptake in *Oct3-/-* embryos versus wild-type embryos, whereas the accumulation of MPP$^+$ in the placenta was not different between groups. These results suggest that OCT3 acts as an uptake transporter for monoamines at the fetoplacental interface. Apart from the distribution of MPP$^+$ into heart and placental transfer of MPP$^+$, no significant phenotypic differences have been observed in *Oct3-/-* mice.

HEREDITARY VARIATION

Much attention has recently been focused on the pharmacogenetics of drug transporters (i.e., the effect of natural human genetic variation in drug transporter genes on drug disposition and drug response). Research in this area has taken primarily a sequence-based approach, beginning with identification of genetic variants likely to produce a functional effect (e.g., amino acid substitutions, or insertions or deletions in the coding region), followed in some cases by functional studies of the variant proteins in heterologous expression systems.

OCT1

Several functionally significant polymorphisms in OCT1 have been described. In a sample of 57 healthy Caucasians, 25 genetic variants were discovered, of which eight resulted in a change in protein sequence. Of these, five (Arg61Cys, Cys88Arg, Phe160Leu, Gly401Ser, and Met420del) were tested for function by measurement of transport activity in *X. laevis* oocytes. Cys88Arg and Gly401Ser showed virtually complete loss of activity toward MPP^+, while Arg61Cys retained approximately 30% of wild-type activity. In contrast to MPP^+, uptake of [^{3}H]serotonin by Cys88Arg and Gly401Ser was detectable (~10% of wild-type), suggesting that these variants influence the substrate selectivity of OCT1. Shu et al. discovered 15 amino acid sequence-altering variants of OCT1 that were identified in a large sample of ethnically diverse healthy subjects, and tested these for function by uptake of MPP^+ in *X. laevis* oocytes.

Variants with significantly reduced or complete loss of function included the Arg61Cys and Gly401Ser described by Kerb et al., as well as Pro341Leu, Gly220Val, and Gly465Arg. An additional variant, Ser14Phe, showed a significant increase in activity; interestingly, this variant corresponds to the likely ancestral allele, as the consensus mammalian OCT1 sequence includes phenylalanine at this position. It was noted that all of the variants with reduced function occurred at evolutionarily conserved amino acid residues, and that variants with reduced function tended to be amino acid substitutions that result in a large chemical change.

Of the variants with significant functional differences from the reference OCT1, five (Ser14Phe, Arg61Cys, Pro341Leu, Gly401Ser, and Gly465Arg) occurred at >1% allele frequency in at least one ethnic group and are attractive candidates for association with drug response phenotypes for OCT1 substrate drugs. Functional studies of several additional OCT1 variants have been performed by two independent groups using microinjected *X. laevis* oocytes or transiently transfected HEK293 cells. Three amino acid sequence variants, Pro283Leu, Arg287Gly, and the previously described Pro341Leu, showed significantly reduced transport activity, with Pro283Leu and Arg287Gly having no appreciable activity toward the model substrates TEA and MPP^+. The reduction in activity of these variants was not explained by reduction in protein expression, as immunofluorescence studies showed all three variants to have membrane expression levels similar to that of the wild-type OCT1.

OCT2

Human genetic variants of OCT2 have been investigated comprehensively by resequencing of the coding region in a large ethnically diverse sample. Eight of the variants identified in this screen result in amino acid substitutions, and one single-nucleotide insertion (134-135insA) leads to a premature stop codon at amino acid position 48. Four of the nonsynonymous variants (Met165Ile, Ala270Ser, Arg400Cys, and Lys432Gln) were polymorphic, with ethnic-specific allele frequencies ≥1%. The remaining four nonsynonymous variants (Pro54Ser, Phe161Leu, Met165Val, and Ala297Gly) as well as the insertion variant were found on only one of 494 chromosomes screened. Of the four polymorphic OCT2 protein sequence variants (Met165Ile, Ala270Ser, Arg400Cys, and Lys432Gln), all retained function as measured by uptake of the prototypical organic cation substrate, MPP^+, in *X. laevis* oocytes expressing the variant transporters.

However, quantitative differences in MPP^+ uptake activity and kinetic differences in interactions with organic cations were observed for the common variants, with Lys432Gln having twofold increased affinity for MPP^+, and the Met 165Ile and Arg400Cys variants having lower maximal transport rates (V_{max}) for MPP^+ than the reference OCT2. In inhibition kinetics studies, Ala270Ser showed an increased K_i value for TBA, while Arg400Cys and Lys432Gln had lower K_i values for TBA inhibition than did the reference OCT2. When the rare variants of OCT2 were expressed in the same system, all of the amino acid substitutions were found to have no effect on OCT2 function. However, as expected, the single nucleotide insertion (frameshift) variant showed a complete loss of function. Additional single-nucleotide polymorphisms (SNPs) in the OCT2 gene were identified by direct sequencing of genomic DNA from 48 unrelated Japanese persons and 116 arrhythmic Japanese patients. In the former study, 27 SNPs and five deletion polymorphisms in the OCT2 gene were identified. Two synonymous variants, also reported by Leabman et al., were identified at amino acid positions 130 and 150.

The remaining SNPs and deletion polymorphisms appeared in introns, 3′ untranslated regions, and 3′ flanking regions of the OCT2 gene. In the latter study, 33 genetic variants, including 14 novel ones, were found. Two nonsynonymous variants were identified (Thr199Ile and Thr201Met) and other SNPs and insertion and deletion polymorphisms were located in exons, the 3′-untranslated region of exon 11, introns, and the 3′-flanking region. The functional effects of these polymorphisms have not been investigated. In addition to direct sequencing of the OCT2 gene, other lines of evidence exist for a role of genetic variation in OCT2 in variable drug response. Based on data from published literature, an estimate of the genetic component contributing to variation in the renal clearance of metformin, which undergoes transporter-mediated secretion, was found to be particularly high (>90%). This finding suggests that variation in the renal clearance of metformin has a strong genetic component and that genetic variation in OCT2 may explain a large part of this pharmacokinetic variability.

OCT3

In a study of genetic variation in 26 membrane transporter genes in a large sample (n = 247) of ethnically diverse human subjects, Leabman et al. discovered 14 nucleotide substitutions in OCT3. The survey region included the exons and flanking intronic region. Of the 14 variants identified, five were located in the coding exonic regions, and three of these (Thr44Met, Ala116Ser, and Thr400Ile) resulted in a change in the amino acid sequence. Only Ala116Ser was polymorphic, with an allele frequency of 1.7% in the African-American subset of the sample. These variants have not been tested for effects on OCT3 function. Genetic variation in OCT3 was investigated independently by resequencing the core promoter, exons, and 3′ untranslated region of the gene in 100 healthy persons of European descent. Six nucleotide substitutions and one single base pair deletion were discovered, including –29A>G in the OCT3 promoter and three synonymous substitutions in the coding region.

The functional consequences of these variants have not been studied. The authors suggest that the synonymous 1233G>A substitution may generate a cryptic 3′-splice acceptor site. Additional data on genetic variation in organic cation transporter genes are available on online public databases. The Pharmacogenetics and Genomics Knowledge Base is a repository for data and information related to all areas of pharmacogenetics. PharmGKB includes well-annotated pages describing genetic variation in OCTs, identified by the Pharmacogenetics of Membrane Transporters project (PMT) in a large collection of DNA samples from ethnically diverse populations. The NCBI single-nucleotide polymorphism database (dbSNP) also includes data on OCT variants. The HapMap project, which aims to take advantage of linkage between SNPs to increase power in genetic association studies, may prove to be a useful tool for studying the genetics of response to OCT substrates.

CONCLUDING REMARKS

The OCTs are a fairly well-studied family of multispecific organic cation transporters, with potential influence for a large number of endogenous and pharmacological compounds. The tissue distribution and subcellular localization of these transporters suggest that their primary role is in the excretion of toxic xenobiotic and endogenous organic cations. OCT1 is ahepatic sinusoidal uptake transporter and appears to be most important for drug distribution into liver, where it aids in presentation of substrate drugs to hepatic metabolizing enzymes or in biliary excretion. OCT2, the kidney-specific organic cation transporter, is primarily responsible for basolateral uptake of organic cations into renal proximal tubules and acts as the first step in active tubular secretion of its substrates.

OCT3 has a more diffuse pattern of expression, as well as a relatively unique substrate selectivity profile, and is thought to be important for extra-neuronal clearance of monoamine neurotransmitters as well as uptake of monoamines into the heart and across the placenta. Knockout mouse models of the OCTs have provided crucial information regarding the importance of these transporters to xenobiotic and endobiotic disposition in vivo. The existence of genetic variants of OCTs with altered function in cellular assays suggests that genetic variation in OCTs may contribute to interindividual variability in drug disposition or drug response; further study of the importance of these variants to human drug disposition may soon allow for clinical pharmacogenetic testing and improvements in rational drug therapy.

4

Endogenous Carriers

Endogenous nucleosides are important precursors of nucleic acid synthesis, a process fundamental to the growth and metabolism in all living systems. The purine nucleoside adenosine is also a key signaling molecule that exerts profound effects on many tissues and organs by activating specific purinergic receptors. Nucleosides are hydrophilic and have low membrane permeability. To facilitate the movement of nucleosides across cell membranes, mammalian cells have evolved two major classes of transporters: *concentrative nucleoside transporters* (CNTs) and *equilibrative nucleoside transporters* (ENTs). At the cellular level, these transporters play key roles in salvaging nucleosides for DNA and RNA syntheses and in regulating adenosine signaling at the receptor sites. From a whole-organism point of view, nucleoside transporters expressed in absorptive and excretory organs are important for maintaining total body homeostasis of nucleosides. The pharmacological significance of nucleoside transporters originates from the wide use of nucleoside analogs in the treatment of cancer, viral infections, and other pathophysiological conditions.

Many of the therapeutic nucleoside analogs rely on nucleoside transporters to enter or exit cells. Consequently, the expression and functional characteristics of these transporters in target or nontarget cells, as well as in absorptive and excretory organs, will have an important impact on the efficacy and toxicity of therapeutic nucleoside analogs. Historically, nucleoside transport activities were first studied by functional and kinetic analyses in mammalian cells and tissue preparations. These early studies established that there exist multiple nucleoside transporters, which can be classified into two major systems: equilibrative and concentrative. The *equilibrative system* consists of some low-affinity facilitated carrier proteins that are further divided into two subtypes, *es* and *ei*, based on their sensitivity to a highly specific inhibitor, nitrobenzylthioinosine [nitrobenzyl-mercaptopurine riboside (NBMPR)]. The *es* (equilibrative inhibitor sensitive) subtype is potently inhibited by NBMPR (K_i, 0.1 to 10 nM), whereas the *ei* (equilibrative inhibitor insensitive) subtype is not affected by NBMPR at concentrations below 1 μM. The *es* and *ei* transport both purine and pyrimidine nucleosides and are widely distributed in different tissues and cell types. The *concentrative system* consists of some high-affinity Na^+-dependent transporters that mediate uphill transport of nucleosides by coupling to the physiologic Na^+ gradient.

Several subtypes with distinct substrate specificity were recognized, with three of them (*cit, cif*, *cib*) well characterized. The *cit* (N2) subtype is pyrimidine selective but also transports adenosine. The *cif* (N1) subtype is mainly purine selective but also transports uridine. The *cib* (N3) subtype is broadly selective, transporting both purine and pyrimidine nucleosides. The Na^+-dependent transporters appear to have relatively limited tissue distributions and are typically found in epithelial tissues such as the intestine, kidney, liver, and choroid plexus. Molecular cloning studies in the past decade led to the isolation of genes encoding individual membrane proteins underlying the foregoing nucleoside transport activities in human and other mammalian species. They belong to two structurally unrelated solute carrier gene families, SLC28 and SLC29. The SLC28 family consists of three members, CNT1 (SLC28A1), CNT2 (SLC28A2), and CNT3 (SLC28A3), which correspond to the previously characterized *cit, cif,* and *cib* subtypes respectively. The SLC29 family consists of four members, ENT1 (SLC29A1), ENT2 (SLC29A2), ENT3 (SLC29A3), and ENT4

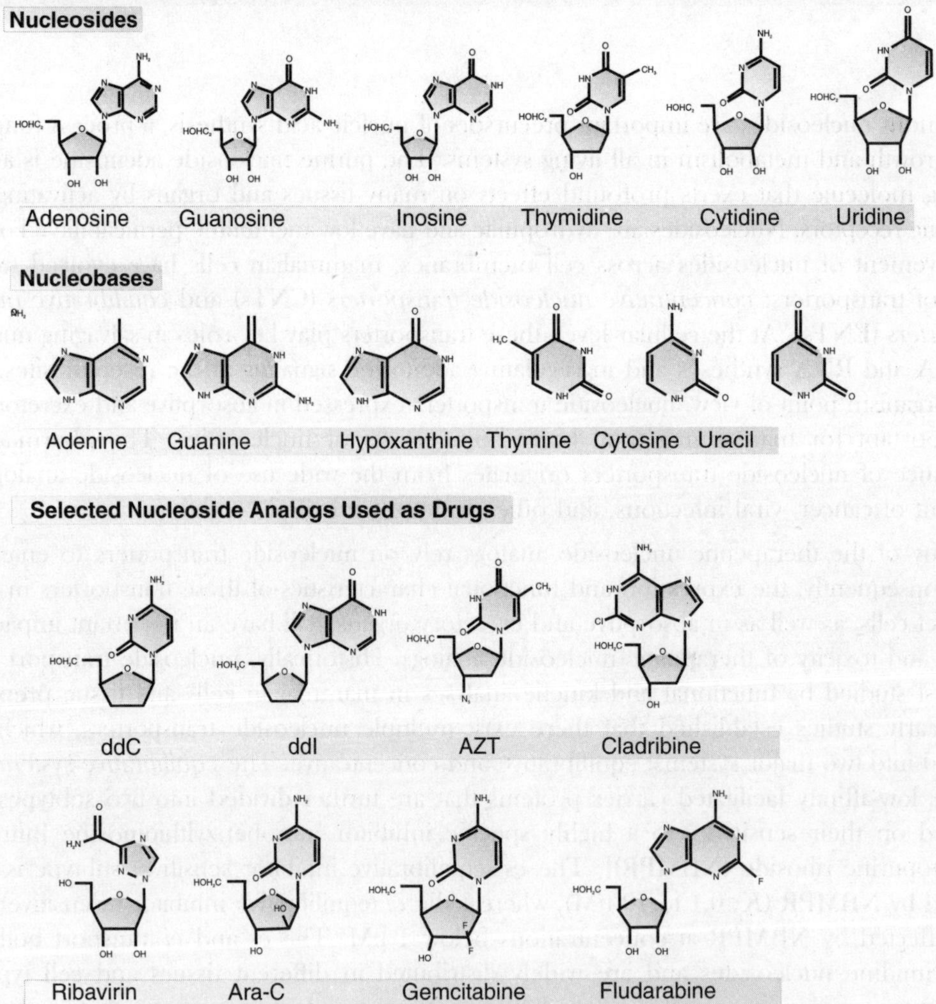

Fig. 4.1. Structures of selected nucleosides and nucleoside analogs.

(SLC29A4). ENT1 underlies the *es* activity, whereas ENT2 is responsible for the *ei* activity. ENT3 and ENT4 are novel members identified from the human genome project. ENT3 is an intracellular low-affinity nucleoside transporter, and ENT4 exhibits novel substrate specificity by functioning as a polyspecific organic cation transporter.

SLC28 AS DRUG CARRIER

CNT1 (SLC28A1). The first cloned Na^+-dependent nucleoside transporter, rCNT1, was isolated from a ratjejunal cDNA library in 1994 by expression cloning in *Xenopus laevis* oocytes. Its human homolog, hCNT1, was subsequently cloned from human kidney. The *hCNT1* gene is localized to chromosome 15q25-26. hCNT1 (649 a.a.) and rCNT1 (648 a.a.) are 85% identical in protein sequence, and are predicted to possess 13 transmembrane (TM) α-helices. A CNT1 ortholog has also been cloned from pig. In humans, hCNT1 mRNA is expressed predominantly in the liver and kidney and exhibits much lower expression in other tissues. There appear to be large interindividual differences of hCNT1 mRNA expression level in the kidney. In rat, rCNT1 mRNA was found in the jejunum, kidney, liver, and various regions of the brain. Western blot further demonstrated the expression of rCNT1 protein in the small intestine, kidney, and liver.

Fluorescence protein-tagged constructs of hCNT1 and rCNT1 were expressed on the apical membranes of polarized LLC-PK1 and MDCK cells. In rat tissues, immuno-histochemistry studies showed that rCNT1 is localized predominantly to the apical membranes of the rat jejunum, renal cortical tubules, and liver parenchymal cells. In rat hepatocytes it was suggested that rCNT1 is targeted to the apical (canalicular) membranes via a transcytotic pathway. Expression of hCNT1 and rCNT1 in *Xenopus* oocytes and transfected cell lines revealed transport properties that are consistent with those of the pyrimidine-selective subtype *cit.* hCNT1 and rCNT1 selectively transport naturally occurring pyrimidine nucleosides (uridine, thymidine, cytidine) as well as the purine nucleoside, adenosine, in a Na^+-dependent manner with high affinities (K_m values in the low-micromolar range). However, both hCNT1 and rCNT1 transport adenosine with much lower capacities (V_{max}) than pyrimidine nucleosides, suggesting that these transporters may function mainly as a pyrimidine-specific nucleoside transporter in vivo.

hCNT1 and rCNT1 interact with a number of pyrimidine and adenosine analogs. hCNT1 transports the anticancer nucleoside analogs gemcitabine (2′,2′-difluorodeoxycytidine) and cytarabine [1-(β-D-arabinofuranosyl)cytosine (AraC)]. 5′-Deoxy-5-fluorouridine (5′-DFUR), an intermediate metabolite of the anticancer drug capecitabine, is a good hCNT1 substrate. hCNT1 also accepts a number of antiviral nucleoside analogs, including zidovudine [3′-azido-3′-deoxythymidine (AZT)], lamivudine [2′,3′-dideoxy-3′- thiacytidine (3TC)], and zalcitabine [2′,3′-dideoxycytidine (ddC)]. Similarly, rCNT1 accepts a number of pyrimidine and adenosine analogs as substrates, such as 5-fluoro-2′-deoxyuridine, 5-fluorouridine, gemcitabine, AZT, ddC, cladribine, and cytarabine.

CNT2 (SLC28A2). rCNT2 cDNA was first isolated from rat liver by expression cloning and was initially named SPNT (sodium-dependent purine nucleoside transporter). The human homolog hCNT2 (also termed hSPNT1) was subsequently cloned by homology cloning from human kidney. The *hCNT2* gene is localized to chromosome 15q15. hCNT2 protein is 72% identical to hCNT1 and shares a similar 13 TM membrane topology. Orthologs of CNT2 were also cloned from rabbit

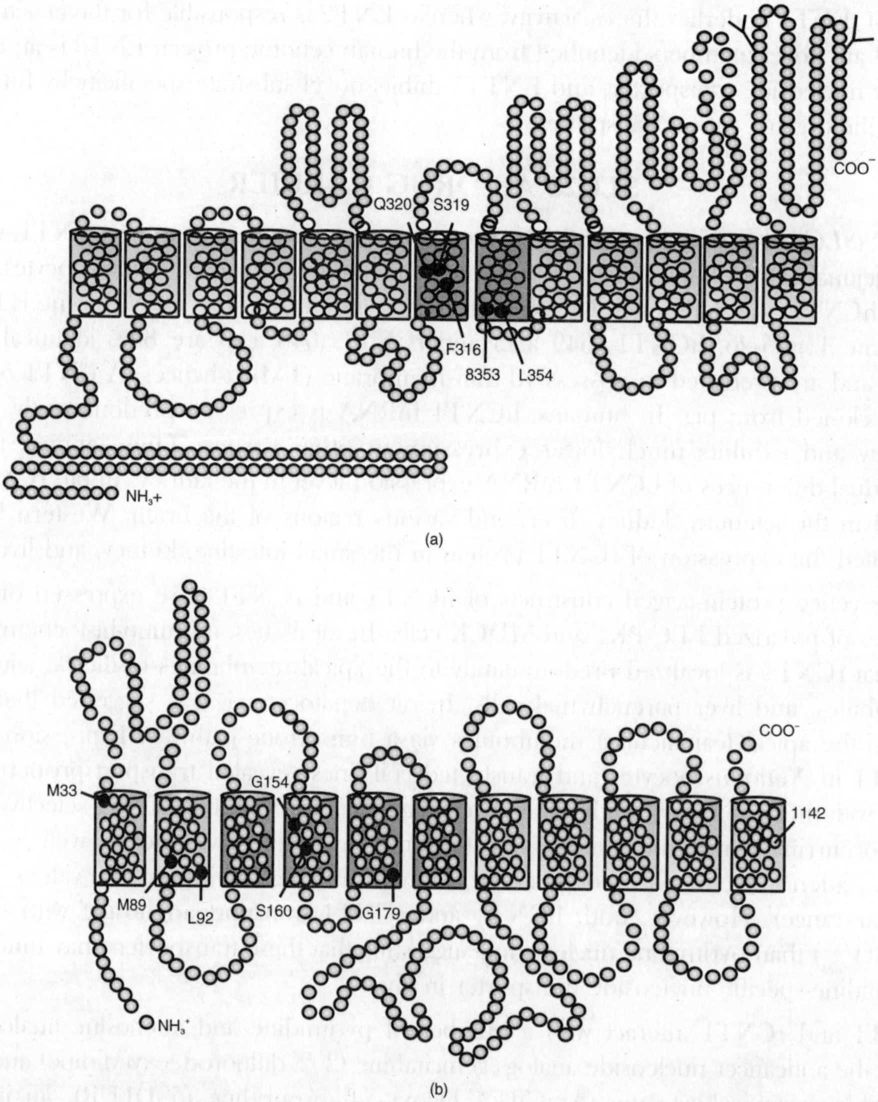

Fig. 4.2. Secondary structures predicted for hCNT1 (a) and hENT1 (b).

(rbCNT2) and mouse (mCNT2). Northern blot studies revealed that hCNT2 is widely distributed in heart, liver, skeletal muscle, kidney, intestine, pancreas, placenta, brain, and lung. hCNT2 is also expressed in normal human leukocytes and a number of neoplastic tissues and cancer cell lines. rCNT2 is expressed in rat liver, jejunum, spleen, heart, skeletal muscle, and various regions of the brain. Interestingly, rCNT2 transcripts were not detected by Northern blot analysis in the rat kidney. rCNT2 was also found in rat brain endothelial and choroid plexus epithelial cells, which form the blood-brain barrier (BBB) and blood-cerebrospinal fluid (CSF) barrier respectively.

Consistent with previous functional data demonstrating *cif* activities in brush border membranes of kidney or intestinal epithelia, green fluorescence protein (GFP)-tagged rCNT2 is predominantly targeted to the apical membranes of polarized MDCK and LLC-PK cells. Interestingly, in liver parenchymal cells, an antibody study by Duflot et al. showed that rCNT2 protein is highly expressed in the basolateral (sinusoidal) membrane and minimally presents at the apical (canalicular) membrane, suggesting that membrane trafficking of rCNT2 in hepatocytes may be different from that in renal or intestinal epithelial cells. The transport characteristics of CNT2 cloned from human, rat, mouse, and rabbit have been studied in *Xenopus* oocytes and transfected cell lines. Substrate specificities of these recombinant CNT2 proteins are consistent with the properties of the classic purine nucleoside-selective subtype *cif.* hCNT2 transports purine nucleosides (adenosine, guanosine, and inosine) and the pyrimidine nucleoside uridine with high affinities and exhibits relatively higher affinities for purine nucleosides (K_m values ranging from 8 to 14 μM for adenosine and inosine) than for uridine (K_m ranging from 80 to 116 μM).

rCNT2, rbCNT2, and mCNT2 exhibited similar transport properties to hCNT2, selectively transporting purine nucleosides and uridine. CNT2 transporters interact with a number of nucleoside analogs. Ribavirin, a broad-spectrum antiviral nucleoside analog used in the treatment of hepatitis C, and clofarabine, an investigational antileukemia drug, were shown to be transported by hCNT2. 5-Fluorouridine is transported by hCNT2 with kinetics similar to those of uridine. Other nucleoside analogs, including gemcitabine, AZT, ddC, ddA, AraA, 5-fluoro-2′-deoxyuridine (FdU), and 5-iodo-2′-deoxyuridine (IdU), are not substrates of hCNT2. There appear to be significant species differences in substrate specificity between the human and rat CNT2 in interacting with nucleoside analogs. For example, ddI is either not a substrate or is poorly transported by hCNT2. In contrast, rCNT2 transports ddI (K_m = 29.2 μM). The anticancer drug cladribine (2-CdA) appears to be a good substrate for rCNT2 but is not transported by hCNT2.

CNT3 (SLC28A3). hCNT3 was cloned from human mammary gland and differentiated human myeloid HL-60 cells. hCNT3 is about 47 to 48% identical in amino acid sequence to hCNT1 and hCNT2, and 57% identical to hfCNT, a broadly selective CNT from the ancient marine prevertebrate hagfish. The *hCNT3* gene is localized to chromosome 9q22.2. The mouse ortholog mCNT3 was isolated from liver, and the protein shares 78% identity with hCNT3. Both hCNT3 and mCNT3 are predicted to possess 13 TMs similar to CNT1 and CNT2. The distribution pattern of hCNT3 was investigated in a human multiple tissue expression RNA array. The highest levels are found in mammary gland, pancreas, bone marrow, trachea, and intestine. Lower levels are found in liver, lung, placenta, prostate, testis, and other tissues, including some regions of the brain and heart. hCNT3 is also found in fetal tissues and various cultured cell lines, including K562, HeLa, and HL-60. A recent study also revealed low expression of hCNT3 in leukocytes and several leukemia cell lines.

The subcellular localization of hCNT3 in polarized epithelial cells has not been investigated directly. When expressed in *Xenopus* oocytes, hCNT3 and mCNT3 exhibit *cib* subtype nucleoside transport properties, with broad selectivities for both purine and pyrimidine nucleosides. hCNT3 transports purine and pyrimidine nucleosides with reported K_m values ranging from 15 to 53 μM in one study and 3 to 11 μM in another. Compared to hCNT1 and hCNT2, hCNT3 is capable of transporting a broader range of synthetic nucleoside analogs. hCNT3 efficiently transports both pyrimidine (5-fluorouridine, 5-fluoro-2′-deoxyuridine, zebularine, and gemcitabine) and purine (cladribine, clofarabine, and fludarabine) nucleoside analogs. Low uptake is also observed for AZT, ddC, and ddI. Recent electrophysiology studies demonstrated further that ribavirin and 3-deazauridine are also hCNT3 substrates. So far, specific high-affinity inhibitors have not been described for hCNT3 and other CNTs.

Method of Transport

Mammalian CNTs are Na^+-coupled transporters which utilize the inwardly directed Na^+ gradient established by Na^+/K^+-ATPase to translocate substrates into cells. Early radio tracer uptake studies in cell and tissue preparations suggested that the Na^+/nucleoside coupling stoichiometry is 1:1 for *cit* (CNT1) and *cif*(CNT2); and 2:1 for *cib* (CNT3). Recent electrophysiological studies have confirmed these coupling stoichiometry for hCNT1 and hCNT3. Electrophysiological analysis of hCNT1 also suggested an ordered simultaneous transport model in which Na^+ first binds to the transporter in order to enhance subsequent binding of nucleoside. Interestingly, while ion coupling of hCNT1 and hCNT2 is Na^+ specific, hCNT3 exhibits unique broad cation specificity and can use H^+ and Li^+ to substitute Na^+ for coupling. Chimeric studies between hCNT1 and hCNT3 suggested that the site for Na^+ ion interaction may reside in the C-terminal half. CNT homologs have also been found in bacteria, yeast, nematodes, insects, and hagfish. CNTs isolated from*Escherichia coli*, *Candida albicans*, and *Caenorhabditis elegans* use a proton gradient as the driving force, whereas the hagfish hfCNT is coupled to Na^+. From the substrate point of view, CNT1-3 specifically transport nucleosides and their analogs. CNT1-3 do not transport nucleobases (i.e., free purine or pyrimidine bases) or nucleotides.

A number of recent studies have examined the structure–activity relationships of the CNTs. Besides isoform-dependent specificity toward the purine or pyrimidine moieties, several common features of the ribose structure are important for CNT–substrate interaction. Generally speaking, the 3′-hydroxyl group of the ribose ring is essential for substrate interaction with the CNTs, and modifications at the C(3′) position are usually not tolerated. CNTs are less sensitive to modifications at the 5′-hydroxyl group and generally tolerate changes at the 2′-hydroxyl group. From the protein standpoint, mammalian CNTs are proposed to have 13 trans- membranes (TMs) with an intracellular N-terminus and an extracellular C-terminus based on antibody and glycosylation studies. hCNT1, hCNT2, and hCNT3 all contain potential N-glycosylation sites and are likely to exist in the glycosylated form. Mutational analysis of the three putative N-linked glycosylation sites in rCNT2 suggests that glycosylation does not affect transporter function or membrane sorting. Protein chimera and site-directed mutagenesis studies have been used to examine the structure–function relationships of CNTs. Domain-swapping studies in rCNT1 and rCNT2 first identified TMs 7 and 8 as the essential sites for substrate recognition and discrimination.

Site-directed mutagenesis studies further identified Ser318 on TM7 of rCNT1 as essential for its pyrimidine selectivity. Replacing Ser318 with glycine transformed rCNT2 into a broadly selective CNT3-like transporter. A Gln319Met mutation augmented this effect by enhancing the apparent affinity for purine nucleosides. Using a similar approach, Loewen et al. confirmed these findings in hCNT1 and hCNT2, and further identified Ser319/Gln320 and Ser353/Leu354 as essential determinants for substrate selectivity. In addition, mutation of a conserved residue Phe316 in TM7 of hCNT1 to tyrosine or alanine increased transporter sensitivity to guanosine. Together, these studies demonstrated that TMs 7 and 8 are essential components of the substrate binding and translocation channel, and that substrate selectivity of the CNTs may be determined by a few key amino acid residues located on TMs 7 and 8. Cysteine-accessibility analysis of hCNT3 suggests that TMs 11 and 12 may also participate in forming the nucleoside translocation pathway. In contrast, TMs 1 to 3 appear not essential for CNT function. When TMs 1 to 3 were truncated from rCNT1 or hCNT1, the resulting transporters retained significant Na^+-dependent transport activity. NupC, a H^+-coupled CNT from *E. coli,* lacks TMs 1 to 3, further suggesting that TMs 1 to 3 may not be required for CNT function.

SLC29 AS DRUG CARRIER

ENT1 (SLC29A1). Purification and N-terminal sequencing of the prototype *es* transporter from human erythrocytes led to the cloning ofhENT1 in 1997. Human ENT1 is a 456-residue glycoprotein with 11 predicted TMs. The *hENT1* gene is localized to chromosome 6p21.1–21.2. ENT1 orthologs were cloned from rat and mouse. rENT1 (457 a.a.) and mENT1.1 (460 a.a.) are about 78 to 79% identical to hENT1. A splice variant, mENT1.2, has been reported for mouse. Studies at both the mRNA and protein levels have revealed that ENT1 is almost ubiquitously distributed in human and rodent tissues, although its abundance varies among tissues. High expression of hENT1 protein or mRNA has been demonstrated and confirmed in many tissues, including erythrocytes, placenta, brain, heart, liver, lung, and colon. However, striking intertissue and interindividual differences in the expression levels of hENT1 were observed. GFP- or YFP (yellow fluorescence protein)-tagged hENT1 is localized predominantly on the basolateral membrane of differentiated MDCK cells, whereas a small amount of hENT1 also appears on the apical membrane.

Besides plasma membrane, Lai et al. showed that hENT1, but not mENT1, is also present in mitochondria, which may contribute to enhanced mitochondrial toxicity of certain antiviral nucleoside analogs. A PEXN motif located in the first extracellular loop of hENT1 was recently identified as an important signal for mitochondrial targeting of hENT1. Recombinant human, rat, and mouse ENT1 transporters transport both purine and pyrimidine nucleosides and exhibit typical *es*-type activities. The K_m values of hENT1 toward endogenous nucleosides range from 50 μM for adenosine to 680 μM for cytidine. Transport is Na^+ independent and inhibited by nanomolar NBMPR. hENT1 does not interact with nucleobases. hENT1 interacts with many nucleoside analogs widely used in the treatment of cancer and viral infections. Cladribine, gemcitabine, fludarabine, cytarabine, clofarabine, and ribavirin have been demonstrated to be transported by hENT1. However, hENT1 does not transport AZT and only weakly transports ddC and ddI. Coronary vasodilators such as dipyridamole, dilazep, and draflazine are potent hENT1 inhibitors (K_i values in nanomolar concentrations).

These compounds bind to the ENT1 protein but are not transported into the cells. There are significant species differences among ENT1 in interacting with coronary drugs. Rat ENT1 is much less sensitive to dipyridamole and dilazep than human ENT1, and mouse ENT1 is only intermediately sensitive to dipyridamole. In addition to NBMPR and coronary vasodilators, a number of protein kinase inhibitors, including the Bcr-Abl tyrosine kinase inhibitor STI-571

(Gleevec) used for treatment of chronic myelogenous leukemia, were shown to be effective in inhibiting ENT1 and/or ENT2, with IC_{50} values ranging from 60 to 1000 nM. Interestingly, Carrier et al. recently showed that plant-derived cannabinoids are potent inhibitors of ENT1 (K_i < 250 nM).

ENT2 (SLC29A2). The human ENT2 transporter was cloned from HeLa cells and human placenta by two independent groups. hENT2 (456 a.a.) is 46% identical in amino acid sequence to hENT1. The most homologous regions between hENT1 and hENT2 are the transmembrane segments, while the least are the hydrophilic termini and loops. The human *hENT2* gene is localized to chromosome 11q13. Two mRNA-spliced variants, encoding nonfunctional truncated transporter proteins, were reported. The rodent orthologs rENT2 and mENT2 were also cloned. The 456-residue mENT2 and rENT2 are 88% identical to hENT2. hENT2 mRNA is particularly abundant in skeletal muscle, but is also expressed in a wide range of tissues, including brain, heart, placenta, thymus, pancreas, prostate, and kidney. In MDCK cells, GFP-tagged hENT2 is localized exclusively to the basolateral membrane. A C-terminal dileucine repeat is implicated in the surface expression of hENT2. Recombinant human and rat ENT2 proteins transport a broad range of purine and pyrimidine nucleosides, and they confer NBMPR-insensitive Na^+-independent nucleoside transport typical to the *ei* system.

Although both hENT1 and hENT2 transporters are broadly selective for purine and pyrimidine nucleosides, hENT2 exhibits 7.7- and 19.3-fold lower affinities for guanosine and cytidine, respectively, but a four fold higher affinity for inosine. More important, ENT2 transports a wide range of purine and pyrimidine nucleobases, whereas ENT1 does not. Specifically, hENT2 and rENT2 transport hypoxanthine, adenine, guanine, uracil, and thymine with low affinities (K_m ranging from 0.7 to 2.6 mM). hENT2, but not rENT2, also transports cytosine. h/rENT2 transports AZT, ddC, and ddI with much higher efficiency than h/rENT1. Gemcitabine and clofarabine have been shown to be transported by hENT2. A number of uridine and cytidine analogs also interact with hENT2.

ENT3 (SLC29A3). The third member of the ENT family, ENT3, was first identified and cloned from human and mouse due to sequence similarity to ENT1 and ENT2. The *hENT3* gene is localized to chromosome 10q22.1. The 475-residue hENT3 and mENT3 proteins are 73% identical in amino acid sequence and exhibit ~30 to 33% identities to ENT1 and ENT2. Structurally, ENT3 differs from ENT1 and ENT2 in possessing a long hydrophilic N-terminal region that contains a dileucine motif (DE)XXXL(LI) for endosomal/lysosomal targeting. ENT3 transcripts and proteins are widely distributed in human and rodent tissues. In humans the highest levels of mRNA were found in the placenta, uterus, ovary, spleen, lymph node, and bone marrow; the lowest levels were found in brain and heart. In contrast to ENT1 and ENT2, the endogenous as well as GFP-tagged hENT3 proteins were found to reside predominantly intracellularly in HeLa cells and to colocalize partially with lysosomal markers.

Truncation of the N-terminal region or mutation of the dileucine motif to alanine (hENT3AA) relocated hENT3 to the plasma membrane in HeLa cells and in *Xenopus* oocytes. Characterization of a cell surface–expressed mutant hENT3AA suggested that hENT3 is a broadly selective and low-affinity nucleoside transporter that transports purine and pyrimidine nucleosides as well as the nucleobase adenine. hENT3AA-mediated transport is Na^+ independent but stimulated strongly by low pH, with the optimum pH being 5.5. At pH 5.5, hENT3AA exhibits about 40- and 10-fold lower affinities for adenosine and uridine than for hENT1. Nucleoside analogs, including cladribine, cordycepin (3′-deoxyadenosine), tubercidin (7-deazaadenosine), fludarabine, zebularine, ddI, ddC, and AZT, are transported by hENT3AA. hENT3AA is relatively insensitive to the classical nucleoside transport inhibitors, such as NBMPR, dipyridamole, and dilazep.

ENT4 (SLC29A4). The fourth member of the human SLC29 family, hENT4, was first cloned and characterized in our laboratory. Gene orthologs of ENT4 are also found in mouse and rat. hENT4 is a 530-residue protein which displays about 86% identity to the 528-residue mouse homolog mENT4. The mENT4 and hENT4 proteins share low but significant sequence identity (about 18 to 20%) to ENT1-3 and possess a similar 11 TM membrane topology. However, hENT4 has a longer hydrophilic N-terminus and exhibits a low overall sequence homology to other ENTs. Phylogenetic analysis suggests that the mammalian ENT4 lineage is evolutionarily distinct from the ENT1/2/3 lineage(s). The gene encoding hENT4 is located at chromosome 7p22.1. Our Northern blot analysis showed that hENT4 mRNA is expressed preferentially in the human brain but is also found in skeletal muscle, kidney, heart, and liver.

In human and mouse brain, hENT4 transcripts are widely distributed in different regions of the CNS. YFP tagging of hENT4 in MDCK cells demonstrated that this protein is localized primarily to the plasma membrane. Although ENT4 was originally suggested to be a nucleoside transporter, extensive functional analysis carried out in this laboratory showed that except for a moderate activity toward adenosine, hENT4 minimally interacts with other nucleosides, nucleobases, or their analogs. Instead, it mediates Na^+-independent, electrogenic transport of monoamine neurotransmitters such as serotonin and dopamine (K_m = 114 and 329 μM, respectively). Functional analysis of the mouse ENT4 ortholog also confirmed that the monoamines are the physiologic substrates of ENT4. Therefore, we previously proposed an alternative functional name, plasma membrane monoamine transporter (PMAT), for ENT4. Our recent analysis showed that hENT4 also transports cationic xenobiotics such as MPP^+ and TEA, and shares striking functional similarities to a genetically unrelated drug transporter family, organic cation transporters (OCTs). Because ENT4 does not typically function as a nucleoside transporter, we will not include it for further discussion.

Method of Transport

ENT1 and ENT2 are facilitated carriers that transport substrates down their concentration gradients. However, in many cell types, cellular uptake of nucleosides and nucleoside analogs are tightly coupled to intracellular metabolism (e.g., phosphorylation). Rapid enzymatic conversion of nucleosides to metabolites (e.g., nucleotides) can thus provide a metabolic "*driving force*" to promote cellular uptake. Efflux occurs only when intracellular concentrations of free nucleoside exceed those outside the cell. The transport mechanism of the lysosome-localized ENT3 is not known yet. ENT3 activity is strongly stimulated by proton, exhibiting maximal activity at pH 5.5 and no activity at pH 8.0. It is unclear whether the pH dependence of ENT3 reflects a proton–nucleoside cotransport mechanism, or an evolutionary adaptation of ENT3 to the acidic environment of lysosomes. Mammalian ENT homologs are also found in fungi, protozoans, nematodes, insects, and plants, but not in bacteria. Interestingly, ENT members from *Leishmania donovani* function as electrogenic proton cotransporters, suggesting that ENT-type transporters are not always "*equilibrative*" and can be electrogenic and concentrative in certain species. Like the CNTs, the 3′-OH of the ribose moiety is critical for substrate interaction with ENT1 and ENT2, whereas the 2′ and 5′ hydroxyl groups are less important. Halogen modifications on most positions at the base are generally accepted.

In general, hENT2 is more tolerant than hENT1 to modification at the ribose ring and even transports nucleobases that lack the ribose moiety. Analysis of a series of uridine analogs with hENT1 and hENT2 suggests that the C(2′)-OH is a structural determinant for uridine-hENT1 but not for uridine-hENT2 interactions. Moreover, hENT2 displayed more tolerance than hENT1 to removal of C(5′)-OH. The changes in binding energies between transporter proteins and the various uridine analogs suggest that hENT1 may form strong interactions with C(3′)-OH and moderate interactions with C(2′)-OH and C(5′)-OH of uridine, whereas hENT2 may form strong interactions with C(3′)-OH, weak interactions with C(5′)-OH, and no interaction with C(2′)-OH. Glycosylation scanning and antibody studies have confirmed the originally proposed 11 TM topology of hENT1. hENT1 is N-glycosylated at a single site and hENT2 at two sites in the large extracellular loop linking TMs 1 and 2. Glycosylation is not required for the transport activity of hENT1 but may slightly affect the binding affinity to transport inhibitors such as NBMPR. Glycosylation does not change hENT2 function but is required for efficient targeting of hENT2 protein to the plasma membrane.

Investigation of human and rat ENT1/2 has begun to identify functionally important domains of these transporters. The human and rat ENT1 transporters exhibit similar binding affinities toward NBMPR, but there is more than a 50-fold difference between hENT1 and rENT1 in interacting with dipyridamole. Using chimeric constructs of hENT1 and rENT1, Sundaram et al. first demonstrated that TMs 1 to 6 of hENT1 are required for interaction with dipyridamole and dilazep, with TMs 3 to 6 being the major site of interaction. Studies with chimeras between rENT1 and rENT2 indicate that TMs 3 to 6 also represent the major site of NBMPR interaction. Additional work revealed that transplantation of a segment encompassing TMs 5 and 6 of rENT2, which transports nucleobases, could enable rENT1 to transport nucleobases. Consistent with these studies demonstrating the importance of TMs 3 to 6, covalent modification with *p*-chloromercuriphenyl sulfonate of Cys140 in TM 4 of rENT2 resulted in the loss of transport activity, and this modification could be protected by coincubation with the substrate uridine.

Mutagenesis of hENT1 and hENT2, which differ in their sensitivities to dipyridamole and dilazep, revealed that residue Met33, located at the end of the predicted TM1, controls sensitivity to these inhibitors. Site-directed mutagenesis of Gly 154 and Gly179 within TM4 and 5 of hENT1 resulted in impaired transport activity and reduced sensitivity to NBMPR, suggesting potential roles for these residues in substrate recognition and NBMPR binding. Ser 160 and Met89 of hENT1 may play a dominant role in conferring sensitivity to dipyridamole and adenosine/guanosine affinity. In addition, Leu92 of hENT1 has been implicated in interacting with inosine and guanosine, and with NBMPR and dilazep. Although it is well accepted that TMs 1 to 6 contain the functional domains of ENTs, a recent study demonstrated that some residues in TM11 of hENT1 (e.g., Leu 442) are also important for interactions with substrates and inhibitors.

CONTROL OF ENDOGENOUS CARRIERS

Nucleoside transport regulation is cell-type dependent, involves multiple mechanisms, and occurs at both transcriptional and posttranscriptional levels. Pathways that stimulate cell proliferation and differentiation influence the type and abundance of nucleoside transporters in cells and tissues. A number of hormones and chemicals, through interaction with cellular signaling pathways, regulate CNTs and ENTs. Regulation of nucleoside transport activities by cell differentiation has been studied in a number of cell culture models. For example, marked changes in nucleoside transport activity during differentiation were observed in human HL-60 promyelocytic leukemia cells. Undifferentiated HL-60 cells exhibit hENT1- and hENT2-mediated *es* and *ei* transport activities predominantly, with a small amount of Na^+-dependent activity. Upon induction of differentiation by the phorbol ester PMA or dimethyl sulfoxide (DMSO), the cells exhibited a decrease in equilibrative Na^+-independent nucleoside transport that was accompanied by an increase in concentrative Na^+-dependent transport of both pyrimidine and purine nucleosides.

The decreased Na^+-independent transport was attributed to *es* and *ei* transporters with a decrease in V_{max} and no change in affinity. For the *es* transporter, the decrease in nucleoside uptake was paralleled by a decline in NBMPR binding sites. The increased Na^+-dependent transport activity was consistent with CNT3. Recent studies showed that hCNT3 mRNA transcripts were substantially increased in differentiated vs. undifferentiated HL-60 cells. An upstream PMA response element was found in the promoter region of the hCNT3 gene, indicating that PMA may directly activate hENT3 transcription in HL-60 cells. A number of cellular signaling pathways have been implicated in the regulation of nucleoside transport. Lipopolysaccharides (LPSs) and phorbol esters (e.g., PMA), which activate B cells, up-regulated CNTs and down-regulated ENTs in lymphocytes from the bone marrow and in a human B-lymphoblast cell line. The induction of the CNT activity and the down-regulation of the *es* system by LPS and PMA were PKC dependent, and activation of CNT-type transporters also involved tumor necrosis factor-α and nitric oxide signaling pathways in activated human B lymphocytes.

Protein and mRNA levels for hENT1 were reduced and transport activity was decreased in human endothelial cells with exposure to D-glucose, a condition that leads to activation of purinergic receptors. A number of studies reported that limited oxygen availability (hypoxia) down-regulated ENT1 mRNA and protein in several cell types. This down regulation of ENT1 is beneficial because it increases extracellular concentrations of adenosine, which initiates a series of innate protective mechanisms under stressful conditions through activation of cell surface receptors. Recent studies suggest that down-regulation of ENT1 is mediated by the hypoxia inducible factor 1 (HIF-1). The hENT1 promoter region contains HIF-1 binding sites, and binding to HIF-1 is responsible for the transcriptional repression of ENT1 observed during hypoxia. Besides transcriptional regulation, nucleoside transporters are also regulated post-translationally. For example, in cultured MCF-7 breast cancer cells and HeLa cells, acute stimulation of PKC caused a rapid increase in hENT1-associated nucleoside uptake, which appeared to involve PKC δ and/or ε, but not α, β, or γ.

This up-regulation of hENT1 activity appears not to involve transporter recruitment or de novo synthesis, but it is not yet clear whether changes in the phosphorylation state of the transporter are involved. Regulation of nucleoside transporters has also been studied in primary cell cultures and in vivo animal models. In rat liver, expression of rCNT1, rCNT2, and rENT 1 was shown to be affected by developmental stages, growth hormones, partial hepatectomy, interleukin-6, and bile acids. In rat small intestine, where CNT1 and CNT2 are coexpressed in the brush border membranes, fasting of the animals resulted in an increased expression of rCNT1 protein with a concomitant increase in Na^+-dependent thymidine and gemcitabine uptake activity. Changes of mRNA levels of CNTs and ENTs were also observed in various tissues from diabetic rats and human umbilical artery smooth muscle cells from subjects with gestational diabetes.

APPLICATIONS

Mammalian cells acquire purines and pyrimidines via two pathways, de novo biosynthesis and salvage pathways. The primary physiological function of ENTs and CNTs in mammalian cells is to facilitate cellular uptake of natural nucleosides, derived from the diet or produced by tissues such as the liver, for nucleotide synthesis in the salvage pathways. This function is particularly important for cells that lack de novo biosynthetic pathways, such as leukocytes, erythrocytes, bone marrow, and certain brain cells. The wide distributions of ENT1 and ENT2 in various cell types and the ability of these transporters to transport both purine and pyrimidine nucleosides suggest that they play a primary role in nucleoside salvage pathways. The ability of ENT2 to transport nucleobases indicates that this transporter may also contribute to cellular uptake of hypoxanthine, a major salvageable purine nucleobase in animals. In addition, ENTs expressed in intracellular compartments may play a role in organelle transport of nucleosides. It has been shown that some hENT1 is also expressed in mitochondria and thus may transport nucleoside into this organelle for mitochondrial DNA synthesis. ENT3, an intracellular transporter expressed in lysosomes, has been postulated to export nucleosides produced by RNA degradation from the lysosomes.

The CNTs are particularly abundant in epithelial cells of the intestine, kidney, and liver. Expressed in the apical membranes, CNTs are thought to work in concert with the basolaterally localized ENTs to mediate vectorial transepithelial flux of nucleosides and therefore regulate total body homeostasis of nucleosides. Another important physiological function of nucleoside transporters is related to the purine nucleoside adenosine, a key signaling molecule that exerts profound effects on many tissues and organs. Through binding to cell surface receptors (A_1, A_{2A}, A_{2B}, and A_3), adenosine regulates a myriad of physiological processes, such as coronary blood flow, vascular tone regulation, neurotransmitter release, platelet aggregation, immuno-suppression during cellular stress, and inflammation. By influencing the concentration of adenosine available to adenosine receptors, nucleoside transporters, particular ENT1, play important regulatory roles in adenosine signaling. For example, studies with rENT1 antibody revealed abundant expression of rat ENT1 protein in the sinoatrial node, cardiac atrial, and ventricular cells, suggesting an important role of this transporter in regulating the cardiovascular activity of adenosine.

In the human brain, hENT1 and A1 receptors colocalize in various brain regions, indicating an important role of hENT1 in the control of neuromodulatory action of adenosine in the CNS. Interestingly, studies of knockout mice lacking the *ENT1* gene suggested that ENT1 has a physiological role in ethanol-associated behaviors via A1 adenosine receptor. The *ENT1*$^{-/-}$ mice appeared phenotypically normal but consumed twice as much alcohol as the wild-type mice. Finally, recent studies have demonstrated that ENT1 is involved in the anti-inflammatory effects of adenosine under stressful conditions such as hypoxia. Down-regulation of ENT1 appears to serve as an innate protective mechanism during hypoxia.

PHARMACEUTICAL APPROACH

The clinical significance of nucleoside transporters can be viewed in several aspects. First, a number of anticancer nucleoside analogs rely on nucleoside transporters to enter cells to reach their cellular targets. As such, the expression level of nucleoside transporters on the target cells is an important determinant for intracellular drug bioavailability, and consequently, responsiveness to therapy. Second, if a drug is a substrate for nucleoside transporters, the distribution of these transporters in various tissues and organs, particularly the absorptive and excretory organs, may influence its pharmacokinetic and toxicological properties. Finally, nucleoside transporters are potential drug targets. By modulating extracellular concentrations of adenosine, nucleoside transport inhibitors can regulate a variety of physiological processes, which can potentially lead to therapeutic benefits. Nucleoside analogs are an important class of drugs used in the treatment of various forms of leukemia (e.g., cytarabine, fludarabine, cladribine) as well as in solid tumors (e.g., gemcitabine). All of the anticancer nucleoside analogs share a similar mechanism of action.

After entering the cells, nucleoside analogs undergo sequential phosphorylation to form the corresponding nucleoside triphosphates, which exert cytotoxicity by interfering with DNA and/or RNA synthesis and metabolisms. Many anticancer nucleoside analogs are hydrophilic and rely on nucleoside transporters, especially ENT1, to enter cells. In this case the number of functional transporters on the target cells becomes an important determinant for intracellular drug bioavailability and, consequently, responsiveness to therapy. Down-regulation of transporter expression or selection of transporter-deficient cells may contribute to clinical resistance to cytotoxic nucleoside analogs. Indeed, many in vitro studies have demonstrated that nucleoside transporters are necessary for many nucleoside analogs to enter cells, and deficiency in transport can result in resistance to cytotoxic nucleoside analogs such as cytarabine, 5-fluorouridine, 5-fluorodeoxyuridine, and gemcitabine. For example, Mackey et al. have shown in a panel of cancer cell lines that treatment of cells with ENT inhibitors such as NBMPR or dipyridamole increases resistance to gemcitabine 39- to 1800-fold. Consistent with the in vitro observations, several recent clinical studies have implicated a role of hENT1 in clinical resistance to cancer chemotherapy.

Galmarini et al. retrospectively analyzed the effects of hENT1 mRNA expression on efficacy of cytarabine treatment of acute myelogenous leukemia in blast cells of 123 patients treated with cytarabine. Decreased expression of hENT1 was associated with an increased risk of early relapse. Stam et al. studied 18 infants and 24 children with acute lymphoblastic leukemia to determine why infants were sensitive to cytarabine. Their results showed that leukemic blasts from infants were threefold more sensitive to cytarabine than blasts from children. Decreased mRNA levels of deoxycytidine kinase (dCK) but increased levels of hENT1 were observed in infants, which was thought to be responsible for the increased sensitivity of infant acute lymphocytic leukemia. In the case of gemcitabine, Spratlin et al. first examined hENT1 and hCNT3 expression in tumor biopsies from 21 pancreatic cancer patients treated with gemcitabine. Their analysis indicated that patients with detectable hENT1 immunostaining had a significantly longer survival time after gemcitabine chemotherapy than did patients without detectable hENT1. More recently, Giovannetti et al. characterized the expression pattern of genes involved in gemcitabine activity in pancreas tumor specimens from 102 gemcitabine-treated patients and its correlation with treatment outcome.

hENT1 expression significantly correlated with clinical outcome; patients with high levels of hENT1 had a significantly longer overall survival. These encouraging results suggest that measurement of transporter abundance may provide a predictive tool for guiding the appropriate use of anticancer drugs in individual patients for treatment optimization. Nucleoside transporters are found in absorptive and excretory organs and thus may influence the systemic pharmacokinetics of nucleoside analogs. For example, in the epithelia of the intestine and kidney, the CNTs and the ENTs are localized to the apical and basolateral membranes, respectively, and are thought to mediate absorption, reabsorption, or elimination of nucleosides and nucleoside analogs. CNTs and ENTs are also asymmetrically expressed in the blood–brain barrier and choroids plexus. How quantitatively these transporters contribute to the bioavailability, clearance, and tissue-specific disposition of therapeutic nucleoside analogs is largely unknown in vivo.

Pharmacokinetic studies using transporter-deficient animal models may help to elucidate the roles of CNTs and ENTs in determining the absorption, distribution, and excretion of nucleoside analogs. Nucleoside transporters themselves can serve as drug targets. As stated above, the endogenous nucleoside adenosine functions as a signaling molecule and regulates a number of physiological processes via binding to specific cell surface receptors. Nucleoside transporter-mediated cellular uptake represents a major mechanism for the termination of adenosine signaling. By enhancing local adenosine concentrations, ENT inhibitors can potentiate the adenosine effect and produce therapeutic benefits.

Coronary vasodilators such as dipyridamole, dilazep, and draflazine are potent hENT1 inhibitors that can substantially increase and prolong the cardiovascular effects of adenosine. Dipyridamole is used clinically as a vasodilator, and draflazine has been shown to exert cardioprotective effect in humans. Adenosine is an endogenous neuroprotective agent. By binding to A1 and A2 receptors, adenosine decreases excitatory amino acid neurotransmission, inhibits inflammation, and promotes vasodilation. Transport inhibitors are thus potentially valuable in ameliorating ischemia- or hypoxia-induced neuronal injury. The neuroprotective effect of propentofylline in acute stroke patients was thought, at least in part, to be due to its ability to inhibit adenosine transport. Recently, Carrier et al. suggested that the immunosuppressive effects of cannabinoids may also be due to their ability to enhance adenosine signaling through inhibition of ENT1.

INTERINDIVIDUAL VARIATIONS

There are larger interindividual variations in nucleoside transporter expression levels and activities in normal and malignant human tissues and cells. This variation has been implicated in interindividual difference in response to nucleoside analog therapies. Genetic polymorphisms in CNT and ENT genes have been hypothesized to be one of the factors underlying the variations in the expression or activity of nucleoside transporters in humans. Recently, Giacomini and colleagues have investigated genetic variations in the coding regions and flanking intronic regions of CNTs and ENTs in a collection of 270 DNA samples from ethnically diverse U.S. populations. Their studies suggest that there are a number of functionally significant protein-altering single-nucleotide polymorphisms (SNPs) in the *hCNT1* gene coding region. In particular, a rare variant (Ser546Pro) and a single base pair deletion variant (1153 del) with an allele frequency of 3% in the African-American population were found to be nonfunctional.

A common CNT1 variant (Val189Ile) with a frequency of 26% had a reduced affinity for the anticancer nucleoside analog gemcitabine. In contrast, few functionally significant coding-region SNPs were identified for hCNT2, hCNT3, hENT1, and hENT2, suggesting that coding region SNPs of these transporters are unlikely to contribute to interindividual differences in response to nucleoside analog drugs. Recently, Myers et al. examined 1.6 kb upstream of the transcription initiation site of *hENT1* in a limited DNA sample collection. Three SNPs (–1345C > G, –1050G > A, and –706G > C) were found, and analysis of four naturally occurring haplotypes suggests that promoter region haplotypes may affect hENT1 gene expression. The regulatory-region SNPs might offer additional clues to the considerable variability in nucleoside transporter abundance observed in normal and tumor tissues.

CONCLUDING REMARKS

In recent years, significant progress has been made in the studies of nucleoside transporters. Molecular cloning of the CNT and ENT genes has promoted studies that have greatly enhanced our understanding of the substrate and inhibitor specificities, tissue and cellular localization, structure–function relationships, and regulation mechanisms of nucleoside transporters. Knowledge obtained from these studies will pave the way for rational design and development of nucleoside analog drugs with improved pharmacokinetic and pharmacodynamic properties. The first nucleoside transporter-deficient animal has recently been generated for ENT1.

Detailed phenotyping and pharmacological analysis in nucleoside-transporter knockout animal models will provide insights into the physiological, pharmacological, and toxicological functions of nucleoside transporters in vivo, in the context of complex intracellular metabolism and the presence of other transporters (e.g., OATs, MRPs) that also transport some nucleoside analogs and/or their active metabolites. Recently, studies have been launched to correlate nucleoside transporter expression levels with clinical outcomes of therapeutic nucleosides. Pharmacogenomic studies have been undertaken to elucidate the mechanisms leading to inter-individual variations in response to nucleoside drugs. Knowledge gained from these studies will undoubtedly benefit the design, development, and clinical optimization of new and existing nucleoside analogs.

5

Magnetic Nanoparticles

Magnetic nanoparticles occur frequently in nature. They are found not only in the mineral world but also in living organisms. Well known examples are magnetotactic bacteria, which are believed to navigate the waters they live in, by using internal magnetic crystals aligned in chains that function as a compass. Higher forms of life, such as humans, also employ iron as an essential metal. In order to ensure a constant supply of iron, the body stores it within the well-defined protein shell ferritin as a 5 to 7 nm hydrous ferric oxide nanoparticle. The use of magnetic powders in medical applications was already conceptualized by ancient Greek and Roman scientists. However, magnetic nanoparticles have only been used since the mid 1970s in the area of biological and medical sciences. A wide range of *in vivo* as well as *in vitro* applications have been or are currently being developed. These applications include magnetic drug delivery, magnetic fluid hyperthermia, magnetic cell separation and extraction when an external magnetic field is applied, and contrast enhancement for diagnostic imaging procedures, as the magnetic nanoparticles' own magnetic field influences their surrounding. From a practical point of view, magnetic nanoparticles are thus versatile tools that enhance yields for many *in vitro* processes such as cell purifications. In addition, in general, no invasive procedures are required when they are used for *in vivo* therapies.

DEVELOPMENT OF NANOPARTICLES

The development of nanoparticles for biomedical applications requires contributions from the basic to the medical sciences. Such interdisciplinary interactions can sometimes lead to communication problems. For example, the term magnetic nanoparticle has a different meaning for a physician and a biochemist, and might have no meaning at all for a physicist. For this reason, definitions satisfying all partners involved in the present research field are required. With the following simplified definitions, we attempt to provide a universal starting point.

Distinguished Properties

The magnetic properties of materials are mainly related to electrons, with all materials showing some kind of magnetic behavior. Materials can be classified according to their response to external magnetic solicitations. Magnetic susceptibility is defined by the initial slope of the magnetic curve, presenting the magnetization "M" (response) as a function of an applied magnetic field "H" (solicitation). The observed behavior of different materials can be explained in terms of their magnetic structure at the atomic level, and can be summarized as diamagnetism, paramagnetism, and ferromagnetism.

Diamagnetic materials consist of atoms with no net magnetic moment. Nevertheless, they tend to oppose any external magnetic field change due to induced dipoles in the material. For this reason, they are characterized by a slight negative magnetic susceptibility. Paramagnetic materials are made of atoms showing a net magnetic moment. The random orientation of these moments is responsible for a slight positive magnetic susceptibility and no magnetization remains when the external magnetic field is switched off. Ferromagnetic materials react strongly to external magnetic fields, unlike dia- and para-magnetic materials. They can be viewed as paramagnetic materials with an organized domain structure. Within a domain, all atomic magnetic moments are parallel. When submitted to an external magnetic field, the different domains, initially in a random orientation, tend to align according to the external field. This alignment requires domain wall motions and results in hysteresis of the magnetization curve. After the external field is switched off, a remaining or remanent magnetization "M_R" can be observed.

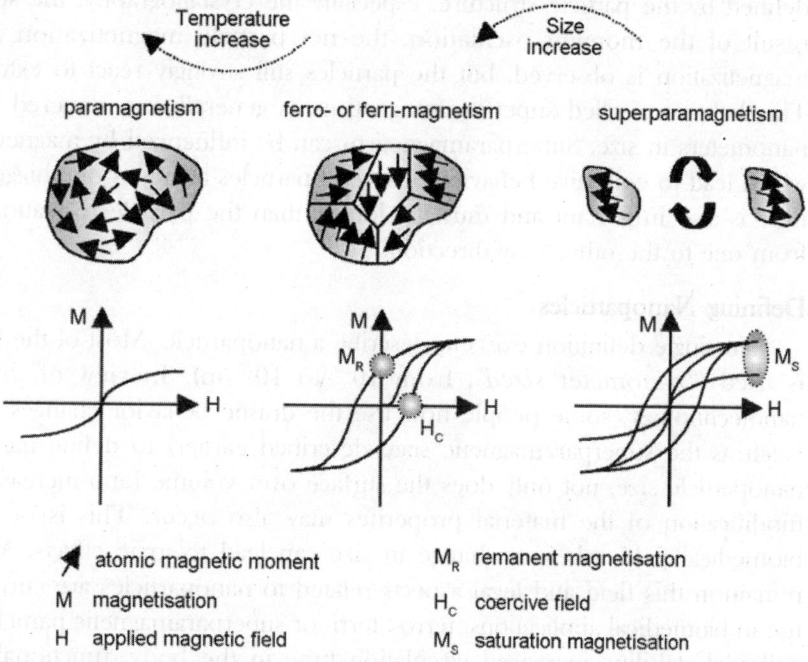

Fig. 5.1. Atomic magnetic moment structure (upper drawings) and corresponding magnetization curves (lower graphs).

Again, in order to achieve a random domain orientation, more energy must be provided by means of an external magnetic field applied in the opposite direction. A coercive field "H_C" is defined as the value of the external field necessary to misalign the domains to a random state. Magnetic materials can be composed of different atoms and ions with various magnetic moments. The most well known example is magnetite (Fe_3O_4), which consists of Fe^{2+} and Fe^{3+} ions. The crystallographic structure of such materials determines whether or not antiferromagnetic or ferrimagnetic properties are present. For magnetic drug targeting, only ferro- and ferri-magnetic materials are of interest, as they react strongly to external magnetic fields due to their non-zero atomic or lattice unit magnetic moment and the domain structure. Temperature also plays a role in the magnitude of magnetic response, as high thermal energy can disturb the atomic moment orientation within the domains, leading to paramagnetic behavior. When ferro- or ferri-magnetic materials are divided, the obtained nanoparticles can become small enough to show single domain structure with a non-zero magnetic moment.

Depending on the particle size, the thermal energy might be high enough to have the particles magnetic moment switch between energetically favorable (or easy) directions. These directions are defined by the particle structure, especially the crystallography, the shape and the surface. As a result of the moment oscillation, the net particle magnetization is zero and no remanent magnetization is observed, but the particles still strongly react to external magnetic solicitations. This behavior, called super-paramagnetism, is generally encountered for particles that are a few nanometers in size. Superparamagnetism can be influenced by magnetic interparticle interactions, which lead to collective behavior of several particles acting as one bigger particle. The observation time is also important and must be longer than the particle relaxation time, necessary to switch from one to the other easy direction.

Defining Nanoparticles

No single definition exists to describe a nanoparticle. Most of the time, an arbitrary size range is used ("*nanometer sized*", from 10^{-9} to 10^{-6} m). In view of the recent developments in nanotechnology, some people now use the drastic behavior changes arising below a critical size (such as the superparamagnetic state described earlier) to define nanoparticles. When reducing nanoparticle size, not only does the surface over volume ratio increase gradually, but a complete modification of the material properties may also occur. This is of primary importance in the biomedical field, where a change in size can lead to toxic effects. Many unanswered questions remain in this field and legal aspects related to nanoparticles are currently under discussion. For use in biomedical applications, ferro-, ferri- or superparamagnetic particles must be coated to ensure colloidal stability, increased circulation time in the body, functional surfaces, and appropriate diagnostic properties.

In this regard, the term "*magnetic nanoparticle*" not only refers to an inorganic core responsible for magnetic properties, but also to a composite structure with one or several cores coated or embedded in a matrix. In addition to a compatible coating, magnetic nanoparticles used in clinical applications must form stable aqueous suspensions. Suspensions are complex dynamic systems. Their equilibrium is influenced by the forces present, including Van der Waals, electrostatic, steric, and magnetic forces, as well as by Brownian motion. On this account, it is crucial to realize that solvent modifications can drastically influence the behavior of the system. The term "*ferrofluid*" is correctly used only in the case of a colloidal stable suspension of single domain nanoparticles.

BIOMEDICAL APPLICATIONS

In general, a single particle type cannot be used for all applications. Instead, the composition, size and production route of synthesized magnetic nanoparticles is determined by the target application. Although superparamagnetic, ferro- and ferrimagnetic particles can all be used for magnetic drug carrier applications, super-paramagnetic particles are favored for biomedical applications, due to the fact that they behave non-magnetically when they are not under the influence of an external magnetic field, thus preventing undesired magnetic agglomeration. To further assist in preventing agglomerations, to optimize bio-interactions with the host environment and to maximize biocompatibility, the choice of appropriate surface chemistries and functionalizations is also important. Many magnetic nanoparticles are available with different surface chemistries.

Synthetic Process

In biomedical applications, the most commonly used magnetic nanoparticles are superparamagnetic magnetite (Fe_3O_4) and maghemite (γ-Fe_2O_3). This is due to their ease of synthesis using chemical or physical approaches, as well as their general bio-compatibility (and FDA approval). Massart's aqueous coprecipitation method, which leads to particles easily dispersible in water, is the most cited method of magnetic nanoparticle preparation. The particle size can be tuned in the 3 to 30 nm size range and the particles usually show an ellipsoidal shape. The stoichiometry ranges from magnetite to maghemite, the two crystallographic structures being very similar. The size distribution is about 10 to 20%. Time consuming size sorting procedures allow for further narrowing of the size distribution to about 5% in the best case. Iron oxide nanoparticles have been synthesized intensively during the past decades, but until recently, phase and size control have been problematic.

A newly developed two-step approach has allowed for much better control over the particle structure. In this approach, metal particles are first obtained and then oxidized in a controlled way. Size distributions of better than 5% can be achieved in the 4 to 16 nm range, as shown by Alivisatos *et al.* and Hyeon *et al.* These particles are, however, often not appropriate for biomedical applications as they do not disperse easily in water. Many other magnetite and maghemite nanoparticle synthesis approaches can be found in the literature, but none are significantly different from the ones presented above. Slightly modified nanoparticles can also be obtained by partly replacing the iron in the magnetite or maghemite structure with cobalt or nickel. This in turn changes the magnetic properties of the particles.

Encapsulation of Cobalt

From a magnetic point of view, particles showing a stronger reaction to magnetic fields are desirable. Cobalt achieves this aim, but its toxicity is a major drawback. One way of preventing or minimizing this toxicity caused by cobalt ion leakage is the inorganic encapsulation of cobalt with, for example, silica.

Coated Iron

Pure iron nanoparticles can be synthesized, but their sensitivity to oxidation is a major drawback for biomedical applications. Thus, a coating, as described for cobalt particles, should be used. Iron has also been coated or alloyed with platinum, cobalt and carbon.

Magnetic Moment

Depending on the application, magnetic nanoparticles may be combined into larger conglomerates to increase the overall magnetic moment. Great care should be paid to interparticle magnetic interactions. The superparamagnetic behavior of a system might, for example, be lost due to such interactions. Also, the magnetic core concentration must be kept constant among the magnetic conglomerates to yield homogeneous magnetic moments, and thus a consistent response to an applied magnetic field. Either a single or a two-step approach can be used to synthesize magnetic particles. In the one-step approach, a "*linker*" is present while synthesizing the magnetic nanoparticles. In the two-step approach, the linker is added subsequently. The linker can be organic or inorganic and is chosen for its chemical and biocompatible characteristics. For biomedical applications, dextran, starch, polyethylene glycol (PEG), polyvinyl alcohol (PVA), silica and gold are among the most common compounds.

Researches on Biocompatibility

One of the first papers to discuss the biocompatibility issues of magnetic particles was published in the early 1970's by Nakamura *et al.* The authors prepared fine carbonyl-iron particles and infused them into different animal species *in vivo*. They concluded that to achieve optimal results, the magnetic particles should be coated with a biocompatible material and be as round as possible. Further research showed that pure magnetic metal particles, such as iron, cobalt and nickel particles, should not be used directly *in vivo* because they oxidize easily and release +2 or +3 charged metal ions that can exert unwanted as well as toxic effects.

Iron ions, for example, are problematic in that they produce and catalyze oxygen radical formation. Cobalt and nickel ions have been found to induce adverse tissue reactions, and to promote infection and metal sensitivity. In contrast to pure magnetic particles, iron oxides and superparamagnetic iron oxide nanoparticles (SPION) coated and stabilized with hydrophilic polymers have been found to be quite thermodynamically stable under physiological conditions, not exerting obvious toxic effects.

In fact, they are similar in size and core composition to the natural non-toxic magnetic nanoparticles found in magnetotactic bacteria and in human tissue. Pharmacokinetic studies of small magnetite nanoparticles destined for magnet resonance imaging have shown that the magnetite nanoparticles are taken up by the cells of the *reticuloendothelial system* (RES) and are transported intracellularly to lysosomes, where they slowly oxidize at low pH and are then recycled by the body. Within 20-40 days, up to 60% of the iron is recovered in the red blood cells, as determined using radiolabeled ^{59}Fe. Recent discussions have centered on the fate and toxic effects of (magnetic) nanoparticles in humans after inhalation.

Rodent models have shown the potential problematic effects of such particles to include the induction of asthma, inflammation, and potentially even cancer. Some of these effects might be due to the fact that particles smaller than 100 nm are not exhaled, but are almost completely retained in the alveoli. For this reason, acute effects can rapidly turn into chronic effects. Another possible cause for concern is the report that small particles have been found behind the blood brain barrier.

Further research needs to clarify if these particles directly crossed the blood brain barrier or via the nose. Care must be taken to relate the effects seen in animal models to the human situation, especially since effects seen in rodents do not seem to develop in humans. Clarification of the short and long term risks of nanoparticle use is the aim of several programs being initiated in 2005 by the European, American and Canadian governments.

MAGNETIC NANOPARTICLES AS DRUG CARRIERS

The following section presents an overview of the use of magnetic nanoparticles sized 1 μm or less for the delivery of drugs. Magnetic microspheres of larger than 1 μm size are also mentioned in a few places, but for a recent and thorough review, as well as for the history of magnetic drug delivery, the reader is advised to consult a more extensive treaty on this matter such as the recent volume of the MML series. In this section, magnetic nanoparticles will be grouped according to their mechanism of action including magnetic hyperthermia; the delivery of magnetic nanoparticles that (slowly) release drugs (tumor treatment, thrombolysis, delivery of antiinfective, antiarthritic, antifungal, and antiscar agents, and local anesthesia or neuroblocking agents) or act without drug release (radiotherapy or embolization); and the improved delivery of peptides (gene transfer).

Heat Production

Magnetic nanoparticles in an alternating current (AC) magnetic field produce heat by Neel and Brownian relaxation. Heat production above a person's normal body temperature is called hyperthermia and can be medically used for the eradication of cancer cells. Temperatures above 56°C lead to thermoablation. Once magnetic nanoparticles have successfully reached certain organs or tissues, especially tumors, magnetic hyperthermia can be induced. Normal tissue nearby, not containing the magnetic nanoparticles, remains at body temperature and is thus spared. One of the first to examine this effect was Gilchrist who published a seminal paper in 1956 on the selective inductive heating of lymph nodes, after injection of maghemite particles sized between 20 and 100 nm diameter directly into the lymph nodes near surgically removed canine tumors. Using 5 mg of Fe_2O_3 per gram of lymph node and a magnetic field strength of 240 Oe, a maximum temperature rise of 14°C was reached within 3 minutes.

To prevent a reoccurrence of the cancer, hyperthermia is normally combined with a second treatment modality such as chemotherapy or irradiation. Twenty years later, Rand *et al.* showed that ferrosilicone can induce heat after being infused into a tumor's blood supply and placed under the influence of a strong magnetic alternating field. Rand's ferromagnetic silicone microspheres were based on Turner *et al.'s* research in 1973, in which magnetic particles of unknown but probably larger size in a silicone fluid were infused into and then clogged (embolized) the capillary bed of several targeted organs. With this technique, the researchers successfully embolized the blood supplies of different tumors. No side effects were reported in the 7 patients who had brain tumors, pheochromocytomas, a tongue tumor and a hypernephroma. Rand's so-called "*magnetic field induced hyperthermia*" was then further developed by Sako *et al.*, who showed that heating was reproducible and proportional to the amount of iron used. The use of single domain, dextran-coated magnetite nanoparticles for tumor hyperthermia was developed by Jordan and Chan, and is currently undergoing *in vivo* and clinical testing.

Jordan reported the optimal nanoparticle core diameter to be in the 10 nm range, although type of coating and coating stability also seemed to be important. Using these nanoparticles, 5 mg of material per gram of tumor was sufficient to increase the tumor temperature by 10°C to cell toxic levels. Jordan is currently conducting a clinical phase II trial combining magnetic hyperthermia and radiation therapy. He recently presented the first clinical results from 8 patients at the 5th

International Conference on the Scientific and Clinical Applications of Magnetic Carriers in Lyon, France. The patients were treated for cervix, rectal, and prostate carcinoma, as well as for a chondrosarcoma, rhabdomyosarcoma, and liver metastasis. During the 1-hour sessions, after local injection of the magnetic particles, the tumor temperature increased to 43-50°C under the influence of a magnetic field of 3 to 9.5 kA/m and a frequency of 100 kHz. While no additional nanoparticle injections were necessary, the hyperthermia treatment was repeated from 2-11 times.

The magnetic fluid hyperthermia was well tolerated. Two patients showed complete remission 9 and 14 months after treatment, while the other six patients showed local control with no recurrent growth of the tumors. These results are very promising. Another group in Germany led by Hilger is working on circumventing the drawback of having to directly inject the particles into a tumor, by using antibody- bound magnetic nanoparticles which are able to target breast cancer, followed by magnetic field hyperthermia. Although their particles are taken up extensively by tumor cells and show a specific heating power of up to 170 W/g, there is still more work needed to increase the number of particles in the tumor and to reach a homogeneous tumor distribution.

Magnetic hyperthermia is also possible with large microspheres that contain magnetic nanoparticles. As an example, Moroz *et al.* incorporated 100 nm maghemite particles into 32 µm biocompatible plastic particles and then embolized the arterial blood supply of liver tumors with them. In an animal study with 10 rabbits, the VX2 tumor volumes decreased significantly within 2 weeks. The development of maghemite nanoparticles with very high AC losses is ongoing. Hergt *et al.* are in the process of characterizing the largest, but still superparamagnetic particles, optimizing the coatings such as carboxydextran or polyethylene glycol, and investigating the exact mechanism of heat production in an AC magnetic field. Magnetic hyperthermia is an exciting cancer treatment possibility and is profiting from ongoing research into its mechanism of action and from improved magnetic materials. The proof of principle has advanced to the clinical stage with the construction and clinical testing of Jordan's magnetic field therapy system. The targeted (cancer) cell uptake of sufficient amounts of magnetic nanoparticles from a patient's blood supply could make magnetic hyperthermia the method of choice for many different kinds of tumors.

Drug Targeting

Magnetic drug delivery is able to concentrate drugs in a tumor if the tumor is accessible through the arterial system and has a good supply of blood. Magnetic drug delivery thus promises to deliver highly effective anticancer drugs with fewer side effects, and with shorter and less toxic treatments. Most drug release from magnetic particles occurs passively, by desorption from and diffusion out of the particle matrix. The main driving forces are pH, osmolarity and concentration differences between particles and the blood/tissue. Widder and Senyei were the first to successfully illustrate this concept with the chemotherapeutic drug doxorubicin encapsulated into albumin-coated magnetite particles sized around 1-2 µm. Targeting a distinct area of a rat's tail, they were able to deliver 200 times more of the drug than intravenous application of the same amount of free drug could achieve. Taking it a step further, they treated Yoshida rats with sarcomas in their tails and attained complete remission in 77% of the rats. The magnetic albumin microspheres were never tested clinically, likely because the magnetophoretic mobility (overall magnetic responsiveness to a magnetic field) was considered too low for deeper applications. This changed with the introduction of iron-carbon particles originally developed in Russia and then brought to clinical trial by the company FeRx. FeRx's irregularly-shaped carbon-coated iron particles, of 0.5 to 5 μm in diameter and with a very high magnetic susceptibility, were loaded with doxorubicin and showed promising results and very low therapy-related toxicity in the treatment of inoperable liver cancer.

Unfortunately, FeRx ceased to exist in 2004 when a preliminary analysis of their ongoing clinical trial failed to convince investors of the method's superiority over other treatment methods. Not only doxorubicin, but also many other chemotherapeutic drugs can be and have been adsorbed to magnetic nanoparticles made from many different matrix materials. Examples of chemotherapeutic magnetic nanoparticles tested *in vivo* include polyalkylcyanoacrylate nanoparticles of 220 nm diameter filled with (adsorbed) dactinomycin, chitosan nanoparticles of 530 nm diameter loaded with oxantrazole, solid lipid nanoparticles of 450-570 nm diameter loaded with methotrexate, and ferro carbon of 100 nm diameter loaded with carminomycine. Each of these nanoparticles has been tested in animal experiments with positive results.

Specifically, after intravenous injection, the drug concentration tripled in the target organ when a magnet was placed above it, compared with a control without an applied magnet. It seems that the intravenous injection of magnetic nanoparticles, even very close to the target region, is not optimal. This was well documented in a clinical cancer therapy trial performed by Lubbe *et al.* in 14 patients. They used magnetic nanoparticles of 100 nm in diameter loaded with 4′-epidoxorubicin for the treatment of advanced solid cancer. The phase I study clearly showed accumulation of magnetic nanoparticles in the target area without toxic effects. MRI measurements, however, indicated that more than 50% of the magnetic nanoparticles were deposited in the liver. This was likely due to the particles' low magnetic susceptibility and small size, which limited their ability to be held at the target organ. Intraarterial injection into the blood supply that leads to the target region might be much more effective for magnetic drug targeting for this reason. The above examples of magnetic drug targeting with magnetic nanospheres are only a subset of all the magnetic drug delivery attempts.

Influence of Magnetic Field

Under the influence of a magnetic field, magnetic particles align in chains and eventually agglomerate. Depending on particle size and shape, this can lead to embolization (clogging) of the blood vessels and especially of the small capillaries of 7 to 10 μm in diameter. This accumulation of particles can be used on its own to starve the target tissue of oxygen, produce hypoxia and induce necrosis. The magnetic particles used for this approach are generally larger, such as the "*iron sponge*" of 10-30 μm used by Sako *et al.*, but can also consist of nanoparticles in a more lipophilic solvent such as the ferrosilicone employed by Turner *et al.* Turner added a catalyst to their ferrosilicone suspension, which resulted in vulcanization of the viscous slurry 14 min after injection into 7 patients with diverse solid tumors. Magnetic particles can also be used for tumor treatment without releasing any drugs. For this purpose, magnetic particles can incorporate radioisotopes either in the matrix or bound to their surface, and then deliver tumor cell-toxic radiation doses wherever they accumulate.

External magnetic fields or internal magnetizable wires can be used to accumulate radioactive magnetic particles and hold them at the target site. The particles irradiate the area within the specific treatment range of the isotope. Initial experiments in mice showed that intraperitoneally injected radioactive poly(lactic acid) based magnetic microspheres (10-20 μm m in diameter) could be concentrated near a subcutaneous tumor in the belly area above which a small magnet had been attached. The dose-dependent irradiation from the β-emitter ^{90}Y-containing magnetic particles resulted in the complete disappearance of more than half of the tumors. Iron carbon-based smaller radioactive particles of around 1 μm have been radiolabeled with different diagnostic (^{99m}Tc, ^{111}In) and therapeutic (^{188}Re, ^{90}Y) radioisotopes. Targeting studies to distinct liver regions in swine by our group showed that more than 90% of the injected radioactive magnetic particles were accumulated underneath a strong NdFeB-magnet.

The radioactive particles stayed in the target region for at least 3 days, even after the removal of the magnet. Magnetic nanoparticles of much smaller diameters are being used clinically for diagnostic purposes, mainly as contrast agents. The accumulation of these magnetic particles is, however, based on non-specific properties such as the tissue- specific pore size (fenestration) or enhanced permeability and retention effect (EPR) seen in tumor tissues, but not on magnetic targeting. Recent examples of nonspecific targeting are the internalization into cells of the positively charged at peptide bound to therapeutic agents, such as radioactively complexed ^{99m}Tc and ^{188}Re, and superparamagnetic iron oxides known as tat-CLIO (tat-cross linked iron oxides). Magnetic drug delivery is also able to deliver other types of drugs such as highly potent antiinfective, blood clot-dissolving, anti-inflammatory, anti-arthritic, photodynamic therapy and paralysis-inducing drugs, among many others.

A good example of these applications was reported in 1988. In this study, Torchilin *et al.* surgically induced a thrombus in both carotid arteries of a dog, fixed a permanent magnet near one of them, and 1 hr later, intravenously injected 1 μm-sized dextran microspheres with covalently bound streptokinase. The side without the magnet completely occluded within 4 hrs, while the magnet side returned to initial blood flow conditions after about 30 min and appeared completely open at histological examination. Torchilin noted that these results were achieved using doses 10 times lower than those used when streptokinase is directly injected. Similar results were obtained in the same year by another group using 30-60 nm PEGylated magnetite particles containing urokinase. In both cases, the thrombolytic activity remained at background levels outside the targeted region.

Newest Application

The newest application of magnetic nanoparticles is for targeted and enhanced gene delivery in potential applications such as wound repair and the treatment of cancer, eye disease, and cystic fibrosis. Magnetically enhanced gene transfer may be able to overcome the current lack of selectivity of the existing vectors and low efficiency of gene transfer. The mechanisms by which magnetic nanoparticles can improve on transfection rates are by magnetically forced contact and by increasing the plasmid concentration magnetically. Magnetofection was first described in a Japanese patent by Harata *et al.* who used magnetic liposomes to transfect cells. Bergemann *et al.* described the first experiments of transfecting cytokine-induced killer cells (CIK-cells) with plasmid DNA carrying distinct interleukin genes. However, their magnetic nanoparticles were only used as plasmid carriers, not as the driving force, which was provided by electroporation. Using a magnet as the driving force was then described a couple of *years* later by Plank *et al.* For successful *in vitro* and *in vivo* transfections, only very small amounts of plasmid were necessary, and the transfection occurred in a matter of just a couple of minutes. This speedy and efficient transfection at low vector doses is the main advantage of magnetofection. Furthermore, remotely controlled vector targeting *in vivo* seems possible.

CONCLUDING REMARKS

The technological advancements in the material and engineering sciences, and especially in the nanotechnology revolution, with its increasing molecular approach to the synthesis, derivatization, combination, self-assembly, and manipulation of materials, will guide improvements in all aspects of magnetic nano- and micropartides. These advancements include the synthesis of higher magnetic nanophases; the increasing availability of stronger magnets and engineered magnetic fields; the ability to prepare more uniform particles; and the rendering of these particles biocompatible and ultimately biodegradable without toxicity. For real therapeutic breakthrough, however, a few challenges in the field of magnetic drug delivery still need to be addressed. One of the challenges is the difficulty involved in generating the focused field profiles needed to target magnetic nanoparticles deep within the body, due to the speed with which the magnetic fields drop off as their distance from the source increases (intensity = $1/r^3$). Another challenge centers on attaining homogeneous particle distributions given that blood flow in the target region can vary from very fast (100-200 cm/s) to very slow (0.05 cm/s).

A final challenge is the optimization of magnetic nanoparticles in terms of magnetization and size uniformity. All these problematic areas are currently being addressed by multidisciplinary groups worldwide. The potential of magnetic drug delivery is great. In addition to their magnetic responsiveness, magnetic nanoparticles carry an innate signal that can be used for magnetic resonance imaging. Furthermore, other imaging modalities such as radioisotope or fluorescence imaging can be used after derivatizing the particle surface. There is no limitation to the kind of drugs that can be encapsulated or bound to magnetic nanoparticles. Also, current pharmaceutical techniques allow for the development of drug release profiles for a large group of drugs and diseases.

Magnetic targeting devices such as the recently FDA approved Niobe system improve on the precise manipulation of magnetic forces. Anatomical and physiological conditions in a patient are, however, complicated, and successful therapies will have to be specifically adapted to each disease and ideally applied under imaging control. The treatment of cancer is an especially good target for magnetic drug delivery. However, any disease that could benefit from precise control over the delivery of highly effective, but potentially toxic substances would also be a good candidate. For example, antiangionic drugs could be delivered to the back of the eye to prevent blindness in patients with age-related macular degeneration. Attempts to develop such a drug delivery system are ongoing.

6

Pharmacokinetics

It has long been recognized that improving one or more of the intrinsic adsorption, distribution, metabolism, and excretion (ADME) properties of a drug is a critical step in developing more effective drug therapies. As early as 1906, Paul Ehrlich proposed altering drug distribution by conjugating toxic drugs to "*magic bullets*" (antibodies) having high affinity for cancer cell-specific antigens, in order to both improve the therapeutic efficacy of cancer while decreasing its toxicity. Since then, it has become clear that directly improving intrinsic ADME through modifications of the drug is limited or precluded by structural requirements for activity. In other words, low molecular mass drugs are too small and have only limited number of atomic groups that can be altered to improve ADME, which often adversely affects drug pharmacological activity. In turn, the modifications of many low molecular mass drugs, aimed to increase their pharmacological activity, often adversely affect their ADME properties. For example, the potency and specificity of drugs can be improved by the addition of hydrophobic groups.

The associated decrease in water solubility, however, increases the likelihood of drug aggregation, leading to poor absorption and bioavailability during oral administration or lowered systemic bioavailability, high local toxicity, and possible pulmonary embolism during intravenous administration. Although there have been considerable difficulties for improving some existing drugs through chemical modifications, the problem became even more obvious with the development of high-throughput drug discovery technologies. Almost half of lead drug candidates identified by high-throughput screening have poor solubility in water, and are abandoned before the formulation development stage. In addition, newly synthesized drug candidates often fail due to poor bioavailability, metabolism and/or undesirable side effects, which together decrease the therapeutic index of the molecules. Furthermore, a new generation of biopharmaceuticals and gene therapy agents are emerging based on novel biomacromolecules, such as DNA and proteins.

The use of these biotechnology-derived drugs is completely dependent on efficient delivery to the critical site of the action in the body. Therefore, drug delivery research is essential in the translation of newly discovered molecules into potent drug candidates and can significantly improve therapies of existing drugs. Polymer-based drugs and drug delivery systems emerged from the laboratory bench in the 1990s as a promising therapeutic strategy for the treatment of certain devastating human diseases. A number of polymer therapeutics are presently on the market or undergoing clinical evaluation to treat cancer and other diseases. Most of them are low molecular weight drug molecules or therapeutic proteins that are chemically linked to water-soluble polymers to increase drug solubility, drug stability, or enable targeting to tumors. Recently, as a result of rapid development of novel nanotechnology-derived materials, a new generation of polymer therapeutics has emerged, using materials and devices of nanoscale size for the delivery of drugs, genes, and imaging molecules.

These materials include polymer micelles, polymer-DNA complexes ("*polyplexes*"), liposomes, and other nanostructured materials for medical use that are collectively known as nanomedicines. Compared with first generation polymer therapeutics, the new generation nanomedicines are more advanced. They entrap small drugs or biopharmaceutical agents such as therapeutic proteins and DNA, and can be designed to trigger the release of these agents at the target site. Many nanomedicines are constructed using self-assembly principles such as the spontaneous formation of micelles or interpolyelectrolyte complexes, driven by diverse molecular interactions (hydrophobic, electrostatic, etc.). This chapter considers polymeric micelles as an important example of the new generation of nanomedicines, which is also perhaps among the most advanced approach toward clinical applications in diagnostics and the treatment of human diseases.

SELF-ASSEMBLED POLYMER MICELLES

Self-assembled polymer micelles are created from amphiphilic polymers that spontaneously form nanosized aggregates when the individual polymer chains ("*unimers*") are directly dissolved in aqueous solution (dissolution method) above a threshold concentration (*critical micelle concentration* or CMC) and solution temperature (*critical micelle temperature* or CMT). Amphiphilic polymers with very low water solubility can alternatively be dissolved in a volatile organic solvent, then dialyzed against an aqueous buffer (dialysis method). Amphiphilic di-block (hydrophilic-hydrophobic) or tri-block (hydrophilic-hydrophobic-hydrophilic) copolymers are most commonly used to prepare self-assembled polymer micelles for drug delivery, although the use of graft copolymers has been reported. For drug delivery purposes, the individual unimers are designed

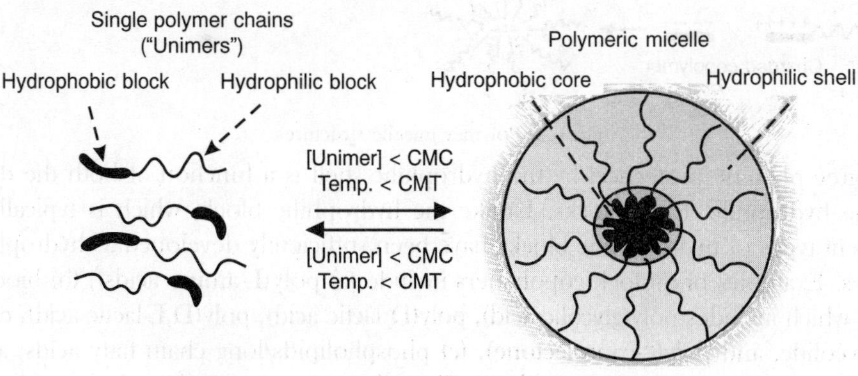

Fig. 6.1. Self-assembly of block copolymer micelles.

to be biodegradable and/or have a low enough molecular mass (< ~40 kDa) to be eliminated by renal clearance, in order to avoid polymer buildup within the body that can potentially lead to toxicity.

The most developed amphiphilic block copolymers assemble into spherical core-shell micelles approximately 10 to 80 nm in diameter, consisting of a hydrophobic core for drug loading and a hydrophilic shell that acts as a physical ("*steric*") barrier to both micelle aggregation in solution, and to protein binding and opsonization during systemic administration. The most common hydrophilic block used to form the hydrophilic shell is the FDA-approved excipient poly(ethylene glycol) (PEG) or poly(ethylene oxide) (PEO). PEG or PEO consists of the same repeating monomer subunit CH_2-CH_2-O, and may have different terminal end groups, depending on the synthesis procedure, e.g. hydroxyl group HO-$(CH_2$-CH_2-$O)_n$-H; methoxy group CH_3O-$(CH_2$-CH_2-$O)_n$-H, etc. PEG/PEO blocks typically range from 1 to 15 kDa. In addition to its FDA approval, PEG is extremely soluble and has a large excluded volume. This makes it especially suitable for physically interfering with intra-micelle interactions and subsequent micelle aggregation. PEG also blocks protein and cell surface interactions, which greatly decreases nanoparticle uptake by the reticulo-endothelial system (RES), and consequently increases the plasma half life of the polymer micelle.

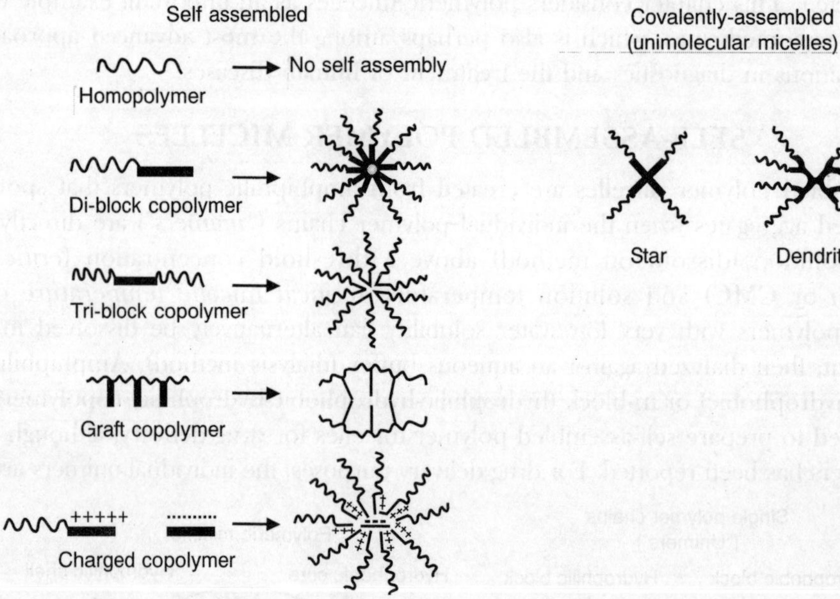

Fig. 6.2. Polymer micelle structures.

The degree of steric protection by the hydrophilic shell is a function of both the density and length of the hydrophilic PEG blocks. Unlike the hydrophilic block, which is typically PEG or PEO, different types of hydrophobic blocks have been sufficiently developed as hydrophobic drug loading cores. Examples of diblock copolymers include (a) poly(L-amino acids), (b) biodegradable poly(esters), which includes poly(glycolic acid), poly(D lactic acid), poly(D,L-lactic acid), copolymers of lactide/glycolide, and poly(ε-caprolactone), (c) phospholipids/long chain fatty acids; and for tri-block copolymers, (d) polypropylene oxide (in Pluronics/poloxamers). The choice of hydrophobic block is largely dictated by drug compatibility with the hydrophobic core (when drug is physically loaded, as described later) and the kinetic stability of the micelle.

The self-assembly of amphiphilic copolymers is a thermodynamic and, consequently, a reversible process that is entropically driven by the release of ordered water from hydrophobic blocks; it is either stabilized or destabilized by solvent interactions with the hydrophilic shell. As such, the structural potential of amphiphilic copolymer unimers to form micelles is determined by the mass ratio of hydrophilic to hydrophobic blocks, which also affects the subsequent morphology if aggregates are formed. If the mass of the hydrophilic block is too great, the copolymers exist in aqueous solution as unimers, whereas, if the mass of the hydrophobic block is too great, unimer aggregates with non-micellar morphology are formed. If the mass of the hydrophilic block is similar or slightly greater than the hydrophobic block, then conventional core shell micelles are formed.

An important consideration for drug delivery is the relative thermodynamic (potential for disassembly) and kinetic (rate of disassembly) stability of the polymer micelle complexes, after intravenous injection and subsequent extreme dilution in the vascular compartment. This is because the polymer micelles must be stable enough to avoid burst release of the drug cargo, as in the case of a physically loaded drug, upon systemic administration and remain as nanoparticles long enough to accumulate in sufficient concentrations at the target site. The relative thermodynamic stability of polymer micelles (which is inversely related to the CMC) is primarily controled by the length of the hydrophobic block.

An increase in the length of the hydrophobic block alone significantly decreases the CMC of the unimer construct (i.e. increases the thermodynamic stability of the polymer micelle), whereas an increase in the hydrophilic block alone slightly increases the CMC (i.e. decrease the thermodynamic stability of a polymer micelle). Although the CMC indicates the unimer concentration below which polymer micelles will begin to disassemble, the kinetic stability determines the rate at which polymer micelle disassembly occurs. Many diblock copolymer micelles possess good kinetic stability and only slowly dissociate into unimers after extreme dilution. Thus, although polymer micelles are diluted well below typical unimer CMCs (10^{-6}–10^{-2} M) after intravenous injection, their relative kinetic stability might still be suitable for drug delivery. The kinetic stability depends on several factors, including the size of a hydrophobic block, the mass ratio of hydrophilic to hydrophobic blocks, and the physical state of the micelle core. The incorporation of hydrophobic drugs may also further enhance micelle stability.

Unimolecular Micelles

Unimolecular micelles are topologically similar to self-assembled micelles, but consist of single polymer molecules with covalently linked amphiphile chains. For example, copolymers with star-like or dendritic architecture, depending on their structure and composition, can either aggregate into multimolecular micelles, or exist as unimolecular micelles. Dendrimers are widely used as building blocks to prepare unimolecular micelles, because they are highly-branched, have well-defined globular shape and controled surface functionality. For example, unimolecular micelles were prepared by coupling dendritic hypercores of different generations with PEO chains. The dendritic cores can entrap various drug molecules. However, due to the structural limitations involved in the synthesis of dendrimers of higher generation, and relatively compact structure of the dendrimers, the loading capacity of such micelles is limited. Thus, to increase the loading capacity, the dendrimer core can be modified with hydrophobic block, followed by the attachment of the PEO chains. For example, Wang *et al.* recently synthesized an amphiphilic 16-arm star polymer with a polyamidoamine dendrimer core and arms composed of inner lipophilic poly(ε-caprolactone) block and outer PEO block.

These unimolecular micelles were shown to encapsulate a hydrophobic drug, etoposide, with high loading capacity. Multiarm star-like block copolymers represent another type of unimolecular micelles. Star polymers are generally synthesized by either the arm-first or core-first methods. In the arm-first method, monofunctional living linear macromolecules are synthesized and then cross-linked either through propagation, using a bifunctional comonomer, or by adding a multifunctional terminating agent to connect precise number of arms to one center. Conversely, in the core-first method, polymer chains are grown from a multifunctional initiator. One of the first reported examples of unimolecular micelles, suitable for drug delivery, was a three-arm star polymer, composed of mucic acid substituted with fatty acids as a lipophilic inner block and PEO as a hydrophilic outer block. These polymers were directly dispersible in aqueous solutions and formed unimolecular micelles. The size and solubilizing capacity of the micelles were varied by changing the ratio of the hydrophilic and lipophilic moieties. In addition, star-copolymers with poly-electrolyte arms can be prepared to develop pH-sensitive unimolecular micelles as drug carriers.

Multimolecular Micelles

The multimolecular micelles structure can be reinforced by the formation of cross- links between the polymer chains. These resulting cross-linked micelles are, in essence, single molecules of nanoscale size that are stabile upon dilution, shear forces and environmental variations (e.g. changes in pH, ionic strength, solvents etc.). There are several reports on the stabilization of the polymer micelles by cross-linking either within the core domain or throughout the shell layer. In these cases, the cross-linked micelles maintained small size and core-shell morphology, while their dissociation was permanently suppressed. Stable nanospheres from the PEO-b-polylactide micelles were prepared by using polymerizable group at the core segment. In addition to stabilization, the core polymerized micelles readily solubilized rather large molecules such as paclitaxel, and retained high loading capacity even upon dilution.

Formation of interpenetrating network of a temperature-sensitive polymer (poly-N-isopropylacrilomide) inside the core was also employed for the stabilization of the Pluronic micelles. The resulting micelle structures were stable against dilution, exhibited temperature-responsive swelling behavior, and showed higher drug loading capacity than regular Pluronic micelles. Recently, a novel type of polymer micelles with cross-linked ionic cores was prepared by using block ionomer complexes as templates. The nanofabrication of these micelles involved condensation of PEO-b-poly(sodium methacrylate) diblock copolymers by divalent metal cations into spherical micelles of core-shell morphology. The core of the micelle was further chemically cross-linked and cations removed by dialysis. Resulting micelles represent hydrophilic nanospheres of core-shell morphology.

The core comprises a network of the cross-linked polyanions and can encapsulate oppositely charged therapeutic and diagnostic agents, while a hydrophilic PEO shell provides for increased solubility. Furthermore, these micelles displayed the pH- and ionic strength-responsive hydrogel-like behavior, due to the effect of the cross-linked ionic core. Such behavior is instrumental for the design of drug carriers with controled loading and release characteristics.

METHODS FOR LOADING DRUGS

In general, there are three major methods for loading drugs into polymer micelle cores: (i) chemical conjugation, (ii) physical entrapment or solubilization, and (iii) polyionic complexation (e.g. ionic binding).

Chemical Drug Linkage

Drug incorporation into polymer micelles via chemical conjugation was first proposed by Ringsdorf's group in 1984. According to this approach, a drug is chemically conjugated to the core-forming block of the copolymer via a carefully designed pH- or enzyme-sensitive linker, that can be cleaved to release a drug in its active form within a cell. The polymer-drug conjugate then acts as a polymer pro-drug which self assembles into a core-shell structure. The appropriate choice of conjugating bond depends on specific applications. The nature of the polymer–drug linkage and the stability of the drug conjugate linkage can be controled to influence the rate of drug release, and therefore, the effectiveness of the prodrug. For instance, recent work by Kataoka's group proposed pH-sensitive polymer micelles of PEO-b-poly(aspartate hydrazone doxorubicin), in which doxorubicin was conjugated to the hydrophobic segments through acid-sensitive hydrazone linkers that are stable at extracellular pH 7.4, but degrade and release the free drug at acidic pH 5.0 to 6.0 in endosomes and lysosomes.

The original approach developed by this group used doxorubicin conjugated to the poly(aspartic acid) chain of PEO-b-poly(aspartic acid) block copolymer through an amide bond. Adjusting both the composition of the block copolymer and the concentration of the conjugated doxorubicin, led to improved efficacy, as evidenced by a complete elimination of solid tumors implanted in mice. It was later determined that doxorubicin physically encapsulated within the micellar core was responsible for antitumor activity. This finding led to the use of PEO-b-poly(aspartate doxorubicin) conjugates as nanocontainers for physically entrapped doxorubicin.

Physical Drug Linkage

The physical incorporation or solublization of drugs within block copolymer micelles is generally preferred over micelle-forming polymer-drug conjugates, especially for hydrophobic drug molecules. Indeed, many polymers and drug molecules do not contain reactive functional groups for chemical conjugation, and therefore, specific block copolymers have to be designed for a given type of drug. In contrast, a variety of drugs can be physically incorporated into the core of the micelles, by engineering the structure of the core-forming segment. In addition, molecular characteristics (i.e. molecular weight, composition, presence of functional groups for active targeting) within a homologous copolymer series can be designed to optimize the performance of a drug for a given drug delivery situation. This concept was introduced by our group in the late 1980s and was initially termed "*micellar microcontainer*", but is now widely known as a "*micellar nanocontainer*".

Haloperidol was encapsulated in Pluronic block copolymer micelles, the micelles were targeted to the brain using brain-specific antibodies or insulin, and enhancement of neuroleptic activity by the solubilized drug was observed. During the last 25 years, a large variety of amphiphilic block copolymers have been explored as nanocontainers for various drugs. Different loading methods can be used for physical entrapment of the drug into the micelles, including but not limited to dialysis, oil in water emulsification, direct dissolution, or solvent evaporation techniques. Depending on the method, drug solubilization may occur during or after micelle assembly. The loading capacity of the polymer micelles, which is frequently expressed in terms of the micelle-water partition coefficient, is influenced by several factors, including both the structure of core-forming block and a drug, molecular characteristics of the copolymer such as composition, molecular weight, and the solution temperature. Many studies indicate that the most important factor related to the drug solubilization capacity of a polymer micelle is the compatibility between the drug and the core-forming block. For this reason, the choice of the core-forming block is most critical.

One parameter that can be used to assess the compatibility between the polymer and a drug is the Flory-Huggins interaction parameter, χ_{sp}, defined as $\chi_{sp} = (\delta_s - \delta p)^2 V_s /kT$, where δ_s and δ_p are Scatchard-Hildebrand solubility parameters, and V_s is the molecular volume of the solubilizate. It was successfully used as a correlation parameter for the solubilization of aliphatic and aromatic hydrocarbons in block copolymer micelles. Recently, Allen's group elegantly demonstrated that the calculation and comparison of partial solubility parameters of polymers and drugs could be used as a reliable means to predict polymer-drug compatibility and to guide formulation development. Polymer micelles, possessing core-forming blocks predicted to be compatible with the drug of interest (Ellipticine), were able to increase the solubility of the drug up to 30,000 times, compared with its saturation solubility in water. The degree of compatibility between the drug and the core-forming block has also been shown to influence the release rate of the drug from the micelles. When the environment within the core of the micelle becomes more compatible with the drug, it results in a considerable decrease in the rate of drug release. For a given drug, the extent of incorporation is a function of factors that also control the micelle size and/or aggregation number. Such factors include the ratio of hydrophobic to hydrophilic block length and the copolymer molecular weight. For example, the loading capacity of Pluronic micelles was found to increase with the increase in the hydrophobic PPO block length.

This effect is attributed to a decrease in CMC, and therefore, an increase in aggregation number and micelle core size. Also, but to a lesser extent, the hydrophilic block length affects the extent of solubilization, such that an increase in percentage of PEO in Pluronic block copolymers results in a decrease in the loading capacity of the micelles. For a given ratio of PPO-to-PEO, higher molecular weight polymers form larger micelles, and therefore, show a higher drug loading capacity. Therefore, the total amount of loaded drug can be adjusted as a function of the micellar characteristics as clearly was demonstrated by Nagaradjan and Kozlov *et al.* Several studies indicate that both the copolymer concentration as well as the drug to polymer ratio upon loading, have a complex effect on the loading capacity of polymer micelles. In general, more polymer chains provide more absorption sites. As a result, solubilization is increased with polymer concentration. However, the solubilization capacity was found to reach a saturation level with an increase of polymer concentration.

The maximum loading level is largely influenced by the interaction between the solubilizate and core-forming block, and stronger interactions enable saturation to be reached at lower polymer concentration. It was also demonstrated in the studies by Hurter and Hatton that the loading capacity of micelles formed from copolymers with high hydrophobic content was independent of the polymer concentration. In addition, the location of the incorporated molecules within polymer micelles (micelle core or the core-shell interface) determines the extent of solubilization, as well as the rate of drug release. It has been found that more soluble compounds are localized at the core-shell interface or even in the inner shell, whereas more hydrophobic *molecules* have a tendency to solubilize in the *micelle* core.

The release rate of drug localized in the shell or at the interface appears to account for the "*burst release*" from the micelles. In general, for drugs physically incorporated in polymer micelles, release is controled by the rate of diffusion of the drug from the micellar core, stability of the micelles, and the rate of biodegradation of the copolymer. If the micelle is stable and the rate of polymer biodegradation is slow, the diffusion rate of the drug will be mainly determined by the abovementioned factors, i.e. the compatibility between the drug and core forming block of copolymer, the amount of drug loaded, the molecular volume of drug, and the length of the core forming block. In addition, the physical state of the micelle core and drug has a large influence on release characteristics. It was demonstrated that the diffusion of incorporated molecules from the block copolymer micelles with glassy cores is slower, in comparison to the diffusion out of the cores that are more mobile.

Charged Therapeutic Agents

Charged therapeutic agents can be incorporated into block copolymer micelles, through electrostatic interactions with an oppositely charged ionic segment of block copolymer. Since it was being proposed independently by Kabanov and Kataoka in 1995, this approach is now widely used for the incorporation of various polynucleic acids into block ionomer complexes, for developing non-viral gene delivery systems. Ionic block lengths, charge density, and ionic strength of the solution affect the formation of stable block ionomer complexes, and therefore, control the amount of drug that can be incorporated within the micelles. The pH- and salt-sensitivity of such block ionomer micelles provide a unique opportunity to control the triggered release of the active therapeutic agent. Furthermore, block ionomer complexes can participate in the polyion interchange reactions which are believed to account for the release of the therapeutic agent and DNA in an active form inside cells.

Several comprehensive reviews can be found in the literature that focus on block ionomer micelles as drug and gene delivery systems. In addition, physicochemical aspects of the DNA complexes with cationic block copolymers have also been recently reviewed. As an example, the metal-complex formation of ionic block copolymer, PEO-b-poly(L-aspartic acid), was explored to prepare polymer micelles incorporating cis-dichlorodiamminoplatinum (II) (CDDP); a potent chemotherapeutic agent widely used in the treatment of a variety of solid tumors, particularly, testicular, ovarian, head and neck, and lung tumors. The CDDP-loaded micelles had a size of approximately 20 nm. These micelles showed remarkable stability upon dilution in distilled water, while in physiological saline, they displayed sustained release of the regenerated Pt complex over 50 hrs, due to inverse ligand exchange from carboxylate to chloride. The release rate was inversely correlated with the chain length of poly(L-aspartic acid) segments in the block copolymer. The stability of CDDP-loaded micelle against salt was shown to be improved by the addition of homopolymer, poly(L-aspartic acid), in the micelles. Recently, CDDP-loaded micelles were newly prepared using another block copolymer, PEO-b-poly(glutamic acid) to improve and optimize the micellar stability, as well as the drug release profile.

The drug loading in the micelles was as high as 39% (w /w), and these micelles released the platinum in physiological saline at 37°C in sustained manner > 150 hrs, without initial burst of the drug. The principle of polyionic complexation can also be used to design new photosensitizers for photodynamic therapy of cancer. The group of Kataoka reported formation of micelles, as a result of mixing of oppositely charged dendrimer porphirin and block ionomer, based on electrostatic assembly or combination of electrostatic and hydrogen bonding interactions. The micelles were stabile at physiological conditions and released the entrapped dendrimers in the acidic pH environment (pH 5.0), suggesting a possibility of pH-triggered drug release in the intracellular endosomal compartments.

Overall, the photodynamic efficacy of the dendrimer porphyrins was dramatically improved by inclusion into micelles. This process resulted in more than two orders of magnitude increase in the photocytotoxicity, compared with that of the free dendrimer porphyrins. In addition, the polyionic complexation has been used to immobilize charged enzymes such as egg white lysozyme or trypsin, which were incorporated in the core of polyion micelles, after mixing with oppositely charged ionic block copolymer. A remarkable enhancement of enzymatic activity was observed in the core of the micelles. Furthermore, the on-off switching of the enzyme activity was achieved through the destabilization of the core domain by applying a pulse electric field. These unique features of the polyion micelles are relevant for their use as smart nanoreactors in the diverse fields of medical and biological engineering. Last, but not the least, a special class of polyion complexes has been synthesized by reacting block ionomers with surfactants of opposite charge, resulting in the formation of environmentally responsive nanomaterials, which differ in sizes and morphologies, and include micelles and vesicles.

These materials contain a hydrophobic core formed by the surfactant tail groups, and a hydrophilic shell formed, for example, by PEO chains of the block ionomer. These block ionomer complexes can incorporate charged surfactant drugs such as retinoic acid, as well as other drugs via solubilization in the hydrophobic domains formed by surfactant molecules. They display transitions induced by changes in pH, salt concentration, chemical nature of low molecular mass counterions, as well as temperature. They can also be fine tuned to respond to environmental changes occurring in a very wide range of conditions that could realize during delivery of biological and imaging agents. The unique self-assembly behavior, the simplicity of the preparation, and the wide variety of available surfactant components that can easily produce polymer micelles with a very broad range of core properties, make this type of materials extremely promising for developing vehicles for the delivery of diagnostic and therapeutic modalities.

BIODIVERSIFIED STUDIES

Incorporation of a low molecular mass drug into polymer micelles drastically alters pharmacokinetics and biodistribution of the drug in the body, which is crucial for the drug action.

Low molecular mass drugs, after administration in the body, rapidly extravasate to various tissues affecting them almost indiscriminately, and then are rapidly eliminated from the body via renal clearance, often causing toxicity to kidneys. Furthermore, many drugs display low stability and are degraded in the body, often forming toxic metabolites. An example is doxorubicinol, a major metabolite of doxorubicin, which causes cardiac toxicity. These impediments to the therapeutic use of low molecular mass drugs can be mitigated by encapsulating drugs in polymer micelles. Within the micelles, the drug molecules are protected from enzymatic degradation by the micelle shell.

The pharmacokinetics and biodistribution of the micelle-incorporated drugs are mainly determined by the surface properties, size, and stability of the micelles, and are less affected by the properties of the loaded drug. The surface properties of the micelles are determined by the micelle shell. The shell from PEO effectively masks drug molecules and prevents interactions with serum proteins and cells, which contributes to prolonged circulation of the micelles in the body. From the size standpoint, polymer micelles fit an ideal range of sizes for systemic drug delivery. On the one hand, micelles are sufficiently large, usually exceeding 10 nm in diameter, which hinders their extravasation in nontarget tissues and prevents renal glomerular excretion. On the other hand, the micelles are not considered large, since their size usually does not exceed 100 nm. As a result, micelles avoid scavenging by the *mononuclear phagocytes system* (MPS) in the liver and spleen. To this end, "stealth" particles whose surface is decorated with PEO are known to be less visible to macrophages and have prolonged half-lives in the blood.

The contribution of the micelle stability to pharmacokintetics and biodistribution is much less understood, although it is clear that micelle degradation should result in a decrease of the size and drug release, perhaps, prematurely. Degradation of the micelles, resulting in the formation of block copolymer unimers, could also be a principal route for the removal of the polymer material from the body. The molecular mass of the unimers of most block copolymers is below the renal excretion limit, i.e. less than ~20 to 40 kDa, while the molecular mass of the micelles, which usually contain several dozen or even hundreds of unimers molecules, is above this limit. Thus, the unimers are sufficiently small and can be removed via renal excretion, while the micelles cannot. A recent study by Batrakova *et al.* determined pharmacokinetic parameters of an amphiphilic block copolymer, Pluronic P85, and perhaps provided first evidence that the pharmacokinetic behavior of a block copolymer can be a function of its aggregation state.

Specifically, the formation of micelles increased the half-life of the block copolymer in plasma and decreased the uptake of the block copolymer in the liver. However, it had no effect on the total clearance, indicating that the elimination of Pluronic P85 was controled by the renal tubular transport of unimers, but not by the rate of micelles disposition or disintegration. Furthermore, the values of the clearance suggested that a significant portion of the block copolymer was reabsorbed back into the blood, probably, through the kidney's tubular membranes. Chemical degradation of the polymers comprising the micelles, followed by renal excretion of the relatively low molecular mass products of degradation, may be another route for the removal of the micelle polymer material from the body. This route could be particularly important in the case of the cross-linked or unimolecular micelles, micelles displaying very high stability, and/or micelles composed from very hydrophobic polymer molecules that can bind and retain considerably biological membranes and other cellular components. The delivery of chemotherapeutic drugs to treat tumors is one of the most advanced areas of research using polymer micelles.

Two approaches have been explored to enhance delivery of drug-loaded polymer micelles to the tumor sites: (i) passive targeting and (ii) vectorized targeting. The passive targeting involves enhanced permeability and retention (EPR) effect. It is based on the fact that solid tumors display increased vascular density and permeability caused by angiogenesis, impaired lymphatic recovery, and lack of a smooth muscle layer in solid tumor vessels. As a result, micellar drugs can penetrate and retain in the sites of tumor lesions. At the same time, extravasation of micellar drugs in normal tissues is decreased, compared with low molecular drug molecules. Among normal organs, spleen and liver can accumulate polymer drugs, but the drugs are eventually cleared via the lymphatic system. The increased circulation time of the micellar drugs should further enhance exposure of the tumors to the micellar drug, compared with the low molecular mass drugs. Along with passive targeting, the delivery of micellar drugs to tumors can potentially be enhanced by the modification of the surface of the polymer micelles with the targeting molecules, vectors that can selectively bind to the surface of the tumor cells.

Potential vectors include antibodies, aptamers and peptides, capable of binding tumor-specific antigens and other molecules diplayed at the surfaces of the tumors. Altered biodistribution of a common antineoplastic agent was demonstrated for CDDP encapsulated in polyionic micelles with PEO-b-poly(glutamic acid) block copolymers. Free CDDP is rapidly distributed to each organ, where its levels peak at about one hr after *i.v.* administration. In contrast, in the case of the CDDP-incorporated micelles, due to their remarkably prolonged blood circulation time, the drug level in the liver, spleen and tumor continued to increase up to at least 24 hrs after injection. Consequently, the CDDP-incorporated micelle exhibited 4-, 39-, and 20-fold higher accumulation in the liver, spleen and tumor respectively, than the free CDDP. At the *same* time, the encapsulation of CDDP into the micelles significantly decreased drug accumulation in the kidney, especially during first hr after administration. This suggested potential for the decrease of severe nephrotoxicity observed with the free drug, which is excreted through the glomerular filtration, thus affecting the kidney. Promising results were also demonstrated for doxorubicin incorporated into styrene-maleic acid micelles. In this case, as a result of drug entrapment into micelles, the drug was redirected from the heart to the tumor, and the doxorubicin cardiotoxicity was diminished.

Complete blood counts and cardiac histology for the micellar drug showed no serious side effects for *i.v.* doses as high as 100 mg/kg doxorubicin equivalent in mice. Similar results were reported for doxorubicin incorporated in mixed micelles of PEO-b-poly(L-histidine) and PEO-b-poly(L-lactic acid) block copolymers. Tissue levels of doxorubicin administered in the micellar formulation were decreased in the blood and the liver, and considerably increased in the solid tumor, compared with the free drug. Further increase in the tumor delivery was achieved by modifying the surface of the micelles with the folate molecules. The accumulated doxorubicin levels observed using folate-modified micelles was 20 times higher than those for free doxorubicin, and 3 times higher than those for unmodified micelles. The first micellar formulation of doxorubicin to reach clinical evaluation stage, used the micelles composed of triblock copolymer, PEO-b-poly(propylene oxide)-b-PEO, Pluronic. Analysis of pharmacokinetics and biodistribution of doxorubicin incorporated into mixed micelles of Pluronics L61 and F127, SP1049C, demonstrated more efficient accumulation of the micellar drug in the tumors, compared with the free drug.

Specifically, the areas under the curves (AUC) in the Lewis lung carcinoma 3LLM-27 solid tumors in C57B1/6 mice were increased about two fold using SP1049, compared with the free doxorubicin. Furthermore, this study indicated that the peak levels of doxorubicin formulated with SP1049 in the tumor were delayed and the drug residence time was increased, in comparison with the free doxorubicin. A clear visualization of drug delivery to the tumor site was shown for doxorubicin covalently incorporated through the pH-sensitive link into polymer micelles of PEO-poly(aspartate hydrazone doxorubicin). A phase-contrast image showed that the tumor blood vessels containing the micelles leaked into extravascular compartments of the tumors, resulting in the infiltration of the micelles into tumor sites.

The micelles circulated in the blood for a prolonged time, and the AUC for micellar doxorubicin was 15-fold greater than the AUC for the free doxorubicin. Furthermore, the AUC values of the micellar doxorubicin in the heart and kidney decreased, compared with the free drug. Thus, the selectivity of drug delivery to the tumor, compared with heart and kidney (AUC_{tumor}/ AUC_{organ}) was increased by 6- and 5-folds respectively. This may result in the reduction of side effects of doxorubicin such as cardiotoxicity and nephrotoxicity. Moreover, the micellar doxorubicin showed relatively low uptake in the liver and spleen, despite very long residence time in the blood. Biodistribution of paclitaxel incorporated into biodegradable polymer micelles of monomethoxy-PEO-b-poly(D,L-lactide) block copolymer, Genexol-PM, was compared with the regular formulation of the drug in Cremophor EL.

Two to three-fold increases in drug levels were demonstrated in most tissues including liver, spleen, kidneys, lungs, heart and tumor, after *i.v.* administration of Genexol-PM, compared with paclitaxel. Nevertheless, acute dose toxicity of Genexol-PM was about 25 times lower than that of the conventional drug formulation, which appears to be a result of the reformulation avoiding the use of Chremophor EL and dehydrated ethanol that are toxic. Selective tumor targeting with paclitaxel encapsulated in micelles, modified with tumor-specific antibodies 2C5 ("*immunomicelles*"), was reported using Lewis lung carcinoma solid tumor model in C57B1/6J mice. These micelles were prepared from PEO-distearyl phosphatidylethanolamine conjugates with the free PEO end activated with the p-nitrophenylcarbonyl group for the antibody attachment.

The amount of micellar drug accumulated in the tumor exceeded that in the non-target tissue (muscles) by more than ten times. It is worth noting that the highest accumulation in the tumor was demonstrated in the micelles containing the longest PEO chains, which also had the longest circulation time in the blood. Furthermore, the immunomicelles displayed the highest amount of tumor-accumulated drug, compared with either free paclitaxel or non-vectorized micelles. It was demonstrated that paclitaxel delivered by plain micelles in the interstitial space of the tumor was eventually cleared after gradual micellar degradation. In contrast, paclitaxel-loaded 2C5 immunomicelles were internalized by cancer cells and the retention of the drug inside the tumor was enhanced. Unexpected results were found using pH-sensitive polymer micelles of N-isopropylacrylamide and methacrylic acid copolymers randomly or terminally alkylated with octadecyl groups. It was demonstrated that aluminium chloride phthalocyanine (AlClPc) incorporated in such micelles was cleared more rapidly and less accumulated in the tumor, than the AlClPc formulated with Cremophor EL.

Furthermore, significant accumulation in the liver and spleen (and lungs for most hydrophobic copolymers) was observed, compared with Cremophor EL formulation. The enhanced uptake of such polymer micelles by the cells of mononuclear phagocyte system (MPS) could be due to micelle aggregation in the blood and embolism in the capillaries. Thus, it attempted to reduce the uptake of the micelles in MPS by incorporating water soluble monomers, N-vinyl-2-pyrrolidone in the copolymer structure. The modified formulation displayed same levels of tumor accumulation and somewhat higher antitumor activity than the Cremophor EL formulation. This work serves as an example reinforcing the need of proper adjustment of the polymer micelle structure, and perhaps the need of using block copolymers to produce a defined protective hydrophilic shell to facilitate evasion of the polymer micelles from MPS.

APPLICATIONS OF POLYMER MICELLES

The studies on the application of polymer micelles in drug delivery have mostly focused on the following areas that are considered below: (i) delivery of anticancer agents to treat tumors; (ii) drug delivery to the brain to treat neurodegenerative diseases; (iii) delivery of antifungal agents; (iv) delivery of imaging agents for diagnostic applications; and (v) delivery of polynucleotide therapeutics.

Pioneering Studies

To enhance chemotherapy of tumors using polymer micelles, four major approaches were employed: (i) passive targeting of polymer micelles to tumors due to EPR effect; (ii) targeting of polymer micelles to specific antigens overexpressed at the surface of tumor cells; (iii) enhanced drug release at the tumor sites having low pH; and (iv) sensitization of drug resistant tumors by block copolymers. A series of pioneering studies by Kataoka's group used polymer micelles for passive targeting of various anticancer agents and chemotherapy of tumors. One notable recent example reported by this group involves polymer micelles of PEO-b-poly(L-aspartic acid) incorporating CDDP. Evaluation of anticancer activity using murine colon adenocarcinoma C26 as an *in vivo* tumor model, demonstrated that CDDP in polymer micelles had significantly higher activity than the free CDDP, resulting in complete eradication of the tumor. A formulation of paclitaxel in biodegradable polymer micelles of monomethoxy-PEO-b-poly(D,L-lactide) block copolymer, Genexol-PM, also displayed elevated activity *in vivo* against human ovarian carcinoma OVCAR-3 and human breast carcinoma MCF7, compared with a regular formulation of the drug in Cremophor EL.

In addition, anthracydine antibiotics, doxorubicin and pirarubicin, incorporated in styrene-maleic acid micelles each revealed potent anticancer effects *in vivo* against mouse sarcoma S-180, resulting in complete eradication of tumors in 100% of tested animals. Notably, animals survived for more than one year, after treatment with the micelle-incorporated pirarubicin at doses as high as 100 mg/kg of pirarubicin equivalent. Complete blood counts, liver function test, and cardiac histology showed no sign of adverse effects for intravenous doses of the micellar formulation. In contrast, animals receiving free pirarubicin had a much reduced survival and showed serious side effects.

Collectively, these studies suggested that various micelle-incorporated drugs display improved therapeutic index in solid tumors, which correlates with enhanced passive targeting of the drug to the tumor sites, as well as decreased side effects, compared with conventional formulations of these drugs.

Tumor-specific targeting of polymer micelles to molecular markers expressed at the surface of the cancer cells has also been explored to eradicate tumor cells. For example, a recent study by Gao's group developed a polymer micelle carrier to deliver doxorubicin to the tumor endothelial cells with overexpressed $\lambda_v\beta_3$ integrins. A cyclic pentapeptide, cRGD was used as a targeting ligand that is capable of selective and high affinity binding to the $\lambda_v\beta_3$ integrin. Micelles of PEO-b-poly(ε-caprolactone) loaded with doxorubicin were covalently bound with cRGD. As a result of such modification, the uptake of doxorubicin-containing micelles in *in vitro* human endothelial cell model derived from Kaposi's sarcoma, was profoundly increased. In addition, folate receptor often overexpressed in cancer cells has been evaluated for targeting various drug carriers to tumors. This strategy has also been evaluated to target polymer micelles. For example, mixed micelles of PEO-b-poly(L-histidine) and PEO-b-poly(L-lactic acid) block copolymers with solubilized doxorubicin or micelles of PEO-b-poly(DL-lactic-co-glycolic acid) block copolymer with covalently attached doxorubicin, were each surface modified by conjugating folate molecules to the free PEO ends. In both cases, *in vitro* and *in vivo* studies demonstrated increased antitumor activity of the micelle-incorporated drug resulting from such modification.

The enhanced delivery of the micellar drugs through the folate receptor, and the enhanced retention of the modified micelles at the tumor sites are possible explanations for the effects of these folate modifications. Micelles conjugated with antibodies or antibody fragments capable to recognize tumor antigens were shown to improve therapeutic efficacy *in vivo* over non-modified micelles. This approach can result in high selectivity of binding, internalization, and effective retention of the micelles in the tumor cells. In addition, recent advances in antibody engineering allow for the production of humanized antibody fragments, reducing problems with immune response against mouse antibodies. For example, micelles of PEO-distearyl phosphatidyl-ethanolamine were covalently modified with the monoclonal antibody 2C5 that binds to nudeosomes, displayed at the surface of many tumor cells. The micelles were then used for incorporating various poorly soluble anticancer drugs including tamoxifen, paclitaxel, dequalinium, and chlorine ε6 trimethyl ester. It was shown that paclitaxel-loaded 2C5-immunomicelles could specifically recognize a variety of tumor types.

The binding of these immunomicelles was observed for all cancer cell lines tested, i.e. murine Lewis lung carcinoma, T-lymphoma EL4, and human breast adenocarcinomas, BT-20 and MCF7. Moreover, paclitaxel-loaded 2C5 immunomicelles demonstrated highest anticancer activity in Lewis lung carcinoma tumor model in mice, compared with plain paclitaxel-loaded micelles and the free drug. The increased antitumor effect of immunomicelles *in vivo* correlated with the enhanced retention of the drug delivered with the immunomicelles inside the tumor. Tumors often display low pH of interstitial fluid, which is mainly attributed to higher rates of aerobic and anaerobic glycolysis in cancer cells than in normal cells. This phenomenon has been employed in the design of various pH-sensitive polymer micelle systems for the delivery of anticancer drugs to the tumors.

One approach consisted in the chemical conjugation of anticancer drugs to the block copolymers through pH sensitive cleavable links that are stable at neutral pH, but are cleavable and release the drug in the mildly acidic pH. For example, several groups used hydrasone-based linking groups, to covalently attach doxorubicin to PEO-b-poly(DL-lactic-co-glycolic acid) block copolymer, PEO-b-block-poly(allyl glycidyl ether) or PEO-b-poly(aspartate hydrazone) block copolymer.

It was suggested that doxorubicin will remain in the micelles in the blood stream, and will be released at tumor sites at lower pH. For example, *in vitro* and *in vivo* studies using PEO-b-poly(aspartate hydrazone doxorubicin) micelles demonstrated that the micelles display an intracellular pH-triggered drug release capability, tumor-infiltrating permeability, and effective antitumor activity with extremely low toxicity. Overall, the animal studies suggested that such polymer micelle drug has a wide therapeutic window due to increased efficacy and decreased toxicity, compared with free doxorubicin. An alternative mechanism for pH-induced triggering of drug release at the tumor sites consists of using pH sensitive polyacids or polybases as building blocks for polymer micelles. For example, mixed micelles of PEO-b-poly(L-histidine) and PEO-b-poly(L-lactic acid) block copolymers incorporate pH-sensitive poly-base, poly(L-histidine) in the hydrophobic core. The core can also solubilize hydrophobic drugs such as doxorubicin.

The protonation of the polybase at acidic conditions resulted in the destabilization of the core and triggered release of the drug. This system was also targeted to the tumors through the folate molecules as described earlier and has shown significant *in vivo* antitumor activity and less side effects, compared with the free drug. Notably, it was also effective *in vitro* and *in vivo* against multidrug resistant (MDR) human breast carcinoma MCF7/ADR that overexpresses P-glycoprotein (Pgp). Pgp is a drug efflux transport protein that serves to eliminate drugs from the cancer cells and significantly decreases the anticancer activity of the drugs. The micelle incorporated drug was released inside the cells, and thus avoided the contact with Pgp localized at the cell plasma membrane, which perhaps contributed to the increased activity of pH sensitive doxorubicin micelles in the MDR cells. A different approach using Pluronic block copolymer micelles to overcome MDR in tumors has been developed by our group. Studies by Alakhov *et al.* demonstrated that Pluronic block copolymers can sensitize MDR cells, resulting in an increased cytotoxic activity of doxorubicin, paclitaxel, and other drugs by 2,3 orders of magnitude.

Remarkably, Pluronic can enhance drug effects in MDR cells through multiple effects including (i) inhibiting drug efflux transporters, such as Pgp and multidrug resistance proteins (MRPs), (ii) abolishing drug sequestration within cytoplasmic vesicles, (iii) inhibiting the glutathione/glutathione S-transferase detoxification system, and (iv) enhancing proapoptotic signaling in MDR cells. Similar effects of Pluronics have also been reported using *in vivo* tumor models. In these studies, mice bearing drug-sensitive and drug-resistant tumors were treated with doxorubicin alone and with doxorubicin in Pluronic compositions. The tumor panel included *i.p.* murine leukemias (P388, P388-Dox), *s.c.* murine myelomas (Sp2/0, Sp2/0-Dnr), *i.v.* and *s.c.* Lewis lung carcinoma (3LL-M27), *s.c.* human breast carcinomas (MCF7, MCF7/ADR), and *s.c.* human oral epidermoid carcinoma (KBv). Using the NCI criteria for tumor inhibition and increased lifespan, Pluronic/doxorubicin has met the efficiency criteria in all models (9 of 9), while doxorubicin alone was only effective in selected tumors (2 of 9).

Results showed that the tumors were more responsive in the Pluronic/doxorubicin treatment groups than in doxorubicin alone. These studies demonstrated improved treatment of drug resistant cancers with Pluronics. The mechanisms of effects of Pluronic on Pgp have been studied in great detail. In particular, exposure of MDR cells to Pluronics has resulted in the inhibition of Pgp-mediated efflux, and this overcomes defects in intracellular accumulation of Pgp-dependent drugs, and abolishes the directionality difference in the flux of these drugs across polarized cell monolayers. The lack of changes in membrane permeability with Pluronics to (i) non-Pgp compounds in MDR cells, and (ii) Pgp probes in non-MDR cells, suggested that Pluronic effects were specific to the Pgp efflux system. These effects were observed at Pluronic concentrations less than or equal to the critical micelle concentration (CMC). Thus, Pluronic unimers rather than the micelles were responsible for these effects. Specifically, Pluronic molecules displayed a dual function in MDR cells. Firstly, they incorporated into the cell membranes and decreased the membrane microviscosity. This was accompanied by the inhibition of Pgp ATPase activity.

Secondly, they translocated into cells and reached intracellular compartments. This was accompanied by the inhibition of respiration, presumably due to Pluronic interactions with the mitochondria membranes. As a result, within 15 min after exposure to select Pluronics, intracellular levels of ATP in MDR cells were drastically decreased. Remarkably, such ATP depletion was not observed in non-MDR cells, suggesting that the Pluronic was "selective", with respect to the MDR phenotype. Combining these two effects, Pgp ATPase inhibition and ATP depletion, resulted in the shut-down of the efflux system in MDR cells. The Pgp remained functionally active when (i) ATP was restored using an ATP supplementation system in the presence of a Pluronic, or (ii) when ATP was depleted, but there was no direct contact between the Pluronic and Pgp (and no ATPase inhibition). Overall, these detailed studies which resulted in the development of a micellar formulation of doxorubicin that is evaluated clinically, reinforce the fact that block copolymers, comprising the micelles, can serve as biological response modifying agents that can have beneficial effects in the chemotherapy of tumors.

Drug Transport to the Nervous System

By restricting drug transport to the brain, the blood brain barrier (BBB) represents a formidable impediment for the treatment of brain tumors and neurodegenerative diseases such as HIV-associated dementia, stroke, Parkinson's and Alzheimer's diseases. Two strategies using polymer micelles have been evaluated to enhance delivery of biologically active agents to the brain. The first strategy is based on the modification of polymer micelles with antibodies or ligand molecules capable of transcytosis across brain microvessel endothelial cells, comprising the BBB. The second strategy uses Pluronic block copolymers to inhibit drug efflux systems, particularly, Pgp, and selectively increase the permeability of BBB to Pgp substrates.

An earlier study used micelles of Pluronic block copolymers for the delivery of the CNS drugs to the brain. These micelles were surface-modified by attaching to the free PEO ends, either polyclonal antibodies against brain-specific antigen, α_2-glycoprotein, or insulin to target the receptor at the lumenal side of BBB. The modified micelles were used to solubilize fluorescent dye or neuroleptic drug, haloperidol, and these formulations were administered intravenously in mice. Both the antibody and insulin modification of the micelles resulted in enhanced delivery of the fluorescent dye to the brain and drastic increases in neuroleptic effect of haloperidol in the animals. Subsequent studies using *in vitro* BBB models demonstrated that the micelles, vectorized by insulin, undergo receptor-mediated transport across brain microvessel endothelial cells.

Based on one of these observations, one should expect development of novel polymer micelles that target specific receptors at the surface of the BBB to enhance transport of the incorporated drugs to the brain. The studies by our group have also demonstrated that selected Pluronic block copolymers, such as Pluronic P85, are potent inhibitors of Pgp, and they have the increased entry of the Pgp-substrates to the brain across BBB. Pluronic did not induce toxic effect in BBB, as revealed by the lack of alteration in paracellular permeability of the barrier, and in histological studies, using specific markers for brain endothelial cells. Overall, this strategy has potential in developing novel modalities for the delivery of various drugs to the brain, including selective anti cancer agents to treat metastatic brain tumors, as well as HIV protease inhibitors to eradicate HIV virus in the brain.

Antifungal Agents

The need for safe and effective modalities for the delivery of chemotherapeutic agents to treat systemic fungal infections in immunocompromised AIDS, surgery, transplant and cancer patients is very high. The challenges to the delivery of anti-fungal agents include low solubility and sometimes high toxicity of these agents. These agents, such as amphotericin B, have low compatibility with hydrophobic cores of polymer micelles formed by many conventional block copolymers. Thus, to increase solubilization of amphotericin B, the core-forming blocks of methoxy-PEO-b-poly(L-aspartate) were derivatized with stearate side chains. The resulting block copolymers formed micelles. Amphotericin B interacted strongly with the stearate side chains in the core of the micelles, resulting in an efficient entrapment of the drug in the micelles, as well as subsequent sustained release in the external environment.

As a result of solubilization of amphotericin B in the micelles, the onset of hemolytic activity of this drug toward bovine erythrocytes was delayed, relative to that of the free drug. Using a neutropenic murine model of disseminated candidas, it was shown that micelle-incorporated amphotericin B retained potent *in vivo* activity. Pluronic block copolymers were used by the same group for incapsulation of another poorly soluble antifungal agent, nystatin. This is a commercially available drug that has shown potential for systemic administration, but has never been approved for that purpose, due to toxicity issues. The possibility to use Pluronic block copolymers to overcome resistance to certain antifungal agents has also been demonstrated. Overall, one should expect further scientific developments using polymer micelle delivery systems for the treatment of fungal infection.

Diagnostic Imaging Agents

Efficient delivery of imaging agents to the site of disease in the body can improve early diagnostics of cancer and other diseases. The studies in this area using polymer micelles as carriers for imaging agents were initiated by Torchilin. For example, micelles of amphiphilic PEO-lipid conjugates were loaded with. In and gadolinium diethylenetriamine pentaacetic acid-phosphatidylethanolamine (Gd-DTPA-PE) and then used for visualization of local lymphatic chain after subcutaneous injection into the rabbit's paw. The images of local lymphatics were acquired using a gamma camera and a magnetic resonance (MR) imager. The injected micelles stayed within the lymph fluid, thus serving as lymphangiographic agents for indirect MR or gamma lymphography. Another polymer micelle system composed of amphiphilic methoxy-PEO-b-poly[epsilon, N-(triiodobenzoy1)-L-lysine] block copolymers, labeled with iodine, was administered systemically in rabbits and visualized by X-ray computed tomography.

The labeled micelles displayed exceptional 24 hrs half-life in the blood, which is likely due to the core-shell architecture of the micelle carriers that protected the iodine-containing core. Notably, small polymer micelles (<20 nm) may be advantageous for bioimaging of tumors, compared with PEG-modified long-circulating liposomes (ca. 100 nm). In particular, the micelles from PEO-distearoyl phosphatidyl ethanolamine conjugates containing. In-labeled model protein were more efficacious in the delivery of protein to Lewis lung carcinoma than larger long-circulating liposomes. Overall, polymer micelles loaded with various agents for gamma, magnetic resonance, and computed tomography imaging represent promising modalities for non-invasive diagnostics of various diseases.

Transport of Polynucleotides

To improve the stability of polycation-based DNA, delivery complexes in dispersion block and graft copolymers containing segments from polycations and non-ionic water-soluble polymers, such as PEO, were developed. Binding of these copolymers with DNA results in the formation of micelle-like block ionomer complexes ("polyion complex micelles"), containing hydrophobic sites formed by the polycation-neutralized DNA and hydrophilic sites formed by the PEO chains. Despite neutralization of charge, complexes remain stable in aqueous dispersion due to the effect of the PEO chains. Overall, the PEO modified polycation-DNA complexes form stable dispersions and do not interact with serum proteins.

These systems were successfully used for intravitreal delivery of an antisense oligonucleotide and the suppression of gene expression in retina in rats. Furthermore, they displayed extended plasma clearance kinetics and were shown to transfect liver and tumor cells, after systemic administration in the body. In addition, there is a possibility targeting such polyplexes to the specific receptors at the surface of the cell, for example, by modifying the free ends of PEO chains with specific targeting ligands. Alternatively, to increase the binding of the complexes with the cell membrane and the transport of the polynucleotides inside the cells, the polycations were modified with amphiphilic Pluronic molecules. One recent study has shown a potential of Pluronic-polyethyleneimine-based micelles for *in vivo* delivery of antisense oligonucleotides to tumors, and have demonstrated sensitization of the tumors to radiotherapy as a result of systemic administration of the oligonucleotide-loaded micelles.

ANTICANCER DRUGS

Three polymer micelle formulations of anticancer drugs have been reported to reach clinical trials. The doxorubicin-conjugated polymer micelles developed by Kataoka's group have progressed recently to Phase I clinical trial at the National Cancer Center Hospital, Tokyo, Japan. The micelle carrier NK911 is based on PEO-b-poly(aspartic acid) block copolymers, in which the aspartic acid units were partially (ca. 45%) substituted with doxorubicin to form hydrophobic block. The resulting substituted block copolymer forms micelles that are further noncovalently loaded with free doxorubicin. Preclinical studies in mice demonstrated higher NK911 activity against Colon 26, M5076, and P388, compared with the free drug. Moreover, NK911 has less side effects, resulting in less animal body and toxic death than the free drug. Clinically, the Pluronic micelle formulation of doxorubicin has been most advanced. Based on the *in vivo* efficacy evaluation, Pluronic L61 was selected for clinical development for the treatment of MDR cancers. The final block copolymer formulation is a mixture of 0.25% Pluronic L61 and 2% Pluronic F127, formulated in isotonic buffered saline. This system contains mixed micelles of L61 and F127, with an effective diameter of ca. 22 to 27 nm and is stable in the serum.

Prior to administration, doxorubicin is mixed with this system, which results in spontaneous incorporation of the drug in the micelles. The drug is easily released by diffusion after dilution of the micelles. The formulation of doxorubicin with Pluronic, SP1049C, is safe, following systemic administration based on toxicity studies in animals. A two-site Phase I clinical trial of SP1049C has been completed. Based on its results, the dose-limiting toxicity of SP1049C was myelosuppression, reached at 90 mg/m^2 (maximum tolerated dose was 70 mg/m^2).

Phase II study of this formulation to treat inoperable metastatic adenocarcinoma of the esophagus is near completion as well. Finally, Phase I studies were reported for Genexol-PM, a Cremophor-free polymer micelle-formulated paclitaxel. Twenty-one patient entered into this study with lung, colorectal, breast, ovary, and esophagus cancers. No hypersensitivity reaction was observed in any patient. Neuropathy and myalgia were the most common toxicities. There were 14% partial responses. The paclitaxel area under the curve and peak of the drug concentration in the blood were increased with the escalating dose, suggesting linear pharmacokinetics for Genexol-PM.

CONCLUDING REMARKS

Approximately two decades have been passed since the conception of the polymer micelle conjugates and nanocontainers for drug delivery. During the first decade, only a few studies were published; however, more recently, the number of publications in this field has increased tremendously. During this period, novel biocompatible and/or biodegradable block copolymer chemistries have been researched, the block ionomer complexes capable of incorporating DNA and other charged molecules have been discovered, the pH and other chemical signal sensitive micelles have been developed.

Many studies focused on the use of polymer micelles for delivery of poorly soluble and toxic chemotherapeutic agents to the tumors to treat cancer. There has been considerable advancement in understanding the processes of polymer micelle delivery into the tumors, including passive and vectorized targeting of the polymer micelles. Notable achievements also include the studies demonstrating the possibilities for overcoming multidrug resistance in cancer, and enhancing drug delivery to the brain using block copolymer micelles systems.

Overall, it is clear that this area has reached a mature stage, reinforced by the fact that several human clinical trials using polymer micelles for cancer drug delivery have been initiated. At the same time, it is obvious that the possibilities for delivery of the diagnostic and therapeutic agents using polymer micelles are extremely broad, and one should expect further increase in the laboratory and clinical research in this field during the next decade.

Targeting polymer micelles to cancer sites within the body will address an urgent need to greatly improve the early diagnosis and treatment of cancer. Capabilities for the discovery and use of targeting molecules will support the development of multifunctional therapeutics that *can* carry and retain antineoplastic agents within tumors. This will also be instrumental in developing novel biosensing and imaging modalities for the early detection of cancer and other devastating human diseases.

7

Nanosized Drug Delivery Systems

Nanosized drug delivery systems have been developed to overcome one or several of the following problems: (i) low and highly variable drug concentrations after peroral administration due to poor absorption, rapid metabolism and elimination (ii) poor drug solubility which excludes i.v. injection of an aqueous drug solution (iii) drug distribution to other tissues combined with high toxicity (e.g. cancer drugs). Several systems, including micelles, liposomes, polymer nanoparticles, nanoemulsions and nanocapsules have been developed. During the last few years, *solid lipid nanodispersions* (SLN) have attracted increased attention. It is the aim of this chapter to discuss the general features of these systems with respect to manufacturing and performance. In the past, solid lipids have been mainly used for rectal and dermal applications. In the beginning of the 80s, Speiser and coworkers developed solid lipid microparticles and "Nanopellets for peroral administration". These Nanopellets were produced by dispersion of melted lipids with high speed mixers or ultrasound.

The manufacturing process was unable to reduce all particles to the submicron size. A considerable amount of microparticles was present in the samples. This might not be a serious problem for peroral administration, but it excludes an intravenous injection. "*Lipospheres*", described by Domb, are close related systems. They are also produced by means of high shear mixing or ultrasound and also often contain considerable amounts of microparticles. The quality of the SLN has been significant improved by the use of high pressure homogenization (HPH) in the early 90s. Higher shear forces and a better distribution of the energy force more effective particle disruption, compared with high shear mixing and ultrasound. Dispersions obtained by this HPH are called *Solid Lipid Nanoparticles* (SLN).

Most SLN dispersions produced by *high pressure homogenization* (HPH) are characterized by an average particle size below 500 nm and a low microparticle content. Other production procedures are based on the use of organic solvents HPH/solvent evaporation or on dilution of microemulsions. The ease and efficacy of manufacturing lead to an increased interest in SLN. Furthermore, it has been claimed that SLN combine the advantages yet without inheriting the disadvantages of other colloidal carriers. Proposed advantages include:

1. Possibility of controlled drug release and drug targeting
2. Increased drug stability
3. High drug pay load
4. Feasibility to incorporate lipophilic and hydrophilic drugs
5. No biotoxicity of the carrier
6. Avoidance of organic solvents
7. No problems with respect to large scale production and sterilization.

However, during the last years, some of these claims have been questioned and it became evident that SLN are rather complex systems which possess not only advantages but also serious limitations.

SOLID LIPID NANOPARTICLES (SLN)

General ingredients include solid lipid(s), emulsifier(s) and water. The term lipid is used generally in a very broad sense and includes triglycerides (e.g. tristearine, hard fat), partial glycerides (e.g. Imwitor), pegylated lipids, fatty acids (stearic acid), steroids (e.g. cholesterol) and waxes (e.g. cetylpalmitate). All classes of emulsifiers (with respect to charge and molecular weight) have been used to stabilize the lipid dispersion. The most frequently used compounds include different kinds of poloxamer, polysorbates, lecithin and bile acids. It has been found that the combination of emulsifiers might prevent particle agglomeration more efficiently. Unfortunately, poor attention has been given by most investigators to the physicochemical properties of the lipid. Fatty acids, partial glycerides and other polar lipids are able to interact with water to much a greater extent, compared with a long chain triglyceride (e.g. they might form liquid crystalline phases). Polar lipids will have much more interaction with stabilizers (e.g. formation of mixed micelles), while more lipophilic lipids will show phase segregation. The author strongly suggests to follow the proposal by Small and to classify lipids according to their interactions with water.

Preparation of SLN

Mechanism of homogenization

High shear homogenization and ultrasound are dispersion techniques which were initially used for the production of solid lipid nanodispersions. Both methods are widespread and easy to handle. However, dispersion quality is often poor due to the presence of microparticles. Furthermore, metal contamination has to be considered if ultrasound is used. Ahlin *et al.* used a rotor-stator homogenizer to produce SLN from different lipids, including trimyristin, tripalmitin, tristearin, partial glycerides and glycerol tribehenate by meltemulsification. They investigated the influence of different process parameters, including emulsification time, stirring rate and cooling conditions on the particle size and the zeta potential.

Poloxamer 188 was used as steric stabilizer (0, 5% w/w). For Witepsol W35 dispersions, the following parameters were found to produce the best SLN quality: stirring 8 min at 20,000 rpm, the optimum cooling conditions 10 min at 5000 rpm at room temperature. In contrary, the best conditions for Dynasan 116 dispersions were 10 min emulsification at 25,000 rpm and 5 min of cooling at 5000 rpm in cool water (T = 16°C). An increased stirring rate did not significantly decrease the particle size, but improved the polydispersity index slightly. No general rule can be derived from differences in the established optimum emulsification and cooling conditions. In most cases, average particle sizes in the range of 100-200 nm were obtained in this study.

High Pressure Homogenization (HPH)

HPH has emerged as a very reliable and probably the most powerful technique for the preparation of SLN. HPH has been used for many years for the production of nanoemulsions for parenteral nutrition. In most cases, scaling up represents zero or limited problems. High pressure homogenizers push a liquid with high pressure (100-2000 bar) through a narrow gap (in the range of few microns). The fluid accelerates on a very short distance to very high velocities. The high shear stress disrupts the particles down to the submicron range. Typical lipid contents are in the range of 5 to 10%. Even higher lipid concentrations (up to 40%) have been homogenized to lipid nanodispersions. Two general approaches of the homogenization step, the hot and the cold homogenization techniques, can be used for the production of SLN. In both cases, a preparatory step involves the drug incorporation into the bulk lipid by dissolving the drug in the lipid melt.

Warm Homogenization

The warm/hot homogenization is carried out at temperatures above the melting point of lipid. Therefore, it is in fact the homogenization of an emulsion. A preemulsion of the drug loaded lipid melt and the aqueous emulsifier phase is obtained by high-shear mixing device (Ultraturrax). The quality of the preemulsion is very important for the final product quality. In general, higher temperatures result in lower particle sizes due to the decrease of the viscosity of the inner phase. However, high temperatures may also increase the degradation rate of the drug and the carrier. The homogenization step can be repeated several times. It should be kept in mind however, that HPH increases the temperature of the sample (approximately 10°C for 500 bar).

In most cases, 3 to 5 homogenization cycles at 500 to 1500 bar are sufficient. Increasing the homogenization pressure or the number of cycles often results in an increase of the particle size due to particle coalescence, which occurs as a result of the high kinetic energy of the particles. It is important to note that the primary product of the hot homogenization is a nanoemulsion due to the liquid state of the lipid. Solid particles are expected to be formed by the following cooling of the sample to room temperature, or to temperatures below. Due to the small particle size and the presence of emulsifiers, lipid crystallization may be highly retarded and the sample may remain as a supercooled melt for several months.

Cool Homogenization

Cool/cold homogenization has been developed to overcome the following three problems of the hot homogenization technique:

1. Temperature induced drug degradation
2. Drug distribution into the aqueous phase during homogenization
3. Complexity of the crystallization step of the nanoemulsion, leading to several modifications and/or supercooled melts

The first preparatory step for cold homogenization is the same as in the hot homogenization procedure and includes the solubilization of the drug in the melt of the bulk lipid. However, the following steps differ. The drug containing melt is rapidly cooled. The high cooling rate favors a homogenous distribution of the drug within the lipid matrix. The solid, drug containing lipid is milled to microparticles. Typical particle sizes obtained by means of ball or mortar milling are in the range of 50 to 100 microns. Low temperatures increase the fragility of the lipid, and therefore favor particle disruption. The solid lipid microparticles are suspended in a chilled emulsifier solution.

The preemulsion is subjected to HPH at or below room temperature. An effective temperature control and regulation is needed in order to ensure the unmolten state of the lipid due to the increase in temperature during homogenization. In general, compared with hot homogenization, larger particle sizes and a broader size distribution are observed in cold homogenized samples. A modified version of this technique has been *recently* published by the group of Miiller-Goymann. They dispersed a solid 1:1 lecithin-hardfat mixture (described as solid reversed micelles) in Tween containing water using high pressure homogenization.

Method of emulsification or evaporation

The solvent emulsification/evaporation processes adapts techniques which have been previously used for the production of polymeric micro- and nanoparticles. The solid lipid is dissolved in a water-immiscible organic solvent (e.g. cyclohexane, or chloroform) that is emulsified in an aqueous phase. Upon evaporation of the solvent, a nanoparticle dispersion is formed by precipitation of the lipid in the aqueous medium. Westesen prepared nanoparticles of tripalmitate by dissolving the triglyceride in chloroform. This solution was emulsified into an aqueous phase by high pressure homogenization.

The organic solvent was removed from the emulsion by evaporation under reduced pressure. The mean particle size ranges from approximately 30 to 100 nm depending on the lecithin/co-surfactant blend. Particles with very small diameters (30 nm) were obtained by using bile salts as co-surfactants. Comparable small particle size distributions were not achievable by melt emulsification of similar composition. The mean particle size depends on the concentration of the lipid in the organic phase. Very small particles could only be obtained with low fat loads (5 w%) related to the organic solvent. With increasing lipid content, the efficacy of the homogenization declines due to the higher viscosity of the dispersed phase.

Process of solvent injection

The solvent injection method has been developed by Fessi to produce polymer nanoparticles. Nanoparticles were only produced with solvents which distribute very rapidly into the aqueous phase (e.g. ethanol, acetone, DMSO), while larger particle sizes were obtained with more lipophilic solvents. According to Fessi, the particle size is critically determined by the velocity of the distribution processes and only water miscible solvents can be used. The solvent injection method can also be used for the production of solid lipid nanoparticles. However, the method is limited to lipids which dissolve in the polar organic solvent. Advantages of the method are the avoidance of elevated temperatures and high shear stress. However, the lipid concentration in the primary suspension will be less compared with High-Pressure-Homogenization. Furthermore, the use of organic solvents clearly represents a drawback of the method.

Technique of microemulsions

SLN preparation techniques which are based on the dilution of microemulsions have been developed by Gasco and coworkers. Unfortunately, there is no common agreement within the scientific community about the definition of a microemulsion. One part of the scientific community understands under microemulsions high fluctuating systems which can be regarded as a critical solution, and therefore do not contain an inner and outer phase. This model has been confirmed by self-diffusion NMR studies of Lindman. In contrast, Gasco and other scientists understand microemulsions as two systems composed of an inner and outer phase (e.g. O/W-microemulsions). They are made by stirring an optical transparent mixture at 65-70°C, typically composed of a low melting lipid fatty acid (e.g. stearic acid), emulsifier (e.g. polysorbate 20, polysorbate 60, soy phosphatidylcholin, taurodeoxycholic acid sodium salt), co-emulsifiers (e.g. Butanol, Na-monooctylphosphate), and water.

The hot microemulsion is dispersed in cold water (2-3°C) under stirring. Typical volume ratios of the hot microemulsion to the cold water are in the range of 1:25 to 1:50. The dilution process is critically determined by the composition of the microemulsion. According to the literature, the droplet structure is already contained in the microemulsion, and therefore, no energy is required to achieve submicron particle sizes. The temperature gradient and the pH-value determine the product quality in addition to the composition of the microemulsion. High temperature gradients facilitate rapid lipid crystallization and prevent aggregation. Due to the dilution step, lipid contents which are achievable are considerably lower, compared with the HPH based formulations. Another disadvantage includes the use of organic solvents. Recent work describes a similar approach to produce SLN. A hot liquid crystalline phase (instead of a microemulsion) is diluted in cold water to yield a solid lipid nanodispersion. This approach avoids the use of high pressure homogenization and organic solvents, and therefore represents an interesting opportunity.

Additional Mechanism

Sterilization

Sterility is required for parenteral formulations. Dry or wet heat, filtration, y-irradiation, chemical sterilization and aseptic production are general, opportunities to achieve sterility. The sterilization should not change the properties of the sample with respect to physical and chemical stability and the drug release kinetics. Sterilization by heat is a reliable procedure which is most commonly used. It was also applied for Liposomes. Steam sterilization will cause the formation of an oil in water emulsion, due to the melting of the lipid particles. The formation of SLN requires recrystallization of the lipids. Concerns are related to temperature induced changes of the physical and chemical stability. The correct choice of the emulsifier is of significant importance for the physical stability of the sample at high temperatures. Increased temperatures will affect the mobility and the hydrophilicity of all emulsifiers, but to a different extent. Schwarz found that Lecithin is preferable to Poloxamer for steam sterilization, as only a minor increase in the particle size and the number of microparticles was observed after steam sterilization.

An increase in particle size for Poloxamer 188 stabilized Compritol-SLN was observed after steam sterilization. It was found that a decrease of the sterilization temperature from 121°C to

110°C can reduce sterilization induced particle aggregation to a large extent. This destabilization can be attributed to the decreased steric destabilization of the Poloxamer. It is well known for PEG-based emulsifiers that increased temperatures lead to dehydration of the ethylenoxide chains, pointing to a decrease of the thickness of the protecting layer. It has been demonstrated by 1H-NMR spectroscopy on Poloxamer stabilized lipid nanoparticles, that even a moderate temperature increase from RT to 37°C decreases the mobility of the ethylenoxide chains on the particle surface.

Results of Freitas *et al.* indicate that the lowering of the lipid content (to 2%), and the surface modification of the glass vials and nitrogen purging might prevent the particle growth to a large extent and avoid gelation. Further studies of Cavalli *et* al. and Heiati demonstrate the possibility of steam sterilization of drug loaded SLN. Filtration sterilization of dispersed systems requires very high pressure and is not applicable to particles >0, 2 μm. As most SLN particles are close to this size, filtration is of no practical use, due to the clocking of the filters. Few studies investigated the possibility of γ-sterilization. It must be kept in mind that free radicals are formed during γ-sterilization in all samples, due to the high energy of the γ-rays. These radicals may recombine with no modification of the sample or undergo secondary reactions which might lead to chemical modifications of the sample.

The degree of sample degradation depends on the general chemical reactivity and the molecular mobility and the presence of oxygen. It is therefore not surprising that chemical changes of the lipid bilayer components of liposomes were observed after γ-irradiation. Schwarz investigated the impact of different sterilization techniques [steam sterilization at 121°C (15 min) and 110°C (15 min); γ-sterilization] on SLN characteristics. In comparison to lecithin stabilized systems, Poloxamer stabilized SLN were less stable than steam sterilization. However, this difference was not detected for γ-sterilized samples. Compared with steam sterilization at 121°C, the increase in particle size after γ-irradiation was lower, but comparable to that at 110°C. Unfortunately, most investigators did not search for steam sterilization or irradiation induced chemical degradation. It should be kept in mind that degradation does not always cause increased particle sizes. In contrast, the formation of species like lysophosphatides or free fatty acids could even preserve small particle sizes, but might cause toxicological problems. Further studies with more focus on chemical degradation products are clearly necessary to permit valid statements of the possibilities of SLN sterilization.

System of drying

SLN are thermodynamic unstable systems, and therefore, particle growth has to be minimized. Furthermore, SLN ingredients and incorporated drugs are often unstable, hydrolyzing or oxidizing. The transformation of the aqueous SLN-suspension in a dry, redispersible powder is therefore often a necessary step to ensure storage stability of the samples. Lyophilization is widely used and is a promising way to increase chemical and physical SLN stability over extended periods of time. Lyophilization also offers principle possibilities for SLN incorporation into pellets, tablets or capsules. Two additional transformations are necessary which might be the source of additional stability problems. The first transformation, from aqueous dispersion to powder, involves the freezing of the sample and the evaporation of water under vacuum. Freezing of the sample might cause stability problems due to the freezing out effect which results in the changes of the osmolarity and the pH.

The second transformation, resolubilization, involves situations at least in its initial stages which favor particle aggregation (i.e. low water and high particle content, high osmotic pressure). The protective effect of the surfactant can be compromised by lyophilization. It has been found that the lipid content of the SLN dispersion should not exceed 5%, so as to prevent an increase in the particle size.

Direct contact of lipid particles are decreased in diluted samples. Furthermore, diluted SLN dispersions will also have higher sublimation velocities and a higher specific surface area. The addition of cryoprotectors (e.g. Sorbitol, Mannose, Trehalose, Glucose, and Polyvinylpyrrolidon) will be necessary to decrease SLN aggregation and to obtain a better redispersion of the dry product. Schwarz *et al.* investigated the lyophilization of SLN in detail. Best results were obtained with the cryoprotectors, Glucose, Mannose, Maltose and Trehalose, in the concentration range between 10% and 15%. The observations come into line with the results of the studies on liposome lyophilization, which indicated that Trehalose was the most sufficient substance to prevent liposome fusion and the leakage of the incorporated drug. Encouraging results obtained with unloaded SLN cannot predict the quality of drug loaded lyophilizates. Even low concentrations of 1% Tetracain or Etomidat caused a significant increase in particle size, excluding an intravenous administration. Westesen investigated the lyophilization of tripalmitate-SLN using glucose, sucrose, maltose and trehalose as cryoprotective agents. Handshaking of redispersed samples was an insufficient method, but bath sonification produced better results.

Average particle sizes of all lyophilized samples with cryoprotective agents were 1.5 to 2.4 times higher than the original dispersions. One year storage caused increased particle sizes of 4 to 6.5 times compared with the original dispersion. In contrast to the lyophilizates, the aqueous dispersions of tyloxapol/phospholid stabilized tripalmitate SLN exhibited remarkable storage stability. The instability of the SLN lyophilizates can be explained by the sintering of the particles. TEM pictures of tripalmitate SLN show an anisometrical, platelet-like shape of the particles. Lyophilization changes the properties of the surfactant layer due to the removal of water, and increases the particle concentration which favors particle aggregation. Increased particle sizes after lyophilization (2.1 to 4.9 times) were also reported by Cavalli. Heiati compared the influence of four cryoprotectors (i.e. trehalose, glucose, lactose and mannitol) on the particle size of azidothymidine palmitate loaded SLN lyophilizates. In agreement to other reports, Trehalose was found to be the most effective cryoprotectant.

The freezing procedure will affect the crystal structure and the properties of the lyophilizate. Literature data suggest that the freezing process needs to be optimized to a particular sample size. Schwarz recommended rapid freezing in liquid nitrogen. In contrast, other researchers observed the best results after a slow freezing process. Again, best results were obtained with samples of low lipid content and with the cryoprotector trehalose. Slow freezing in a deep freeze (–70°C) was superior to rapid cooling in liquid nitrogen. Furthermore, introduction of an additional thermal treatment of the frozen SLN dispersion (2 hr at –22°C; followed by 2 hr temperature decrease to –40°C) was found to improve the quality of the lyophilizate. Lately, lyophilization has been used to stabilize retinoic acid loaded SLN. An interesting alternative to lyophilization has been recently suggested by Gasco's group. Drying with a nitrogen stream at low temperatures of 3 to 10°C has been found to be superior. Compared with lyophilization, the advantages of this process are the avoidance of freezing and the energy efficiency resulting from the higher vapor pressure of water. Spray drying has been scarcely for SLN drying, although it is cheaper compared with lyophilization. Freitas obtained a redispersable powder with this method, which meets the general requirements of i.v.-injections, with regard to the particle size and the selection of the ingredients. Spray drying might potentially cause particle aggregation due to high temperatures, shear forces and partial melting of the particles. Freitas recommends the use of lipids with high melting points >70°C to avoid sticking and aggregation problems. Furthermore, the addition of carbohydrates and low lipid contents favor the preservation of the colloidal particle size in spray drying.

FEATURES OF SLN

The characterization of SLN is a necessity and a great challenge. Lipid characterization itself is not trivial as the statement by Laggner shows: "Lipids and fats, as soft condensed material in general, are very complex systems, which not only in their static structures but also with respect to their kinetics of supramolecular formation, Hysteresis phenomena or supercooling can gravely complicate the task of defining the underlying structures and boundaries in a phase diagram". This is especially true for lipids in the colloidal size range. Therefore, possible artifacts caused by sample preparation (removal of emulsifier from particle surface by dilution, induction of crystallization processes, changes of lipid modifications) should be kept in mind. For example, the contact of the SLN dispersion with new surfaces (e.g. a syringe needle) might induce lipid crystallization or modification, and sometimes result in the spontaneous transformation of the low viscous SLN-dispersion into a viscous gel.

The most important parameters of SLN include particle size and shape, the kind of lipid modification and the degree of crystallization, and the surface charge. *Photon correlation spectroscopy* (PCS) and *Laser diffraction* (LD) are the most powerful techniques for routine measurements of particle size. It should be kept in mind that both methods are not "*measuring*" particle sizes. Rather, they detect light scattering effects which are used to calculate particle sizes. For example, uncertainties may result from nonspherical particle shapes. Platelet structures commonly occur during lipid crystallization and are very often described in the SLN literature. The influence of the particle shape on the measured size is discussed by Sjostrom. Further difficulties arise both in PCS and LD measurements for samples which contain several populations of different size. Therefore, additional techniques might be useful. For example, light microscopy is recommended although it is not sensitive to the nanometer size range. It gives a fast indication about the presence and the character of microparticles.

Electron Microscopy provides, in contrast to PCS and LD, direct information on the particle shape. *Atomic force microscopy* (AFM) has attracted increasing attention. A cautionary note applies to the use of AFM in the field of nanoparticles, as an immobilization of the SLN by solvent removal is required to assess their shape by the AFM tip. This procedure is likely to cause substantial changes of the molecular structure of the particles. Zur Miihlen demonstrated the ability of AFM to image the morphological structure of SLN. The sizes of the visualized particles are of the same magnitude, compared with the results of PCS measurements. The AFM investigations revealed the disk- like structure of the particles. Dingler investigated cetylpalmitate SLN (stabilized by polyglycerol methylglucose distearate) by electron microscopy and AFM and found an almost spherical form of the particles.

The usefulness of cross flow *Field-flow-fractionation* (FFF) for the characterization of colloidal lipid nanodispersions has been recently demonstrated. Lipid nanodispersions with constant lipid content, but different ratios of liquid and solid lipids did show similar particle sizes in dynamic light scattering. However, retention times in FFF were remarkably dissimilar due to the different particle shapes (i.e. spheres vs. platelets). Anisotropic particles such as platelets will be constrained by the cross flow much more heavily compared with the spheres of similar size. The very high anisometry of the SLN particles has been confirmed by electron microscopy, where very thin particles of 15 nm thickness and the length of several hundred nanometers became visible.

The measurement of the zeta potential allows predictions about the storage stability of colloidal dispersions. In general, particle aggregation is less likely to occur for charged particles (i.e. high zeta potential) due to electric repulsion. However, this rule cannot strictly apply to systems which contain steric stabilizers, because the adsorption of steric stabilizer will decrease the zeta potential due to the shift in the shear plane of the particle. Particle size analysis is just one aspect of SLN quality. The same attention has to be paid on the characterization of lipid crystallinity and modification, because these parameters are strongly correlated with drug incorporation and release rates. Thermodynamic stability and lipid packing density increase, and drug incorporation rates decrease in the following order:

$$\text{supercooled melt} < \alpha\text{-modification} < \beta'\text{-modification} < \beta\text{-modification}$$

In general, it has been found that melting and crystallization processes of nanoscaled material can differ considerable from that of the bulk material. The thermodynamic properties of material having small nanometer dimensions can be considerably different, compared with the material in bulk form (e.g. the reduction of melting point). This occurs because of the tremendous influence of the surface energy. This statement is also valid for SLN, where lipid crystallization and modification changes might be highly retarded, due to the small size of the particles and the presence of emulsifiers. Moreover, crystallization might not occur at all and has been shown that samples which were previously described as SLN (solid lipid particles) were in fact supercooled melts (liquid lipid droplets). The impact of the emulsifier on SLN lipid crystallization has been shown by Bunjes. The same group demonstrated also a size dependent melting of SLN. Differential Scanning Calorimetry (DSC) and X-ray scattering are most commonly applied to asses the status of the lipid. DSC uses the fact that different lipid modifications possess different melting points and melting enthalpies.

By means of X-ray scattering, it is possible to assess the length of the long and short spacings of the lipid lattice. It is highly recommended to measure the SLN dispersion themselves, because solvent removal will lead to modification changes. Sensitivity problems and long measurement times of convential X-ray sources might be overcome by synchrotron irradiation. In addition, this method permits to conduct time resolved experiments and allows the detection of intermediate states of colloidal systems which will be non detectable by convential X-ray methods. Recent work shows that SLN might form superstructures by parallel alignment of SLN platelets. These reversible particle self-assemblies were observed by Illing *et al.* in tripalmitin dispersions when the lipid concentration exceeds 40 mg/g.

Higher lipid concentrations did enhance particle self-assembly. The tendency to form self-assemblies has been found to depend on the particle shape, the lipid and the surfactant concentration. Infrared and Raman Spectroscopy are useful tools to investigate structural properties of lipids and they might give complentary information to X-ray and DSC. Raman measurements on SLN show that the arrangement of lipid chains of SLN dispersions changes with storage. Rheometry might be particularly useful for the characterization of the viscoelastic properties of SLN dispersions. The rheological properties are important with respect to the dermatological use of SLN, but they also provide useful information about the structural features of SLN dispersions and their storage dependency. Studies of Lippacher show that the SLN dispersion posses higher elastic properties than emulsions of comparable lipid content. Furthermore, a sharp increase of the elastic module is observed at a certain lipid content. This point indicates the transformation from a low viscous lipid dispersion to an elastic system, with a continuous network of lipid nanocrystals.

Illing and Unruh did compare the rheological properties of trimyristic, tripalmitic and tristearic SLN suspensions. The results indicate that the viscosity of triglyceride suspensions increases with the lipid chain length and an increased anisotropy of the particles. Souto *et al.* used rheology to study the influence of SLN addition on the rheological properties of hydrogels. The co-existence of additional colloidal structures (micelles, liposomes, mixed micelles, nanodispersed liquid crystalline phases, supercooled melts, drug-nanoparticles) has to be taken into account for all SLN dispersions. Unfortunately, many investigators neglect this aspect, although the total amount of surface active compounds is often comparable to the total amount of the lipid. The characterization and quantification are serious challenges due to the similarities in size. In addition, the sample preparation will modify the equilibrium of the complex colloidal system.

Dilution of the original SLN dispersion with water might cause the removal of surfactant molecules from the particle surface and induce further changes such as crystallization or the changes of the lipid modifications. It is therefore highly desirable to use methods which are sensitive to the simultaneous detection of different colloidal species, which do not require preparatory steps such as Raman, NMR and ESR spectroscopy. NMR active nuclei of interest are ^{1}H, ^{13}C, ^{19}F and ^{35}P. Due to the different chemical shifts, it is possible to attribute the NMR signals to particular molecules or their segments. For example, lipid methyl protons give signals at 0.9 ppm, while protons of the polyethylenglycole chains give signals at 3.7 ppm. Simple ^{1}H-spectroscopy permits an easy and rapid detection of supercooled melts, due to the low linewidths of the lipid protons. This method is based on the different proton relaxation times in the liquid and semisolid/ solid state. Protons in the liquid state give sharp signals with high signal amplitudes, while semisolid/solid protons give very broad or invisible NMR signals under these circumstances.

NMR has been used to characterize calixarene SLN and hybrid lipid particles (NLC), which are composed of liquid and solid lipids. Protons from solid lipids are not detected by standard NMR, but they can be visualized by solid state NMR. A drawback of solid state NMR is the rapid spinning of the sample that might cause artifacts. A recent paper describes the use of this method to monitor the distribution of Q10 in lipid matrices. Unfortunately, the authors did use "drying of the sample to constant weight" as a preparatory step, which will cause significant changes of the sample characteristics. ESR requires the addition of paramagnetic spin probes to investigate SLN dispersions. A large variety of spin probes is commercially available. The corresponding ESR spectra give information about the microviscosity and micropolarity. ESR permits the direct, repeatable and non-invasive characterization of the distribution of the spin probe between the aqueous and the lipid phase. Experimental results demonstrate that storage induced crystallization of SLN leads to an expulsion of the probe out of the lipid into the aqueous phase. Furthermore, using an ascorbic acid reduction assay, it is possible to monitor the time scale of the exchange between the aqueous and the lipid phase. The transfer rates of molecules between SLN and liposomes or cells have been determined by ESR.

ALTERNATIVE SLN MODELS

Lipid nanoemulsions are composed of a liquid oily core and a surfactant layer (lecithin). They are widely used for the parenteral delivery of poorly soluble drugs. The original idea of SLN was to achieve a controlled release of incorporated drugs by increasing the viscosity of the lipid matrix. Therefore it is not surprising that in original model, SLN is being described as *"frozen emulsions"*. However, lipids are known to crystallize very frequently in anisotropic platelet shapes and anisotropic. Sjostrom *et al.* described in 1995 that the particle shape of Cholesterylacetate SLN did strongly depend on the emulsifier. Platelet shaped particles have been detected for lecithin stabilized particles, while PEG-20-sorbitanmonolaurate stabilized particles preserved their spherical shape.

Anisotropic particles have been found in numerous other SLN dispersions. Based on the experimental results, a platelet shaped SLN model can be proposed as an alternative. In the year 2000, Westesen questioned the frozen emulsion droplet model with the following statement: "Careful physicochemical characterization has demonstrated that these lipid-based nanosuspensions (solid lipid nanoparticles) are not just emulsions with solidified droplets. During the development process of these systems, interesting phenomena have been observed, such as gel formation on solidification and upon storage, unexpected dynamics of polymorphic transitions, extensive annealing of nanocrystals over significant periods of time, stepwise melting of particle fractions in the lower-nanometer-size range, drug expulsion from the carrier particles on crystallization and upon storage, and extensive supercooling. Her comment highlights the complex behavior and changes of SLN dispersions. In addition, the presence of competing colloidal structures (e.g. micelles, liposomes,

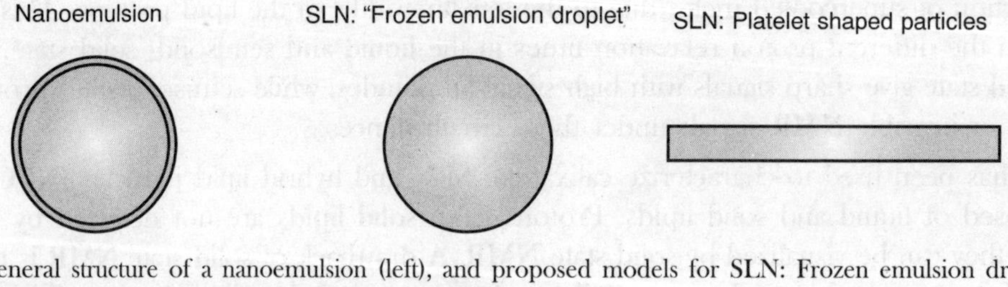

Fig. 7.1. General structure of a nanoemulsion (left), and proposed models for SLN: Frozen emulsion droplet model (middle) and platelet shaped SLN model (right)

mixed micelles, nanodispersed liquid crystalline phases, supercooled melts and drug-nanoparticles) should be considered. Additional colloids might have an impact on very different aspects, including the correct measurement of particle size, drug incorporation and toxicity. A recent study shows that the cell toxicity of the SLN dispersion was reduced by dialysis due to the removal of water soluble components.

NANOSTRUCTURED LIPID CARRIERS (NLC)

Nanostructured lipid carriers (NLC) have been recently proposed as a new SLN generation with improved characteristics. The general idea behind the system is to improve the poor drug loading capacity of SLN by "mixing solid lipids with spatially incompatible lipids leading to special structures of the lipid matrix", while still preserving controlled release features of the particles. Three different types of NLC have been proposed (NLC I: The imperfect structured type, NLC II: The structureless type and NLC III: The multiple type). Unfortunately, these structural proposals have not been supported by experimental data. They assume a spherical shape and they are not compatible with lipid platelet structures. For example, NLC III structures should contain small oily droplets in a solid lipid sphere.

Detailed analytical examination of NLC systems by Jores *et al.* demonstrate that "nanospoon" structures are formed, in which the liquid oil adheres on the solid surface of a lipid platelet. Jores *et al.* did conclude that "Neither SLN nor NLC lipid nanoparticles showed any advantage with respect to incorporation rate or retarded accessibility to the drug, compared with conventional nanoemulsions. The experimental data concludes that NLCs are not spherical solid lipid particles with embedded liquid droplets, but rather, they are solid platelets with oil present between the solid platelet and the surfactant layer". Very similar structures have been found on Q10 loaded SLN by Bunjes *et al.*

Advantages

Proposed advantages of SLN, compared with nanoemulsions, include increased protection capacity against drug degradation and controlled release possibilities due to the solid lipid matrix. The general low capacity of crystalline structures to accommodate foreign molecules is a strong argument against the proposed rewards. It is therefore necessary to distinguish between drug association and drug incorporation. Drug association means that the drug is associated with the lipid, but it might be localized in the surfactant layer or between the solid lipid and the surfactant layer (similar to the oil). Drug incorporation would mean the distribution of the drug within the lipid matrix. Another limiting aspect comes from the fact that the platelet structure of SLN, which is found in many systems, leads to a tremendous increase in surface area and the shortening of the diffusion lengths.

Furthermore, additional colloid structures present in the sample are alternatives for drug localization the SLN for drug incorporation as it was pointed out by Westesen: "The estimation of drug distribution is difficult for dispersions consisting of more than one type of colloidal particle. Depending on the type of stabilizer and on the concentration ratio of stabilizer to matrix material significant numbers of particles such as liposomes and/or (mixed) micelles may coexist with the expected type of particles". The detailed investigation of drug localization is very difficult and only a few studies exist. Parelectric spectroscopy has been used to investigate the localization of glucocorticoids. The results indicate that the drug molecules are attached to the particle surface, but not incorporated into the lipid matrix. With Betamethasonvalerate, the loading capacity of the particle surface was clearly below the usual concentration of 0.1%. Lukowski used Energy Dispersive X-ray Analysis and found that the drug Triamcinolone, Dexamethasone and Chloramphenicol are partially stored at the surface of the individual nanoparticles. The importance of the emulsifier is reflected in a study from Danish scientists. They produced gamma-cyhalothrin (GCH) loaded lipid micro- and nanoparticles. GCH had only limited solubility in the solid lipid and was expulsed during storage.

The appearance of GCH crystals was strongly dependent from the solubility of the GCH in the emulsifier solutions. Emulsifier with high GCH solubility provoked rapid crystal growth. This observation is in accordance with a mechanism of crystal growth according to Ostwald ripening. Slovenian scientist found that ascorbylpalmitate was more resistant against oxidation in non-hydrogenated soybean lecithin liposomes, compared with SLN. It shows that liposomes might have a higher protection capacity compared with SLN. Fluorescence and ESR studies have been used by Jores *et al.* to monitor the microenvironment and the mobility of model drugs. The results indicate that even highly lipophilic compounds are pushed into a polar environment during lipid crystallization. Therefore, the incorporation capacity of SLN is very poor for most molecules. A nitroxide reduction assay gave results in accordance with the results of the distribution. Compared with nanoemulsions, nitroxides were more accessible in SLN and NLC to ascorbic acid, localized in the aqueous environment. Therefore, nanoemulsions were more protective than SLN and NLC systems. Drug release from SLN and NLC could be either controlled by the diffusion of the drug or the erosion of the matrix.

The original idea was to *achieve* a controlled release of SLN due to the slowing down of drug diffusion to the particle surface. This idea is, however, questionable due to drug expulsion during lipid crystallization. In addition, very short diffusion lengths in nanoscaled delivery systems lead to short diffusion times, even in highly viscous or solid matrices. In most cases, the delivery of the drug will be controlled by the slow dissolution rate in the aqueous environment. Drug release rate will be highly dependent on the presence of further solubilizing colloids (e.g. micelles), which are able to work as a shuttle for the drug and the presence or absence of a suitable acceptor compartment. Many investigators studied only the release in buffer media. A controlled release pattern under such conditions is not surprising, as it is caused by low solubilization kinetics due to the poor solubility of the drug. *In vivo,* acceptor compartments will be present (e.g. lipoproteins, membranes) and will speed up release processes significantly.

Whenever possible, drug loaded SLN should be compared with nanosuspensions to separate the general features of the drug and the influence of the lipid matrix. Results by Kristl *et al.* indicate that lipophilic nitroxides diffuse between SLN and liposomes. The diffusion kinetics was strongly dependent on the nitroxide structure. In contrast, uptake of nitroxides in cells was similar between lipophilic nitroxides, suggesting endocytosis as the main mechanism. The detailed mechanisms of drug release *in vivo* are poorly understood. *In vitro* data by Olbrich demonstrate that SLN are degraded by lipases. Degradation by lipase depends on the lipid and strongly on the surfactant.

Steric stabilization (e.g. by poloxamer) of SLN and NLC are less accessible because lipase needs an interface for activation. It is also known that highly crystalline lipids are poorly degraded by lipase.

DIFFERENT ROUTES

SLN and NLC can be administered at different routes, including peroral, dermal, intravenously and pulmonal. Peroral administration of SLN could enhance the drug absorption and modify the absorption kinetics. Despite the fact that in most of the SLN, the drug will be associated but not incorporated in the lipid, SLN might have advantages due to enhanced lymphatic uptake, enhanced bioadhesion or increased drug solubilization by SLN lipolysis products such as fatty acids and monoglycerides. A serious challenge represents the preservation of the colloidal particle in the stomach, where low pH values and high ionic strengths favor agglomeration and particle growth. Zimmermann and Willer studied the stability of different SLN formulations in artificial gastric juice. The main findings of this study are that (i) some SLN dispersions preserve their particle size under acidic conditions, and (ii) there is no general lipid and surfactant which are superior to others.

The particular interactions between lipid and stabilizer are determining the robustness of the formulation. Therefore, the suitable combination of ingredients has to be determined on a case by case basis. Several animal studies show increased absorption of poorly soluble drugs. The efficacy of orally administrated Triptolide free drug and Triptolide loaded SLN have compared in the carrageenan-induced rat paw edema by Mei *et al.* Their results suggest that SLN can enhance the anti inflammatory activity of triptolide and decrease triptolide-induced hepatotoxicity. The usefulness of SLN to increase the absorption of the poorly soluble drug all-trans retinoic acid has been shown by Hu *et al.* on rats. Gascos group investigated the uptake and distribution of Tobramycin loaded SLN in rats. They observed an increased uptake into the lymph, which causes prolonged drug residence times in the body of the animals. Furthermore, AUC and clearance rates did depend on the drug load.

The same group described also enhanced absorption of Idarubicin-loaded solid lipid nanoparticles (IDA-SLN), in comparison to the drug solution. Furthermore, the authors described that SLN were able to pass the blood-brain barrier and concluded that duodenal administration of IDA-SLN modifies the pharmacokinetics and tissue distribution of idarubicin. Parenteral administration of SLN is of great interest too. To avoid the rapid uptake of the SLN by the RES system, stealth SLN particles have been developed by the adoption of the stealth concept from liposomes and polymer nanoparticles. Reports indicate that Doxorubicin loaded stealth SLN circulate for long period of time in the blood and change the tissue distribution. Therefore, SLN could be alternatives to marketed stealth-liposomes, which can decrease the heart toxicity of this drug due to changed biodistribution. Long circulation times have also been observed for Poloxamer stabilized SLN with Paclitaxel. The dermal application is of particular interest and it might become the main application of SLN.

SLN pose occlusive properties which are related to the solid structure of the lipid. Human *in vivo* results of the group of Miller demonstrate that SLN can improve skin hydration and viscoelasticity. SLN have also UV protection capacity due to their reflection of UV light. Furthermore, data by Schafer-Korting suggest SLN can be used to decrease drug side effects due to SLN mediated drug targeting to particular skin layers. Further reports describe additional applications of SLN as well as gene delivery, delivery to the eye, pulmonary delivery, and drug targeting of anticancer drugs. Studies of the different groups also propose the use of SLN for brain targeting to deliver MRI contrast agents or antitumour drugs.

CONCLUDING REMARKS

SLN and NLC are now investigated by many scientists worldwide. In contradiction to early proposals, they certainly do not combine all the advantages of the other colloidal drug carriers and avoid the disadvantages of them. SLN are complex colloidal dispersions, not just "*frozen emulsions*". SLN dispersions are very susceptible to the sample history and storage conditions. Disadvantages of SLN include gel formation on solidification and upon storage, unexpected dynamics of polymorphic transitions, extensive annealing of nanocrystals over significant periods of time, stepwise melting of particle fractions in the lower-nanometer-size range, drug expulsion from the carrier particles on crystallization and upon storage, and extensive supercooling. The anisotropic shape of many SLN dispersions increases the surface area significantly, decreases the diffusion lengths to the surface and changes the rheological behavior dramatically (e.g. gel formation). Furthermore, the presence of alternative colloidal structures (micelles, liposomes) has to be considered to contribute to drug localization.

In most cases, the drug will be associated with the lipid and not incorporated. Studies demonstrate that SLN and NLC might have no advantages compared with submicron emulsions, in regard to protection from the aqueous environment. On the other side, animal data suggest that SLN can change the pharmacokinetics and the toxicity of drugs. In many cases, drug incorporation might not be required and drug association with the lipid can be sufficient for lymphatic uptake. Clearly, more detailed studies are necessary to get a deeper understanding of the *in vivo* fate of these carriers. Whenever possible, SLN and NLC systems should be compared directly with alternative colloidal carriers (e.g. liposomes, nanoemulsions, nanosuspensions) to evaluate their true potential.

8

Drug Nanocrystals

Since the last ten years, the number of poorly soluble drugs is steadily increasing. According to estimates, about 40% of the drugs in the pipelines have solubility problems. The increased use of high throughput screening methods leads to the discovery of more drugs being poorly water soluble. In the literature, figures are quoted that about 60 percent of the drugs coming directly from synthesis are now-a-days poorly soluble. Poor solubility is not only a problem for the formulation development and clinical testing, it is also an obstacle at the very beginning when screening new compounds for pharmacological activity. From this, there is a definite need for smart technological formulation approaches to make such poorly soluble drugs bioavailable. Making such drugs bioavailable means that they show sufficiently high absorption after oral administration, or they can alternatively be injected intravenously. There is quite a number of formulation approaches for poorly soluble drugs which can be specified as "*specific approaches*". These approaches are suitable for molecules having special properties with regard to their chemistry (e.g. solubility in certain organic media) or to the molecular size or conformation.

Of course it would be much smarter to have a "*universal formulation approach*" applicable to any molecule. Such a universal formulation approach to increase the oral bioavailability is micronization, meaning the transfer of drug powders into the size range between typically 1-10 µm. However, nowadays many drugs are so poorly soluble that micronization is not sufficient. The increase in surface area, and thus consequently in dissolution velocity, is not sufficient to overcome the bioavailability problems of very poorly soluble drugs of the biopharmaceutical specification class II. A consequent next step was to move from micronization to nanonization. Since the beginning of the 90s, the company Nanosystems propagated the use of nanocrystals (instead of microcrystals) for oral bioavailability enhancement, and also to use nanocrystals suspended in water (nanosuspensions) for intravenous or pulmonary drug delivery.

The solution was simple; in general, simple solutions possess the smartness that they can be realized easier than complex systems and introduction to the market is faster. Nevertheless, it took about ten years before the first nanocrystals in a tablet appeared on the market, the product

Rapamune by the company Wyeth in 2000. Compared with liposomes developed in 1968 with the first products on the market around 1990 (e.g. Alveofact, a lung surfactant), this was still relatively fast. What were the reasons that it took about one decade for nanocrystals to enter the market? From our point of view, pharmaceutical companies prefer to use formulation technology already established with know how available in the company. In addition, if formulation technologies are established, a company also has the possibility for production of the final product. Therefore, all the traditional formulation approaches were exploited to solve a formulation problem.

In addition, formulation approaches were preferred, being even simpler than nanocrystals. For example, production of drug-containing microemulsions administered in a capsule is, in many cases, even simpler. Another reason for the reluctance of pharmaceutical companies at the beginning was the lack of large scale production methods. These were not available at the very beginning of the development of the nanocrystal technology. Meanwhile, this has changed and the major pharmaceutical companies try to secure or have already secured their access to nanocrystal technology. Access to nanocrystal technology is possible either by licencing in or alternatively by the attempt to develop one's own production technologies for the nanocrystals, which do not depend on already existing intelectual property (IP). This chapter discusses the physicochemical properties of nanocrystals which make them interesting for drug delivery, reviews and discusses briefly the various production methods available and highlights the opportunities for improved drug delivery using different application routes.

MEANING OF DRUG NANOCRYSTALS

Drug nanocrystals are crystals with a size in the nanometer range, meaning that they are nanoparticles with a crystalline characters. There are discussions about the definition of a nanoparticle, referring to the size of a particle to be classified as a nanoparticle. Depending on the discipline, e.g. in colloid chemistry, particles are only considered as nanoparticles when they are in sizes below 100 nm or even below 20 nm. Based on the size unit, in the pharmaceutical area, nanoparticles should be defined as having a size between a few nanometers and 1000 nm (1 μm); thus, microparticles possess consequently a size 1-1000 micrometer. A further characteristic is that drug nanocrystals are composed of 100% drug; there is no carrier material as in polymeric nanoparticles. Dispersion of drug nanocrystals in liquid media leads to "*nanosuspensions*", in contrast to "*microsuspensions*" or "*macrosuspensions*". In general, the dispersed particles need to be stabilized, e.g. by surfactants or polymeric stabilizers. Dispersion media can be water, aqueous solutions or non-aqueous media [e.g. liquid polyethylene glycol (PEG), oils]. Depending on the production technology, processing of drug microcrystals to drug nanoparticles can lead to either a crystalline or to an amorphous product, especially when applying precipitation. In the strict sense, such an amorphous drug nanoparticle should not be called nanocrystal. However, one often refers to "nanocrystals in the amorphous state".

GENERAL PROPERTIES

The reason for micronization is to increase the surface area, thus consequently according to the Noyes-Whitney equation, increasing the dissolution velocity. Therefore, micronization can be succesfully employed if the dissolution velocity is the rate-limiting step for oral absorption (drugs of BSC II). Of course, by moving one dimension further to smaller particles, the surface area is further enlarged and consequently, the dissolution velocity is further enhanced. In most cases, a low dissolution velocity is correlated with a low saturation solubility.

Saturation Solubility

The general textbook statement is that the saturation solubility c_s is a constant depending on the compound, the dissolution medium and the temperature. This is valid for powders of daily life with a size in the micrometer range or above. However, below a critical size of 1-2 µm, the saturation solubility is also a function of the particle size. It increases with decreasing particle size below 1000 nm. Therefore, drug nanocrystals possess an increased saturation solubility. This has two advantages:

1. According to Noyes-Whitney, the dissolution velocity is further enhanced because dc/dt is proportional to the concentration gradient (c_s - cx)/h (c_x – bulk concentration, h – diffusional distance).
2. Due to the increased saturation solubility, the concentration gradient between gut lumen and blood is increased, consequently, the absorption by passive diffusion.

The interesting question very often asked is "How manyfold is the increased saturation solubility?". Data published in the literature or available to us from discussions range from 2-14 fold. What are the factors affecting the increase in saturation solubility? The factors can be identified when looking at the theoretical background. The Kelvin equation describes the increase in the vapor pressure of droplets in a gas medium as a function of their particle size, i.e. as a function of their curvature:

$$\ln\left(\frac{P}{P_0}\right) = \frac{-\gamma - {}^*V_L{}^* \cos\theta}{r_K{}^* RT}$$

P = vapor pressure

P_0 = equilibrium pressure of a flat liquid surface

γ = surface tension

V_L = molar volume

cos (θ) = contact angle

r_K = radius of droplet

R = universal gas constant

T = absolute temperature (K)

The vapor pressure increases with increasing curvature of the surface, that means decreasing particle size. Each liquid has its compound specific vapor pressure, thus the increase in vapor pressure will be influenced by the available compound-specific vapor pressure. The situation of a transfer of molecules from a liquid phase (droplet) to a qas phase is in principal identical to the transfer of molecules from a solid phase (nanocrystal) to a liquid phase (dispersion medium). The vapor pressure is equivalent to the dissolution pressure. In the state of saturation solubility, there is an equilibrium of molecules dissolving and molecules recrystallizing. This equilibrium can be shifted in case the dissolution pressure increases, thus increasing the saturation solubility. Identical to liquids with different vapor pressures under normal conditions (micrometer droplet size), each drug crystal has a specific dissolution pressure in micrometer size. The important question is how the dissolution pressure changes, depending on the specific dissolution pressure of each compound and on the particle size. Model calculations were performed applying the Kelvin equation to compounds with different vapor pressures (droplets) as a function of droplet size. Liquids with low medium and high vapor pressure were selected, such as oleic acid as an oil, water and ether. The important result for a drug formulation was:

1. The increase in vapor pressure is more pronounced for compounds having *a priori* a low vapor pressure. Applied to solid compounds, increase in dissolution pressure will be more pronounced for compounds having *a priori* a low dissolution pressure, i.e. the relative increase is highest for poorly soluble drugs.
2. The increase in vapor pressure is exponential, with a very pronounced increase occurring at droplet sizes below 100 nm.

Effect of Size

Transferring this to drug nanocrystals means that really smart crystals with highest increase in saturation solubility should have a size of e.g. 50 nm or 20-30 nm. From this, it can be concluded that the slogan "*size matters*" is correct regarding the increase in saturation solubility, and consequently, the increase in dissolution velocity caused by a higher c_s. It needs to be kept in mind which blood profile is anticipated with a certain drug. In many cases, too fast a dissolution is not desired (creation of high plasma peaks, reduction of t_{max}). There is the request to combine drug nanocrystals with traditional controlled release technology (e.g. coated pellets) to avoid too fast a dissolution, too high plasma peaks, too early a t_{max} and to reach prolonged blood levels. To summarize, the optimal drug nanocrystal size will depend on:

1. Required blood profile
2. Administration route

In the case of i.v. injected nanocrystals, the size should be as small as possible in case the pharmacokinetics of a solution should be mimicked. In the event that a targeting is the aim (e.g. to the brain by PathFinder technology, the drug nanocrystals should possess a certain size to delay dissolution and to give them the chance to reach the blood-brain barrier (BBB) for internalization by the endothelial cells of the BBB.

Effect of Amorphous Drugs

It is well known that amorphous drugs possess a higher saturation solubility, compared with crystalline drug material. A classical example from the literature is chloramphenicol palmitate. The polymorphic modification I has a solubility of 0.13, the high energy modification II a solubility of 0.43 and the amorphous material of 1.6 mg/mL. The same is valid for drug nanoparticles, amorphous drug nanoparticles possess a higher saturation solubility, compared with equally sized drug nanocrystals in the crystalline state. Therefore, to reach highest saturation solubility increase, a combination of nanometer size and amorphous state is ideal. However, prerequisite for exploitation in pharmaceutical products is that the amorphous state can be maintained for the shelf life of the product.

METHODS OF PRODUCTION

Precipitation Methods

Hydrosol technique

The Hydrosol technology was developed by Sucker and the intellectual property owned by the company Sandoz, now known as Novartis. It is basically a classical precipitation process known to pharmacists under the term "via humida paratum" (v.h.p.). This v.h.p. process was employed to prepare ointments containing finely dispersed, precipitated drugs. The drug is dissolved in a solvent, the solvent added to a non-solvent leading to the precipitation of finely dispersed drug nanocrystals. A problem associated with this technology is that the formed nanoparticles need to be stabilized to avoid growth in micrometer crystals. In addition, the drug needs to be soluble at least in one solvent. This creates problems for the newly synthesized or discovered drugs, being poorly soluble in water and simultaneously in organic media. Lyophilization is recommended to preserve the particle size. To our knowledge, this technology has not been applied to a product to date.

NanoMorph

Depending on the precipitation methodology, drug nanoparticles can be generated which are in the amorphous state. A nice example are carotine nanoparticles in food industry. A solution of the carotinoid, together with a surfactant and a digestible oil, are admixed into an appropriate solvent at a specific temperature. The solution is mixed with a protective colloid. This tranforms the hydrophilic solvent components into the water phase and the hydrophobic phase of the carotinoid forms a monodisperse phase. X-ray analysis after subsequent lyophilization shows that approximately 90% of the carotinoid is in the amorphous state. Amorphous precipitation technology is used by the company Soliqs and the technology is advertised under the tradename NanoMorph. The preservation of the amorphous state could be achieved successfully for food products. To exploit the amorphous technology for pharmaceutical products, the stricter requirements for pharmaceuticals need to be met.

Process of Homogenization

Method of microfluidizer

The previous Canadian company RTP employed the microfluidizer to homogenize drug suspensions. The microfluidizer is a jet stream homogenizer of two fluid streams collied frontally with high velocity (up to 1000 m/sec) under pressures up to 4000 bar. There is a turbulant flow, high shear forces, particles collied leading to particle diminution to the nanometer range. The high pressure applied and the high streaming velocity of the lipid can also lead to cavitation additionally, contributing to size diminution. The patent describes examples requiring up to 50 passes through the microfluidizer to obtain a nanosuspension. Sometimes, up to 100 cycles are required when applying the microfluidizer technology. This does not pose any problem on the small lab scale, but it is not production friendly for larger lab scale. The dispersion medium is water.

Method of dissocubes

In 1994, Mueller *et al.* developed a high pressure homogenization method based on piston-gap homogenizers for drug nanosuspension production. Dispersion medium of the suspensions was water. A piston in a large bore cylinder creates pressure up to 2000 bar. The suspension is pressed through a very narrow ring gap. The gap width is typically in the range of 3-15 micrometer at pressures between 1500-150 bar. There is a high streaming velocity in the gap according to the Bernouli equation. Due to the reduction in diameter from the large bore cylinder (e.g. 3 cm) to the homogenization gap, the dynamic pressure (streaming velocity) increases and simultaneously decreases the static pressure on the liquid.

The liquid starts boiling, and gas bubbles occur which subsequently implode, when the suspension leaves the gap and is again under normal pressure (cavitation). Gas bubble formation and implosion lead to shock waves which cause particle diminution. The patent describes cavitation as the reason for the achieved size diminution. Piston-gap homogenizers which can be used for the production of nanosuspensions are e.g. from the companies APV Gaulin, Avestin or Niro Soavi. The technology was aquired by Skyepharma PLC at the end of the 90s and employed in its formulation development.

Nanosuspensions

For oral administration, the drug nanosuspensions themselves are, in most cases, not the final products. For patient's convenience, the drug nanocrystals should be incorporated in traditional dry dosage form, e.g. tablets, pellets and capsules. An elegant method to obtain a final formulation directly is the production of nanocrystals in non-aqueous homogenization media. Drug nanocrystals dispersed in liquid polyethylene glycol (PEG) or oils can be directly filled as drug suspensions into gelatine or HPMC capsules. The non-aqueous homogenization technology was established against the teaching that cavitation is the major diminution force in high pressure homogenization. Efficient particle diminution could also be obtained in non-aqueous media. To prepare tablets or pellets, the dispersion medium of the nanosuspension needs to be removed, i.e. in general, evaporated. Evaporation is faster and possible under milder conditions when mixtures of water with water miscible liquids are used, e.g. water-ethanol. To obtain isotonic nanosuspensions for intravenous injection, it is beneficial to homogenize in water-glycerol mixtures. The IP owned by Pharmasol covers, therefore, water-free dispersion media (e.g. PEG, oils) and also water mixtures.

Co-operative Process

Nanoedge technology

The Nanoedge technology by the company Baxter covers a combination of precipitation and subsequent application of high energy shear forces, preferentially high pressure homogenization with piston-gap homogenizers. As outlined above, the precipitated particles have a tendancy to grow. According to the patent by Kipp *et al.,* treatment of a precipitated suspension with energy (e.g. high shear forces) avoids particle growth in precipitated suspensions (= annealing process). The relative complex patent description can be summarized in a simplified way that the subsequent annealing stabilizes the obtained particle size by precipitation.

As described above, precipitated particles can be amorphous or partially amorphous. This implies the risk that during the shelf life of a product, the amorphous particles can recrystalize, leading subsequently to a reduction in oral bioavailability or a change in pharmacokinetics after intravenous injection. The annealing process by Baxter converts amorphous or partially amorphous particles to completely crystalline material.

Scaling up

An important criteria for a technology is its scaling up ability and the possibility to produce on large scale, applying "normal" production conditions. The number of 50-100 passes through a homogenizer as partially required for the microfluidizer technology is not production friendly. Piston-gap homogenizers proved to be more efficient, typically between 10-20 homogenization cycles are sufficient to obtain a nanosuspension. However, it would of course be desirable to apply even less homogenization cycles, reducing production time, potential product contamination by wearing of the machine and production costs. Pharmasol developed a new combination process, Nanopure XP (Xtended Performance) leading to:

1. Identical particle sizes compared with high pressure homogenization in water, but at half the cycle numbers or less.
2. Lower particle sizes at identical cycle numbers.

The process is again a combination technology, a pre-treatment step is followed by a high pressure homogenization step, typically performed with a piston-gap homogenizer. The code for this homogenization technology is H42. Applying the novel H42 technology leads to distinctly smaller crystals after just 15 cycles, compared with the "old" technology of applying 40 cycles.

COMMERCIAL UTILIZATION

Through Mouth

Most attractive regarding regulatory and commercial aspects is the oral administration route. Compared with parenteral administration, the regulatory hurdles are much lower. In addition, the patient prefers an oral dosage form, that is why oral products possess the largest percentage of the pharmaceutical market. However, for the oral administration route, it is generally necessary to transfer the liquid nanosuspension into a solid dosage form. Aqueous nanosuspensions can be used as a granulation fluid for producing tablets or as a wetting agent for pellet production. In addition, spray drying can be performed in order to obtain a product which can subsequently be processed to oral products. The first nanosuspension product in the market was Rapamune, introduced in 2000 by the company Wyeth.

Rapamune is available on the market as oral solution, and alternatively as tablet. The tablet is more user-friendly. Comparing the oral bioavailabilities of solution and nanocrystal tablet, the bioavailability of the nanocrystals is 21% higher compared with the solution. The oral single dose of Rapamune is 1 or 2 mg, the total tablet weight being 364 mg for 1 mg formulation and 372 mg for the 2 mg formulation, meaning that it contains a very low percentage of its total weight as nanocrystals. An important point is that the drug nanocrystals are released from the tablet as ultrafine suspension. In the event that crystal aggregation takes place to a pronounced extent, the dissolution velocity, and subsequently, the oral bioavailability of the BSC II drugs will be reduced. Therefore, there is an upper limit to load tablets with nanocrystals.

In case the limit is exceeded and nanocrystals get in contact with each other within the excipient mixture of the tablet, the nanocrystals might fuse to larger crystals under the compression pressure during tablet production. For drugs with a low oral single dose such as Sirolimus in Rapamune, incorporation into tablets causes little issues. A total nanoparticle load of less than 1% is well below the percentage being critical. The second product on the market was Emend, introduced in 2001 by the Company Merck. The drug Aprepiptant is for the treatment of emesis (single dose is either 80 or 125 mg). Aprepiptant will only be absorbed in the upper gastrointestinal tract. Bearing this in mind, nanoparticles proved to be ideal in overcoming this narrow absorption window. The large increase in surface area due to nanonization leads to rapid *in vivo* dissolution, fast absorption and increased bioavailability.

The formulation of a tablet from micronized bulk powder made higher doses necessary, leading to increased side effects. The drug nanocrystals are contained within the hard gelatin capsules as pellets. Aprepiptant was formulated as capsules for it to be user friendly by healthcare providers and patients, and on the other hand, to make it applicable as pellets via a stomach tube. Currently, studies are being undertaken to evaluate the change in pharmaco-kinectics (if any) between the pellets and the capsules. All nanocrystals in these first two products were produced using the pearl mill technology by Nanosystems/Elan. The prerequiste was the bioavailability of sufficient large scale production facilities for the respective product.

In general, the candidates of first choice for nanosuspension technology are drugs with a relatively low dose. It is interesting that drugs such as Naproxen are formulated as nanosuspension (e.g. for fast action onset and reduced gastric irritancy), however, it requires more sophisticated formulation technology to ensure the release of the drug nanocrystals as fine suspension when incorporated in a tablet in a relatively high concentration of a single dose of 250 mg. The tablet size (weight) has to be acceptable for the patient and that a dosing with two tablets should be avoided, for reasons of patient's compliance and marketing purposes. Alternatively, to aqueous nanosuspensions, nanosuspensions in nonaqueous media can be produced by the Nanopure technology (Pharmasol).

Nanocrystals dispersed in liquid PEG or oil can be directly filled into gelatine or HPMC capsules. It saves the step of water removal and subsequent dispersion of the powder in a liquid capsule filling medium. The Nanopure technology also allows production of nanocrystals in melted PEG (at 60°C). After solidification of the PEG nanosuspension, the drug nanocrystals are fixed (and kept seperated) in the solid PEG matrix. The solidified drug nanocrystal containing PEG can either be milled into powders and filled into the capsules, or alternatively, the hot liquid PEG nanosuspension can be directly filled into the capsules. Instead of using aqueous nanosuspensions as fluids for the wet granulation process or extrusion of pellet mass, the nanosuspensions can be converted into a dry powder which is subsequently further processed into a tablet or a capsule. It also appears attractive to package such powders in sachets for redispersion in water or soft drinks prior to oral administration.

Spray-drying is the only feasable cost-effective way to produce such powders. An attractive approach is the production of so-called "*compounds*" as described in the direct compress technology. The term "*compound*" does not mean a chemical compound; in powder technology, "*compounds*" are defined as freely flowable granulate powders. In the direct compress technology, water-insoluble polymeric particles (e.g. Eudragit RSPO, ethyl-cellulose) are dispersed in the aqueous drug suspension, and lactose is dissolved. The mixture is a freely flowable compound yielded by spray-drying. The lactose is responsible for the good flowing properties. The water-insoluble polymeric particles also contribute to the formation of flowable granules, while at the same time allowing the compound to be compressed in a direct compaction process into tablets. The polymers form the matrix structure of the tablet.

Depending on the percentage of the insoluble polymeric particles added, the resulting tablets may disintegrate fast or present a prolonged release system. A prolonged release of dissolving nanocrystal is desired in the case of high plasma that peaks at very early times (short t_{max}) and a targeted sustained blood level. Alternatively, the drug nanocrystal compound can be filled into hard gelatine capsules. Due to the presence of lactose and surfactant from the original nanosuspension, the compounds disperse relatively fast in liquids. As outlined above, efficient release and redispersion of the drug nanocrystals in a fine, nonaggregated state is a prerequisite for benefiting fully from the drug nanocrystal features.

Intravenous Administration

Intravenous administration is the second frequently investigated route. The company Baxter, with its technology NANOEDGE, is presently focusing on intravenous nano suspensions. They investigated Itraconazole nanosuspensions intensive. It could be nicely shown that the side effects of the commercial product Sporanox could be distinctly reduced by the administration of a nanosuspension. The nephrotoxicity of Sporanox is not caused by the drug, but by the excipient used for solubilizing the drug, the hydroxypropyl-β-cyclodextrin. The itraconzole nanosuspension was stabilized with Tween 80 surfactant being well tolerated intravenously. Administration of nanosupsensions into body cavities is also of great interest, e.g. to increase the tolerability of the drug, to achieve a local treatment or to have a depot with slow release (e.g. into the blood). It could be shown that intraperitonal administration of a nanosuspension was well tolerated, whereas administration of a macrosuspension leads to irritancy [azodicarbonamide (ADA), unpublished data]. Intraperitonal administration can be used for local treatment or to obtain a depot with prolonged release into the blood.

Interesting therapeutic targets include local inflammations, e.g. in joints. For instance, arthritic joint inflammations are caused by secretion products of activated macrophages. An interesting approach is therefore the administration of a corticoid nanosuspension directly into the joint capsule. The drug particles will be phagocytosed, the drug dissolves and reduces the hyperactivity of the macrophages. This concept is not new, being adopted by the company Boots in the 80s in an attempt to incorporate the corticoid prednisolone into polymeric nanoparticles made from PLA-

GA-copolymer. The particle load (polymer load) required to achieve a therapeutic drug level was being calculated. However, incubating macrophage cell cultures with the required particle concentration lead to cytotoxicity. The concept could not be realized, as it cannot occur with drug nanocrystals since no carrier polymer to required and present. Producing parenteral products with drug nanocrystals has to meet higher regulatory hurdles and product quality standards distinctly.

The produced drug nanosuspensions need to be terminally sterilized or alternatively produced in an aseptic process. In principal, sterilization is possible by autoclaving. However, the increase in temperature can reduce hydration of steric stabilizers, thus leading to some aggregation during the sterilization process. Gamma irradiation is *a priori* a non-preferred process by the industry due to the necessary analytics (i.e. proof of absence of toxic irradiation products). In addition, it was also observed that irradiation can cause aggregation not by directly interacting with the drug nanocrystals, but with the stabilizing surfactant.

Irradiation of Tarazepide nanosuspension leads to aggregation; simultaneously, a decrease in zeta potential also occurred during the irradiation process. A decrease in zeta potential, i.e. electrostatic repulsion, was considered as the cause for the aggregation process. It can be concluded that the production of drug nanosuspensions in an aseptic, controlled process has to be preferred, compared with the terminal sterilization by irradiation. The aseptic production process can be validated and documented relatively easy, therefore, being simpler to handle as an irradiation sterilization with accompanied analytics.

Other Intake Methods

Oral and parenteral/intravenous routes are the ones in which developments are focusing, clearly due to the commercial background and the relation between the development costs for a market product versus its potential annual sales. However, drug delivery could also be improved when using drug nanocrystals for pulmonary and ophthalmic adminstration or dermal application. Poorly soluble drugs could be inhaled as drug nanosuspension. The drug nanosuspension can be nebulized using commercially available nebulizers. Disposition in the lungs can be controlled via the size distribution of the generated aerosol droplets. Compared with microcrystals, the drug is more evenly distributed in the droplets when using a nanosuspension. The number of crystals are higher, consequently, the possibility that one or more drug crystals are present in each droplet is higher. It could be shown that nanoparticles possess a prolonged retention time in the eye, most likely due to their adhesive properties. From this, poorly soluble drugs could be administered as a nanosuspension. However, the major obstacles are the commercial considerations.

In many cases, the sales volume do not justify the costs for the development of a new market product. This is especially the case when a company has already a drug formulation which might be less efficient, but is already a product on the market. The price achievable with an improved product is not sufficiently high to cover the development costs of this new product. An additional major obstacle for the development of such improved products is the cost reduction policy of the healthcare systems worldwide. A longer treatment time with a less efficient product might still be less expensive for the healthcare system than a shorter treatment time with a more efficient, but distinctly more expensive product. The same is valid for dermal products. Sales per product are lower compared with e.g. oral products, as the dermal market is smaller.

Dermal nanosuspensions are mainly of interest if conventional formulation technology fails or if it is distinctly less efficient. Dermal drug nanosuspensions lead to a supersaturated system because of their increased saturation solubility. The higher concentration gradient between topical formulation and skin can improve drug penetration into the skin. In addition, because of their small size, drug nanocrystals could target the hair follicle by protruding into the gap around the hairs. This was illustrated in solid lipid nanoparticles of a similar size. Adhesive properties of drug nanocrystals are also an area of interest. Adherence to the skin reduces the "loss" of drug to the environment/third persons. This is especially so in the event that highly active compounds are applied, e.g. hormones. For this reason, the drug estradiole was incorporated into solid lipid nanoparticles to better localize it on the skin.

FINAL PRODUCTS

As described above, nanosuspensions can be produced such that nanocrystals appear in final products. Alternatively, drug nanosuspensions can be used as intermediate product, i.e. the drug nanocrystals do not appear in the final product. Recently, the SolEmuls technology was developed to produce drug-loaded emulsions for intravenous injection, i.e. localizing poorly soluble drugs in the interfacial layer of lecithin emulsions. The applicability of the technology has been proven for several drugs including amphotericin B, itraconazole, ketoconazole, and carbamazepine, among others. The drug Amphotericin B is on the market as a solution (Fungizone), but also in liposomes (Ambisome); the latter having the benefit of reduced nephrotoxicity. Liposomes are relatively expensive (daily treatment costs approximately EUR 1000-2000), therefore Amphotericin B was incorporated into parenteral emulsions. These emulsions can also reduce nephrotoxicity, but for their production, it was necessary to use organic solvents.

Egg lecithin and amphotericin B were dissolved in an organic solvent, the solvent evaporated and the obtained drug-lecithin mixture was used to produce an o/w emulsion. In these emulsions, amphotericin B was located in the interfacial lecithin layer as Amphotericin B is simultaneously poorly soluble in water and in oils. There were also attempts to incorporate amphotericin B in the emulsion by simply adding Amphotericin B powder to the emulsion and subsequently shaking

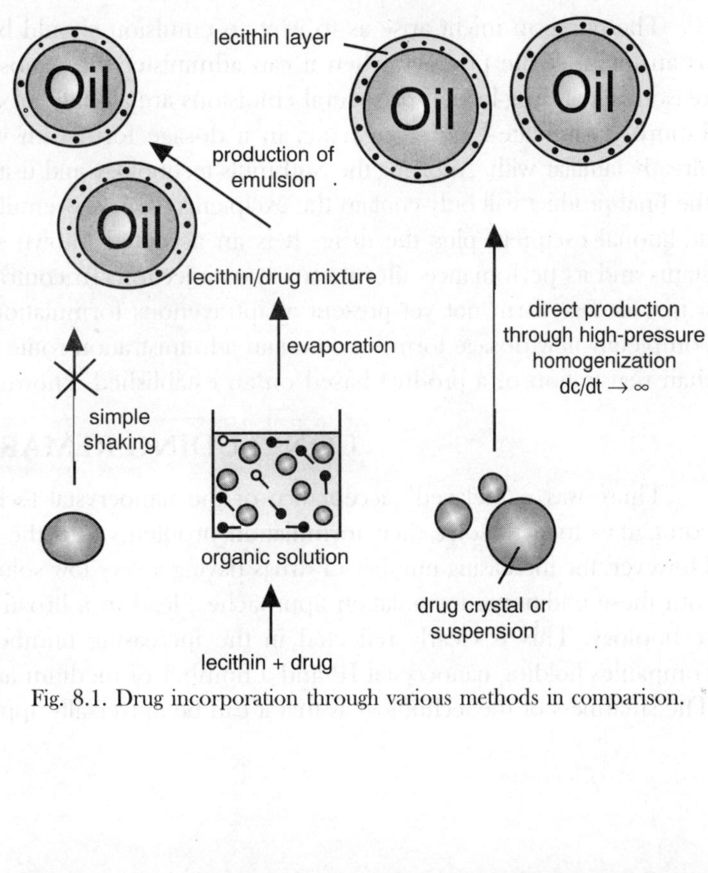

Fig. 8.1. Drug incorporation through various methods in comparison.

it. However, even shaking for 18 hours with 1800 rph was unable to completely dissolve the Amphotericin B. The reason was simply due to its low solubility in the water, and the dissolution velocity was also extremely low, i.e. the process of dissolution and redistribution into the lecithin layer takes too long for it to be used in pharmaceutical production. The problem was solved by the SolEmuls technology, i.e. simple co-homogenization of oil droplets and micro- crystals. For a de novo production, a coarse pre-emulsion of lecithin stabilized oil droplets in water is prepared, the drug powder is admixed under stirring and the obtained hybrid suspension subsequently homogenized at 600 bar (pressure being in the range to be used in pharmaceutical production lines).

The high streaming velocities in the homogenization process lead to fast dissolution of the drug microcrystals and the re-distribution into the interfacial lecithin layer. Depending on the size of the drug crystals, 5-10 homogenization cycles are required. The number of homogenization cycles can be reduced when adding the drug not as microcrystals, but as nanocrystals in the form of a nanosuspension. A concentrated nanosuspension is prepared (e.g. 20-30% solid content) and added to the pre-emulsion. Ideally the nanosuspension is also stabilized by lecithin, i.e. the same emulsifier for the suspension and the emulsion. Alternatively, intravenously accepted stabilizers such as Tween 80 or Poloxamer 188 can be used. They are accepted intravenously without posing any regulatory issues. In addition, mixing the emulsion and nanosuspension at a ratio of 10:1 or higher will dilute the stabilizer concentration used in the nanosuspension by at least a factor of 10, meaning that in the final product, the nanosuspension surfactant concentration is typically 0.1 or 0.01%.

The question might arise as to why an emulsion should be prepared using a nanosuspension as an intermediate product, when it can administer the nanosuspension itself intravenously? The reason is that drug-loaded parenteral emulsions are already products on the market (e.g. Diazepam-Lipuro, Etomidate-Lipuro, etc.), i.e. in a dosage form with which the regulatory authorities are already familiar with. Applying the SolEmuls technology and using lecithin-stabilized nanosuspension, the final product will only contain the excipients of an o/w emulsion for parenteral nutrition, without additional excipient plus the drug. It is an accepted known system with regard to the excipient status and its perfomance after intravenous injection. In contrast, drug nanosuspensions represent a new dosage form not yet present as intravenous formulations on the market. Registration of a completely new dosage form for a certain administration route is just more complicated and timely than registration of a product based on an established, known technology.

CONCLUDING REMARKS

There was a "delayed" acceptance of the nanocrystal technology in the 90s. Pharmaceutical companies tried to solve their formulation problems with the traditional formulation approaches. However, the increasing number of drugs having a very low solubility, and not able to be formulated with these traditional formulation approaches, lead to a broad acceptance of the drug nanocrystal technology. This is clearly reflected in the increasing number of licensing agreements between companies holding nanocrystal IP and a number of medium and large pharmaceutical companies. The smartness of the technology is that it can be universally applied to practically any drug. Identical to micronization, it is a universal formulation principle, but limited to BSC drugs class II. The time between the beginning of intensive research in the drug nanocrystal technology and the first products on the market was relatively short, about one decade.

The value of a formulation principle or technology can be clearly judged by looking at the number of products on the market, in the clinical phases, and /or the time of entry into the market. Based on these criteria, the drug nanocrystal technology is a successful emerging technology. Meanwhile, "Big Pharma" also realized the drug nanocrystal value. In combination with the further increasing number of poorly soluble drugs, a distinct increase in drug nanocrystal-based products on the market can be expected. In many cases, oral products will dominate because of the market share, higher sales volumes and less regulatory hurdles and quality requirements, compared with parenteral products.

9

Clinical Drug Carriers

Large protein structures (in nanometer range) may be utilized as pharmaceutical carriers of drugs and DNA for targeted and other specialized delivery in biological systems. Lipoproteins are such structures which function as natural biological carriers and transport various types of lipids in blood circulation. There are many studies suggesting that lipoproteins can serve as efficient carriers for anticancer drugs, gene or other type of compounds. Previous results showed that hydrophobic cytotoxic drugs could be incorporated into lipoproteins, without changing the integrity of native lipoprotein structure.

Lipoproteins as drug carriers offer several advantages. Firstly, they are endogenous components and do not trigger immunological response. They have a relatively long half-life in the circulation. Secondly, they have small particle *size* in the nanometer range, allowing the diffusion from vascular to extravascular compartments. Thirdly, lipoproteins can potentially serve as the carriers for targeted drug delivery through specific cellular receptors. For example, *low density lipoprotein* (LDL)-drug complexes may target cancer cells which, in many cases, have higher LDL-receptor expression than normal cells. Fourthly, the lipid core of lipoprotein provides a suitable compartment for carrying hydrophobic drugs. As a result of these advantages, lipoproteins have received wide attentions in recent years in the development of drug-targeting strategies to use them as specialized delivery vehicles. This review intends to provide an overview of the development and the specialized utilization of lipoproteins for drug delivery purpose. After briefly introducing the structure and the basic biological functions of lipoproteins, we will focus on four classes of lipoproteins, namely, chylomicron, *very low-density lipoprotein* (VLDL), *low-density lipoprotein* (LDL), and *high-density lipoprotein* (HDL), as the carriers for various drug compounds. Cholesterol-rich emulsions (LDE) and artificial lipoproteins as drug carriers will also be discussed.

MEMBRANOUS PROTEINS

Lipoproteins, as implied by their names, are biological protein-lipid complexes. Lipoproteins serve the functions of carrying hydrophobic substances in blood circulation and transporting them

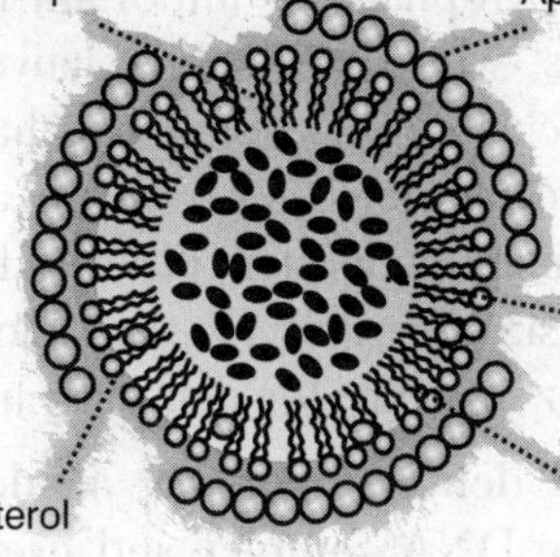

Fig. 9.1. General structure of lipoproteins.

to various biological sites through the protein-receptor interactions. The size of lipoproteins is in the nanometer range and they have a spherical shape with complex physicochemical properties. The hydrophobic core contains water-insoluble substances and is surrounded by a polar shell. The polar shell consists of phospholipids, unesterified cholesterol and different types of apolipoproteins, which bind to various cellular receptors for specific biological functions. Therefore, based on their physicochemical properties, lipoproteins are nanoemulsions with targeting functions provided by the apolipoproteins.

Owing to the unique structure of lipoproteins, they can serve a two-mode function of solubilizing hydrophobic substances, including triglycerides and cholesteryl esters, within the nanoemulsion core and allow themselves to float in blood circulation. Lipoproteins can be classified into five major classes, based on their densities from gradient ultracentrifugation experiments. The lipoprotein classification includes chylomicron, *very low-density lipoprotein* (VLDL), *intermediate-density lipoprotein* (IDL), *low-density lipoprotein* (LDL), and *high-density lipoprotein* (HDL). These classes of lipoproteins have different sizes, different protein to lipid ratios and different types of apolipoproteins. In general, chylomicrons act on transporting dietary triacylglycerols and cholesterol to the adipose tissue and liver, following the absorption of dietary hydrophobic substances from the intestines. Very low density lipoprotein, intermediate density lipoprotein and low density lipoprotein work at different stages to transport triacylglycerols and cholesterol from the liver to various tissues. High density lipoprotein brings endogenous cholesterol from the tissues back to the liver.

Table 9.1 Physicochemical properties of lipoproteins.

Lipoprotein	*Transport route*	*Size (nm)*	*Protein (%)*	*Total lipids (%)*
Chylomicron	Intestines to Liver	75-1200	1.5-2.5	97-99
VLDL	Liver to tissues	30-80	5-10	90-95
IDL	Liver to tissues	25-35	15-20	80-85
LDL	Liver to tissues	18-25	20-25	75-80
HDL	Tissues to liver	5-12	40-55	45-60

LIPOIDAL CARRIER

Chylomicrons are assembled in the intestine from the absorbed dietary lipids and transported by lymphatic system. Although most of the drugs administered orally are absorbed directly into the portal blood to reach the systemic circulation, an alternative absorption route through the intestinal lymphatics may be available for hydrophobic drugs. It is estimated that a high hydrophobicity (log o/w partition coefficient > 5) of drug molecules is required for intestinal lymphatic transport. Chylomicrons can thus potentially serve as an important natural carrier for hydrophobic drugs to transport through lymphatic system. It is known that targeted drug delivery through the lymphatics is important for anti-viral drug molecules for the protection of B- and T-lymphocytes, which maintain relatively higher concentrations through the lymphatics than the systemic circulation. Chylomicrons have a much larger size than other lipoproteins, and thus can carry more drug molecules from the absorption site. With the presence of food, chylomicrons are the predominant lipoprotein produced by the small intestine to carry dietary lipids efficiently because of its large size. Various types of bioactive molecules have been incorporated into reconstituted chylomicron structure for delivery purposes.

In gene delivery, Hara *et al.* developed reconstituted chylomicron which incorporated a hydrophobic DNA complex and used it as an *in vivo* gene transfer vector. They found that the DNA-incorporated chylomicrons induced a high gene expression in mouse liver after the reconstituted chylomicron was administered through portal vain injection. Furthermore, it was also reported that artificial, protein-free lipid emulsions could be utilized to model the metabolism of lymph chylomicron in rats, not only in the initial partial hydrolysis by lipoprotein lipase, but also in the delivery of a remnant-like particle to the liver. As a targeted therapeutic approach to hepatitis B, anti-viral iododeocyuridine was incorporated into recombinant chylomicrons, resulting in the drug molecules being selectively targeted to the liver parenchymal cells. It has been suggested that chylomicron can serve as a special carrier for liver cell targeting. Due to the targetability, this approach could be further developed as an effective therapy for hepatitis B patients.

ROLE OF VLDL

VLDL particles have a size range of 30-80 nm. They are assembled in the endoplasmic reticulum (ER) and matured in Golgi apparatus of hepatocytes before secretion. After entering into the plasma, VLDL particles are catabolized by a series of biochemical actions, including apolipoprotein exchange of apoC-I, apoC-II, apoC-III, and apoE; lipolysis by triglyceride lipase; and cell-surface receptor-mediated uptake. As lipolysis proceeds, VLDL particles become smaller and are eventually converted to IDL. Some of the IDL particles are rapidly taken up by hepatocytes via a receptor-mediated mechanism while others undergo further hydrolysis before being converted to LDL. The catabolism route of VLDL suggests the possibility of using VLDL as a drug carrier for targeted delivery. ApoE is a protein ligand present on the surface of VLDL and it is well known that the receptor of apoE is overexpressed on some types of cancer cells. Therefore, VLDL can potentially serve as an antineoplastic drug carrier. As a drug carrier, VLDL is an interesting candidate because it contains a relatively small amount of proteins (about 5-10% protein) and a large amount of triglycerides (about 50-65% within the emulsion core) which can be used to solubilize hydrophobic substances sufficiently.

By mimicking the compositions and structure of VLDL, Shawer *et al.* developed a VLDL-resembling phospholipid nanoemulsion system that carried a new anti-tumor boron compound for targeted delivery to cancer cells. The nanoemulsion demonstrated sufficient capability to solubilize the hydrophobic compound. The structure of the phospholipid nanoemulsion was verified based on the changes in the molecular surface area and the molecular volume of each component of the nanoemulsion when the particle size is changed (from different size fractions). If certain molecules are located at the core of nanoemulsion, their numbers per overall volume should not be changed when the particle size is increased. If certain molecules are located at the surface of nanoemulsion, their numbers per overall volume should decrease when particle size is increased. This is because the overall surface area decreases when particle size is increased. Similar to the natural lipoproteins, it was demonstrated that phospholipid was predominately located at the surface and the hydrophobic substances, triolein and cholesteryl oleate, were mainly located in the core of the phospholipid nanoemulsion. Recently, a similar nanoemulsion formulation was used to encapsulated *quantum dots* (QD) as a new bioimaging carrier.

Quantum dots (QDs) are semiconductor nanocrystals that are emerging as unique fluorescence probes in biomedicine. When manufactured, most of the quantum dots have organic ligand coating on their surface and are extremely hydrophobic. The research goal was to encapsulate QDs in phospholipid nanoemulsion and to examine the physical stability, size distribution and their interactions with cancer cells. It was found that CdSe QDs can be efficiently encapsulated in the phospholipid nanoemulsion. The QD-encapsulated phospholipid nanoemulsion are stable and interact well with cultured cells to deliver the QDs inside the cells for fluorescence imaging. In other studies, it has been demonstrated that cytotoxic drugs such as 5-fluorouracil (5-FU), 5-iododeoxyuridine (IudR), doxorubicin (Dox), and vindesine can be effectively incorporated into VLDL, and the resultant complexes showed effective cytotoxicity to human carcinoma cells.

ROLE OF LDL

LDL (18-25 nm) is not directly synthesized in human body. Instead, most of them are formed through the VLDL pathway. LDL is the major circulatory lipoprotein for the transport of cholesterol and cholesteryl esters, and it can be internalized by cells via LDL receptor-mediated endocytosis. The internalization process of LDL has been well characterized and the understanding of the mechanism can potentially help the designing of the drug targeting strategy through the LDL receptor. The binding of dephosphorylated adaptor protein to the plasma membrane initiates the formation of coated pits which are covered by the protein clathrin. The receptors from the surrounding regions of the plasma membrane shift towards the binding site for internalization. Apolipoproteins including apo B-100 and apo E are recognized and bound by the LDL receptor on the cell surface to form a complex which is internalized into the coated pits. After internalization of the LDL, the coated pits are pinched off and within a very short time, they shed off their clathrin coating.

The internalized LDL particle is transferred to endocytotic vesicles or endosomes. Due to the acidic pH within the endosomes, LDL dissociates from its receptor. This is followed by the fusion of the endosomes with lysosomes which contain hydrolases. The protein component of LDL is broken into free amino acids, while the cholesteryl ester component is cleaved by lysosomal lipase. The free cholesterol is released and incorporated into the cell membrane. Excess cholesterol is re-esterified by the action of acyl-CoA:cholesterol acyltransferase (ACAT). Among various lipoproteins, LDL has been widely studied as a drug carrier for targeted and other specialized deliveries, because many types of cancer cells show elevated expression of LDL receptors than the corresponding normal cells. Comparing with chylomicron, VLDL, and IDL, LDL also has a longer serum half-life of 2-4 days, making it a desirable drug carrier. Low density lipoprotein was found to be suitable as carriers for cytotoxic drugs to target cancer cells. LDL-drug complexes can be formed through various processes without changing the lipoprotein integrity.

Role of Doxorubicin

Doxorubicin (Dox) is widely used in treating different tumors. Its main side effects are cadiotoxicity and multidrug resistance, especially during prolonged treatment in the patients. LDL has been studied as a target carrier for Dox in nude mice, bearing human hepatoma HepG2 cells. Both *in vitro* and *in vivo* studies indicated that when Dox was incorporated into LDL, the multidrug resistance could be circumvented and the cardiotoxicity could be reduced as well. Kader and Pater used VLDL, LDL and HDL as carriers to deliver four cytotoxic drugs, 5-fluorouracil (5-FU), 5-iododeoxyuridine (IUdR), doxorubicin (Dox) and vindesine. They found that significant drug loading was achieved in all three classes of lipoproteins, consistent with the sizes and hydrophobicity of the drug.

Experiments were carried out to examine the changes in drug cytotoxicity against HeLa cervical and MCF-7 breast carcinoma cells, after the incorporation into lipoprotein. The results demonstrated that VLDL-drug complex did not affect their IC_{50} on both HeLa and MCF-7 cell lines, when compared with free drugs. However, the IC_{50} values of LDL- and HDL-drug complexes were significantly lower compared with free drugs. Their studies further indicated that drugs were incorporated into lipoproteins without disrupting their integrity; drugs remained in their stable forms inside lipoproteins; and human LDL and HDL could be particularly useful in the delivery of antineoplastic drugs.

Anticancer Compounds

Although LDL has been widely studied as a carrier to deliver anticancer compounds, it may also be useful to deliver other types of bioactive compounds. LDL may serve as a carrier for site-specific delivery of drugs to atherosclerotic lesions. When *dexamethasone palmitate* (DP), a steroidal anti-inflammatory drug, was incorporated in LDL, an inhibitory effect of this complex on foam cell formations was demonstrated. The study indicated that LDL could potentially carry DP to atherosclerotic lesions.

Fluorophore-labeled LDL was also used for optical imaging in tumors diagnosis. For example, carbocyanine dyes can be used as near infrared (NIR) optical imaging probes with long wavelength absorption, high extinction coefficients and high fluorescence quantum yield. *In vitro* confocal microscopic study and *ex vivo* low-temperature fluorescent scanning demonstrated that carbocynine-labled LDL probes, DiI-LDL, could be selectively delivered to B16/HepG2 tumor cells and the corresponding animal tumors via the LDL receptor pathway. It was also proposed that DiI is located and oriented in the phospholipid monolayer when it binds to LDL.

Gene in Delivery

LDL has also been investigated as gene delivery carriers. Comparing with viral gene-delivery vectors and some other types of non-viral gene delivery vectors, the LDL system shows certain advantages in transfection efficiency and safety considerations. Several LDL based gene delivery systems have been reported. Kim's group developed a terplex system which comprises LDL, lipidized poly(L-lysine) and plasmid DNA. The complex had a diameter of about 100 nm. The studies showed high efficiency to deliver plasmid DNA to smooth muscle cells and fibroblast cells. In addition, a novel LDL-DNA complex was formulated by Khan *et al.* LDL was cationized using carbodiimide and the modified lipoprotein complex significantly increased the DNA binding capacity with improved stability. The novel delivery system also demonstrated the ability to target cells through LDL receptor.

ROLE OF HDL

Among various lipoproteins, HDL has the smallest size with a diameter of 5-12 nm. It shares common structural characteristics with other lipoproteins. However, its polar shell contributes more than 80% of the total mass. Newly synthesized HDL hardly contains any cholesteryl ester molecules. Cholesteryl esters are gradually added to the particles by lecithin via enzymatic reaction: cholesterol acyltransferase (LCAT), which is a 59-kD glycoprotein associated with HDL. The interaction of HDL with cells appears similar to that of LDL. Although the function of HDL in the human body is not well-defined, it generally transports excess cholesterol and cholesteryl esters from various tissue cells back to the liver.

Comparing with other types of lipoproteins, small size and fast internalization by tumor cells are the major advantages of utilizing HDL for drug delivery and targeting. HDL has mainly been utilized for the delivery of water insoluble anticancer drugs through the targeting function: When the anticancer drug, Taxol, was incorporated into HDL, stable complexes were formed and they were examined for cancer-cell targeting. Reconstituted HDL was explored as a drug carrier system for a lipophilic prodrug, IDU-Ol_2. The studies indicated that the lipophilic prodrug could be efficiently incorporated into reconstituted HDL particles. This approach may also be useful to encapsulate other lipophilic derivatives of water-soluble drugs.

The utilization of HDL for drug targeting may lead to a more effective therapy for infectious diseases, such as hepatitis B, since the HDL-drug complexes were demonstrated to be selectively taken by parenchymal liver cells: Comparing with free drugs in cytotoxicity assays, the IC values of HDL-drug complexes were significantly decreased, about 2.5 to 23-fold lower. Interestingly, it was observed that HDL-drug complex specifically increased the cytotoxicity to carcinoma cells. Earlier studies showed that HDL could increase the sensitivity of HeLa cells to the cytotoxic effects of Dox. Similar to LDL-drug complex, the lipoprotein receptor pathway appears to be involved in the interactions between HDL-drug complex and cancer cells.

ROLE OF LDE

LDE is a lipid based formulation, an emulsion with a lipid structure resembling LDL particle and it is made without protein incorporation. Essentially, it is composed of a cholesteryl ester core surrounded by a monolayer of phospholipids. Comparing with native LDL, LDE is removed from the blood circulation more rapidly. It appears possible that LDE can acquire apoE and other apolipoproteins from native lipoproteins in plasma. ApoE can be recognized by LDL receptors, thus allowing the binding of LDE to the receptors. However, it is known that LDE binds to receptors through apoE, but not through apoB100. The interaction between apoE and the receptor appears stronger than that of apoB100. LDE is considered as a potential carrier for anticancer drugs to deliver chemotherapeutic agents to neoplastic cells. Although there is no protein in the LDE formulations, previous studies showed that the LDL receptor could still play an important role in the cellular uptake of these lipid complexes.

LDE binds to low-density lipoprotein receptors which are upregulated in cancer cells, leading to a higher concentration in neoplastic tissues. LDE-carmustine complex was studied with a neoplastic cell line and its biodistribution was studied in mice. An exploratory clinical study was also conducted. The result showed that the uptake of LDE-carmustine complex by tumor was several fold greater than the uptake by the corresponding normal tissue. The association of carmustine with LDE preserves the tumor-cytotoxicity of carmustine with reduced side effects. Preliminary clinical study was also carried out using LDE-carmustine complex to treat patients with advanced cancers. The results demonstrated that the systemic toxicity of the drug was significantly reduced. Rodrigues *et al.* investigated the formulation of LDE containing antineoplastic compound paclitaxel.

The experiments revealed a 75% incorporation efficiency and the stable complex of the drug molecules incorporated in LDE emulsion. Its LD_{50} was ten-fold greater than that of a commercial formulation of paclitaxel. It was suggested by the authors that the cellular uptake and the cytotoxic activity of LDE-paclitaxel complex might be mediated by the LDL receptors due to the cholesterol moiety in the LDE formulation. In addition to LDE, artificial lipoproteins have been constructed. Several research groups have developed various types of artificial lipoproteins. Most of them constructed the artificial lipoproteins by incorporating natural apoB protein into lipid microemulsion for the purpose of examining the lipoprotein metabolism. Artificial lipoproteins containing polylysine has also been investigated as the DNA carrier for cellular transfection, with the potential to reduce the cytotoxicity and to improve the transfection efficiency.

10

Immunization

Prevention of microbial infections by the use of vaccines is a preferred alternative to treatment. Vaccines have been applied successfully, for example, in the eradication of smallpox as well as against tetanus, diphtheria, whooping cough, polio and measles, thus preventing millions of deaths each year. However, vaccines made of attenuated organisms, which mimick natural infections usually without the disease, can be potentially unsafe. For instance, there is a risk of reversion during replication of live viruses or even mutation to a more pathogenic state. Furthermore, with immunocompromised individuals, some of the attenuated viruses may still provoke disease. On the other hand, with killed virus vaccines, their extracellular localization and subsequent phagocytosis by professional antigen presenting cells (APC) or antigen-specific B cells, lead to MHC-II class restricted presentation and to T helper cell and humoural immunity. However, they do not elicit significant cytotoxic T cell (CTL) responses. Moreover, subunit vaccines produced from biological fluids may not be entirely free of infectious agents. Even with subunit and peptide vaccines produced recombinantly or synthetically (and thus considered safe), immune responses are weak and often not of the appropriate kind.

The great variety of immunological adjuvants that are now available go a long way in rendering subunit and peptide vaccines stronger and more efficient. However, more than seventy years after the introduction of aluminium salts as an adjuvant, only two other adjuvants, liposomes and MF59, have been approved for use in humans. Thus, inspite of considerable progress, the road to the ideal vaccine appears as elusive as ever, until recently. Recent developments have led to a novel and exciting concept, namely *de novo* production of the required vaccine antigen by the host's cells *in vivo,* which promises to revolutionize vaccination especially where vaccines are either ineffective or unavailable. The concept entails the direct injection of antigen-encoding plasmid DNA which, on uptake by cells, localizes to some extent into the nucleus where it transfects the cells episomally. The produced antigen is recognized as foreign by the host and is thus subjected to pathways similar to those observed for antigens of internalized viruses (but without their disadvantages), leading to protective humoural and cell mediated immunity. A series of publications since 1992 first established the ability of plasmid DNA to induce an immune (antibody) response to the encoded foreign protein; in experiments with DNA encoding influenza nucleoprotein, immunity was both humoural and cell-mediated, and also protective in mice challenged with the virus: This was the first demonstration of an experimental DNA vaccine.

Another observation was the induction of humoural and cell-mediated immunity against HIV-1 using plasmids encoding the HIV *rev* and *env* proteins. Similar results were obtained with a gene for the hepatitis B surface antigen (HBsAg). DNA immunization was also found to apply in cancer treatment. For instance, injection of plasmids encoding tumor antigens promoted immune responses which were protective in an animal model. The concept of DNA immunization has now been adopted by vaccinologists worldwide using an ever increasing number of plasmids encoding immunogens from bacterial, viral and parasitic pathogens, and a variety of tumors. In many of these studies, genetic immunization has led to the protection of animals from infection. A number of clinical trials for the therapy of, or prophylaxis against, a variety of infections are in progress.

PLASMID DNA VACCINE

A plasmid DNA vaccine is usually supercoiled and consists of the gene encoding the vaccine antigen (the section of the target pathogen which elicits protective immunity), a promoter sequence which is often derived from *cytomegalovirus* (CMV) or *Rous sarcoma virus* (RSV)), an mRNA stability polyadenylation region at the 3′ end of the insert, and the plasminogen activator gene which controls the secretion of the recombinant product. In addition, there are an origin of replication for the amplification of the plasmid in bacteria, and a gene for antibiotic resistance to select the transformed bacteria. Immunization procedures with DNA vaccines are carried out by the intramuscular and, to a lesser extent, the intraepidermal route. Other routes include the oral, nasal, vaginal, intravenous, intraperitoneal and subcutaneous routes.

Intramuscular injection of DNA vaccines leads to such types of immunity as CTL. This was unexpected because antigen presentation requires the function of professional APC. However, myocytes which were shown[5] to take up the plasmid only to a small extent and with only a fraction of cells participating in the uptake, are not professional APCs. Although myocytes carry MHC class I molecules and can present endogenously produced viral peptides to the $CD8^+$ cells to induce CTLs, they do so inefficiently as they lack vital costimulatory molecules (e.g. the B7-1 molecule). It is thus difficult to accept that antigen presentation, leading to a CTL response, occurs via myocytes. Instead, it was reported that CTL responses occur as a result of the transfer of antigenic material between the myocytes and professional APC to some extent. In parallel, it could also be that plasmid secreted by the myocytes or as such, is taken up directly by APC infiltrating the injected site. Such APC would include dendritic cells which will express and present peptides to $CD8^+$ cells following transport to the lymph nodes or spleen. On the other hand, $CD4^+$ cells may be activated by APCs via MHC class II presentation of antigen secreted by the myocytes (or released from them after their destruction via a Tc response) and captured by the cells. Such events would lead to both cellular (Th 1) and humoural (Th 2) immunity. Indeed, it has been shown that dendritic cells are the essential APC involved in immune responses elicited by intramuscularly given DNA vaccines.

LIPOSOMAL VACCINATION

Vaccination with naked DNA by the intramuscular route is dependent on the ability of myocytes to take up the plasmid. However, some of the DNA may also be engulfed by APC infiltrating the site of injection, or in the lymph nodes following migration of the DNA to the lymphatics. The extent of DNA degradation by extracellular deoxyribonucleases is unknown, but depending on the time of its residence interstitially, degradation could be considerable. Therefore, approaches that protect DNA from the extracellular nucleases and promote DNA uptake by cells more efficiently, or target it to APC, should contribute to the optimal design of DNA vaccines. It has been suggested that as APC are a preferred alternative to muscle cells for DNA vaccine uptake and expression, liposomes (known to be taken up avidly by APC infiltrating the site of injection or in the lymphatics, an event that has been implicated in their immunoadjuvant activity) would be a suitable means of delivery of entrapped DNA to such cells.

Liposomes would also protect their DNA content from deoxyrubonuclease attack. Moreover, the structural versatility of the system would ensure that its tranfection efficiency is further improved by the judicial choice of its structural characteristics or by the co-entrapment of cytokine genes, other adjuvants (e.g. immunostimulatory sequences), or indeed protein antigens (see later) together with the plasmid vaccine. As a number of injectable liposome-based drug formulations, including vaccines against hepatitis A and influenza, have been *already* licensed for clinical use, acceptance of the system clinically would be less problematic than with other systems that are still at an experimental stage.

Quantitative Studies

A variety of plasmid DNAs have been quantitatively entrapped into liposomes by a mild dehydration-rehydration procedure. The procedure consists of mixing preformed *small unilamellar vesicles* (SUV) with a solution of the DNA destined for entrapment, freeze-drying of the mixture, followed by controlled rehydration of the formed powder, and centrifugation to remove non-entrapped material. Formed liposomes are multilamellar. However, when an appropriate amount of sucrose is added to the SUV and DNA mixture prior to dehydration, the resulting liposomes

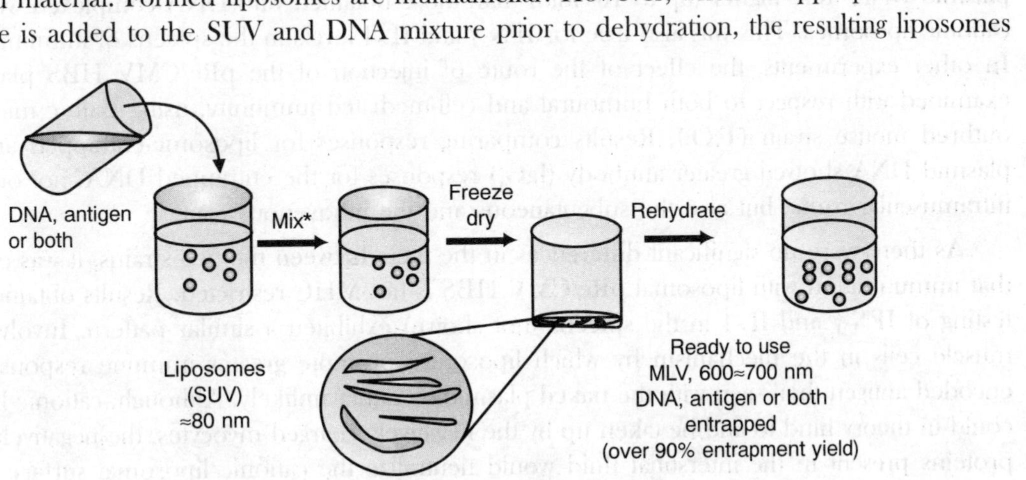

Fig. 10.1. Entrapment of DNA and/or protein into cationic liposomes.

are much smaller (about 100-160 nm in diameter). As expected, DNA incorporation values were higher (up to 90% of the amount used) when a cationic lipid was present in the bilayers. No apparent relationship was observed between amount of DNA used (10-500 μg) and the values of incorporation for the compositions and lipid mass used.

The possibility that DNA was not entrapped within the bilayers of cationic liposomes, but was rather complexed with their surface (as suggested by the high "*incorporation*" values obtained on mixing) was examined by treating liposome-entrapped and liposome-complexed DNA with deoxyribonuclease. Substantially, more liposome-entrapped DNA remained intact than when it was complexed, presumably because of the inability of the enzyme to reach its substrate in the former case. The significant resistance of complexed DNA (despite its accessibility) to the enzyme could be attributed to its condensed state.

Additional evidence that the DNA was entrapped within liposomes was obtained by gel electrophoresis of a mixture of cationic SUV and plasmid DNA before (complexed DNA) and after dehydration-rehydration of the mixture (entrapped DNA). When the anionic sodium dodecylsulphate (SDS) was incorporated in the gel, complexed DNA was dissociated from the SUV, presumably because of ionic competition for the cationic charges. As expected, "*entrapped*" DNA retained its association with the liposomes, suggesting its unavailability to the competing SDS anions.

Antigenic Role

Previously, liposome-entrapped plasmid found to transfect cells *in vitro* regardless of the vesicle surface charge was tested in immunization experiments, using a plasmid (pRc/CMV HBS) encoding the S region of the hepatitis B surface antigen (HBsAg; subtype ayw). Mice (Balb/c) that are repeatedly injected intramuscularly with 5 or 10 μg plasmid entrapped in cationic liposomes, exhibited at all times much greater (up to 100-fold) antibody (IgG_1) responses against the encoded antigen than animals immunized with the naked plasmid. Values of other subclasses (IgG_{2a} and IgG_{2b}) were also greater (up to 10-fold). Moreover, IgG_1 responses for the liposome-entrapped plasmid DNA were higher (up to 10- fold) than those obtained with DNA complexed with similar cationic liposomes. This was also true for IFN-γ and IL-4 levels in the spleens of immunized mice. In other experiments, the effect of the route of injection of the pRc/CMV HBS plasmid was examined with respect to both humoural and cell-mediated immunity, using Balb/c mice and an outbred mouse strain (T.O.). Results comparing responses for liposome-entrapped and naked plasmid DNA showed greater antibody (IgG_1) responses for the entrapped DNA, not only by the intramuscular route, but also the subcutaneous and the intravenous routes.

As there were no significant differences in the titers between the two strains, it was concluded that immunization with liposomal pRc/CMV HBS is not MHC restricted. Results obtained on the testing of IFN-γ and IL-4 in the spleens (not shown) exhibited a similar pattern. Involvement of muscle cells in the mechanism by which liposomes promote greater immune responses to the encoded antigen than seen with the naked plasmid, is rather unlikely. Although, cationic liposomes could in theory bind to and be taken up by the negatively charged myocytes, the negatively charged proteins present in the interstitial fluid would neutralize the cationic liposomal surface and thus interfere with such binding. In addition, vesicle size (about 600-700 nm average diameter) would render access to the cells difficult, if not impossible. It is therefore more likely that cationic liposomes are endocytosed by APC, including dendritic cells, in the lymphatics where liposomes are expected to end up. Uptake of liposomal plasmid DNA is supported in studies where mice were injected intramuscularly or subcutaneously with liposomes entrapping the plasmid (pCMV. EFGP), encoding the enhanced fluorescent green protein or with the naked plasmid.

Fluorescence microscopy of sections of the lymph nodes draining the injected site revealed much more green fluorescence when the plasmid was administered in the liposomal form. It appears that the key ingredient of the DNA-containing liposomes, contributing to enhanced immune responses, is the cationic lipid. The mechanism by which liposomal DNA reaches the nucleus for episomal transfection is poorly understood. It is conceivable, however, that some of the endocytosed liposomal DNA escapes the endocytic vacuoles prior to their fusion with lysosomes (in a way similar to that proposed for vesicle-DNA complexes) to enter the cytosol for eventual episomal transfection and presentation of the encoded antigen. It is perhaps at this stage of intracellular trafficking of DNA, spanning its putative escape from endosomes and access to the nucleus, that the cationic lipid, possibly together within the fusogenic phosphatidylethanolamine (PE) component, plays a significant role.

Syngeneic Method

Immunization studies with liposome-entrapped DNA vaccines were expanded to include the cytotoxic T lymphocyte (CTL) component of the immune response. This was measured by the specific killing of syngeneic target cells pulsed with a recognized CTL epitope peptide derived from the antigen tested. To that end, the type and degree of immune response induced following subcutaneous injection of DNA in cationic liposomes was monitored and compared with that obtained with DNA alone injected by the same route. 6-8 week old, female C57/BL6 (H-2d) mice were injected subcutaneously with one or two doses of 2.5 or 10 µg ovalbumin (OVA)-encoding plasmid DNA (pCI-OVA), either alone or entrapped in liposomes.

Animals immunized subcutaneously with 100 µg of OVA protein complexed with 1 µg of cholera toxin (CT) served as a positive control. Blood samples and spleens were collected from all animals one week after the last injection and tested for anti-OVA total IgG (serum), CTL activity and cytokine release (spleen). After a single dose of antigen, only animals immunized with either protein or 10 µg of liposomal DNA showed significant anti-OVA antibody titres by ELISA. After two doses of antigen, only animals immunized with either protein or liposomal DNA (both 2.5 and 10 µg DNA) showed significant levels of seroconversion and serum antibody titres against OVA by ELISA.

Similarly, no anti-OVA CTL activity was detected in animals immunized with DNA alone. However, animals immunized with two doses of 10 µg of liposomal DNA displayed a CTL response higher (60% cell killing vs 50%) than that obtained in the positive control group immunized with OVA protein and adjuvant (CT). Thus, delivery of a small dose of liposomal plasmid DNA subcutaneously, a route of immunization not normally inducing significant plasmid DNA mediated immune activation, results in a strong antigen-specific cellular response which is greater than that achieved by higher doses of a conventional protein antigen together with a powerful adjuvant (CT).

ANALOGOUS APPROACH

Proteins that are synthesized within a cell (e.g. from plasmid DNA having a mammalian-active promoter) are continuously sampled as peptides by the proteosome/ class-I MHC antigen presenting pathway. Conversely, proteins that are acquired exogenously by antigen-presenting cells are sampled in an analogous way by the endosomal/MHC-class-II pathway. It follows that the delivery of both protein and plasmid-DNA-encoded forms of a protein antigen to the same individual antigen-presenting cell would result in the simultaneous presentation of the antigen via both class-I and class-II pathways, thereby providing an opportunity for synergy in the resulting immune response to the antigen. Several appropriate liposomal formulations were designed to test the "*co-delivery*" hypothesis, exploiting the advantages of the dehydration-rehydration liposome technology that entraps both DNA and protein immunogens efficiently.

The formulations, comprise various test and control permutations of plasmid DNA and protein, either free or entrapped (together or separately) in the liposomal vehicle. Immunization with DNA encoding the influenza haemagglutinin protein has been explored previously with naked or liposomally formulated DNA. Although immune responses elicited by DNA alone were adequate to achieve protective efficacy against influenza virus challenge in preclinical studies, only weak anti-HA antibody responses were elicited. The present "*co-delivery*" concept was designed to rectify this deficiency of DNA-based influenza vaccines. In a series of experiments, plasmid DNA encoding the haemagglutinin (HA) antigen [DNA(ha)] of the influenza virus (A /Sichuan/87 or A /PR/8 strains) was co-entrapped with the corresponding whole inactivated virus (HA) within the same liposomes using the dehydration-rehydration method.

A variety of control preparations including liposomes co-entrapping irrelevant DNA (i.e. plasmid DNA encoding ovalbumin) with HA or irrelevant protein (i.e. ova) with DNA (ha), entrapping DNA(ha) or HA alone, a mixture of the latter two preparations, and a mixture of the naked DNA(ha) and HA were used to immunize mice. Results demonstrate that the "co-delivery" hypothesis formulation (comprising both HA and its corresponding DNA in the same liposomes), elicited a greater response than all other formulations at each time point in the series, and it is by far the strongest response after a single dose. Notably, the formulation "Lip (OVA/ha)", which is a control for the CpG adjuvant effect of plasmid DNA, gave a response which was much lower than that of "*co-delivery*" with the appropriate homologous pair of HA DNA and protein.

Likewise, Lip (HA/ova) (an inappropriate pairing according to the hypothesis), gave a markedly weaker response. Figure also demonstrates that separately entrapped HA DNA and protein (in neighbouring vesicles) gave rise to an inferior response, supporting the hypothesis that delivery of both payloads to the same cell (which is best achieved by co-entrapment in the same liposome) is important in achieving the optimal antibody response. It is also remarkable that, inspite the modest DNA dose (10 μg) and small number of immunizations used, several formulations completely failed to generate an anti-HA response. These included HA DNA alone, and liposomally entrapped HA DNA. These findings serve to emphasize the striking degree of superiority of "*co-delivery*" over previous methods of DNA-based immunization against influenza virus. In conclusion, the present studies demonstrate that very small doses of protein as an additive in DNA immunization can dramatically improve the antibody response to the target protein, provided that the protein and DNA are homologous to one-another (i.e. that the DNA can express the protein), and that the payloads are delivered in the same individual liposomal vehicle.

The simplest hypothesis to explain our observation is that the synergy observed between the appropriately delivered "*homologous pair*" of protein and DNA involves delivery of both payloads to the same antigen-presenting cell. The application of the co-delievery concept to alternative delivery systems, e.g. niosomes, dendimers, PLA/PLGA, chitosans, alginates and other microparticles awaits investigation. It is anticipated that the "*co-delivery*" approach will lead to better DNA-based vaccines for prophylactic and therapeutic use, particularly where vaccines require the elicitation of antibody responses (e.g. influenza vaccines).

11

Cytological Carriers

Microparticle and nanoparticle polymeric systems currently occupy an important place in the field of drug delivery and targeting. Nevertheless, there are biological drug carriers that offer an efficient alternative to such systems. Within the different systems of biological carriers, of great importance are cells and cell ghosts, which are both efficient and highly compatible systems from the biological point of view, capable of providing the sustained release and specific delivery to tissues, organs and cells of drugs, enzymatic systems and genetic material. Cell systems such as bacterial ghosts, erythrocyte ghosts, polymorphonuclear leukocytes, apoptotic cells, tumor cells, dendritic cells, and more recently, genetically engineered stem cells, are all examples of how cell systems of very diverse nature can be suitably manipulated and loaded with drugs and other substances, to permit specific drug delivery *in vivo* with important therapeutic applications. Cell carriers for drug delivery are used in very different applications such as cancer therapy, cardiovascular disease, Parkinson's, AIDS, gene therapy, etc. Table 11.1 shows the classification of biological carriers for drug delivery based on the use of cells and cell ghosts.

PROKARYOTIC GHOSTS

Bacterial ghosts are intact, non-living, non-denatured bacterial cell envelopes devoid of cytoplasmic contents. They are created by lysis of bacteria, but maintain their cellular morphology and native surface antigenic structures, including their bioadhesive properties. Bacterial ghosts allow the encapsulation of drugs and other substances, and their specific attachment to mammalian tissues and cells. This kind of cell carrier acts as a true drug delivery system, allowing the permanency of drugs in the systemic circulation to be increased together with tissue-specific targeting. They are thus a promising alternative to conventional drug delivery systems such as liposomes or nanoparticles. The main advantages of bacterial ghosts as delivery systems are the fact that they are non-living, i.e. they can act as delivery systems of drugs, antigens or DNA; allow specific delivery to different tissues and cell types; and are well captured by phagocytic cells and antigen-presenting cells as dendritic cells. Among the drawback of bacterial ghosts is the possibility that they might revert to being virulent, the possibility of horizontal gene transfer, the stability of the recombinant phenotype, and pre-existing immunity against the carrier used. Usually, bacterial ghosts are produced by protein E-mediated lysis of Gram-negative bacteria.

Table 11.1 Kinds of cells and cell ghosts used for drug and gene delivery.

Cell carrier	*Target*	*Encapsulated substance*
Bacterial ghost	Tissues, macrophages, cells	Drugs, vaccines, genetic material
Erythrocyte ghost	RES, macrophages	Drugs, enzymes, peptides
Engineered stem cells	Tumor cells, T cells, macrophages	
Polymorphonuclear leucocytes	Tissues	Drugs
Apoptopic cells	Tumor cells	Drugs
Tumor cells	Tumor cells	Drugs
Dendritic cells	T cells	Drugs

The production of bacterial ghosts is based on the controlled expression of the bacteriophage PhiX174-derived lysis gene E. Expression of this gene from a plasmid in Gram-negative bacteria leads to the formation of a trans-membrane lysis tunnel structure that penetrates the inner and outer membranes, and is formed by protein E with border values fluctuating between 40-200 nm in diameter. Protein E is a hydrophobic protein localized exclusively in the cell envelope. E-mediated lysis has been achieved in many Gram-negative bacteria. Scanning electron micrographs of E-lysed cells reveal that bacterial ghosts contain only one E hole in a bacterial ghost, although in a few cases, there are two holes. The cytoplasm is expelled as a consequence of the high osmotic pressure inside the cell. The collapse of membrane potential and the release of cytoplasmic components such as proteins, nucleic acids, etc occur simultaneously.

In the case of strains of *E. coli,* this effect occurs within a period of 10 min after the induction of expression. The resulting empty bacterial cell envelope is considered a bacterial ghost. Bacterial ghosts show all the morphological, structural and immunogenic properties of a living cell. Since bacterial ghosts are derived from Gram-negative bacteria that are able to adhere to structures such as firnbriae and lipopolysaccharide, they are used for specific binding to human tissue. Bacterial ghost drug-loading is accomplished by the suspension of lyophilised bacterial ghosts in a buffered medium containing the drug. The ghosts are then subjected to an incubation process varying from 5 to 30 min at 24°C. They are then washed to remove excess drug. In order to prevent rapid leakage of loaded water-soluble drugs or other substances, the bacterial ghosts are *sealed* by fusion of the cell membrane with membrane vesicles at the edges of the lysis pore. For the sealing step, the bacterial ghosts suspension is incubated in the fusion buffer at 28°C for 10 min.

The *in vitro* release of drugs from loaded bacterial ghosts is performed from a suspension of drug-loaded bacterial ghosts that is dialysed through a membrane suitable for excluding the ghosts. Dialysis is performed at 28°C in PBS buffer. The concentrations of drug released into the buffer at preset times are quantified using an appropriate analytical technique. In studies addressing the adherence and capture of loaded bacterial ghosts by target cells such as macrophages, human colorectal adenocarcinoma cells (Caco-2) or dendritic cells, fluorescent markers such as fluorescein isothiocyanate (FITC) are used. These allow adherence to be assessed using fluorescence microscopy and flow cytometry techniques. Macrophages internalize bacterial ghosts to a greater extent than Caco-2 cells. Studies carried out using confocal laser scanning microscopy with *M. haemolytica* ghosts loaded with Doxorubicin have shown that the drug was associated with the ghosts membranes and the inner lumen. Dendritic cells that are professional phagocytic cells displaying the phagocytic capacity of antigens also have a good capacity for capturing bacterial ghosts, allowing the latter to be used as a vehicle for immunization and immunotherapy.

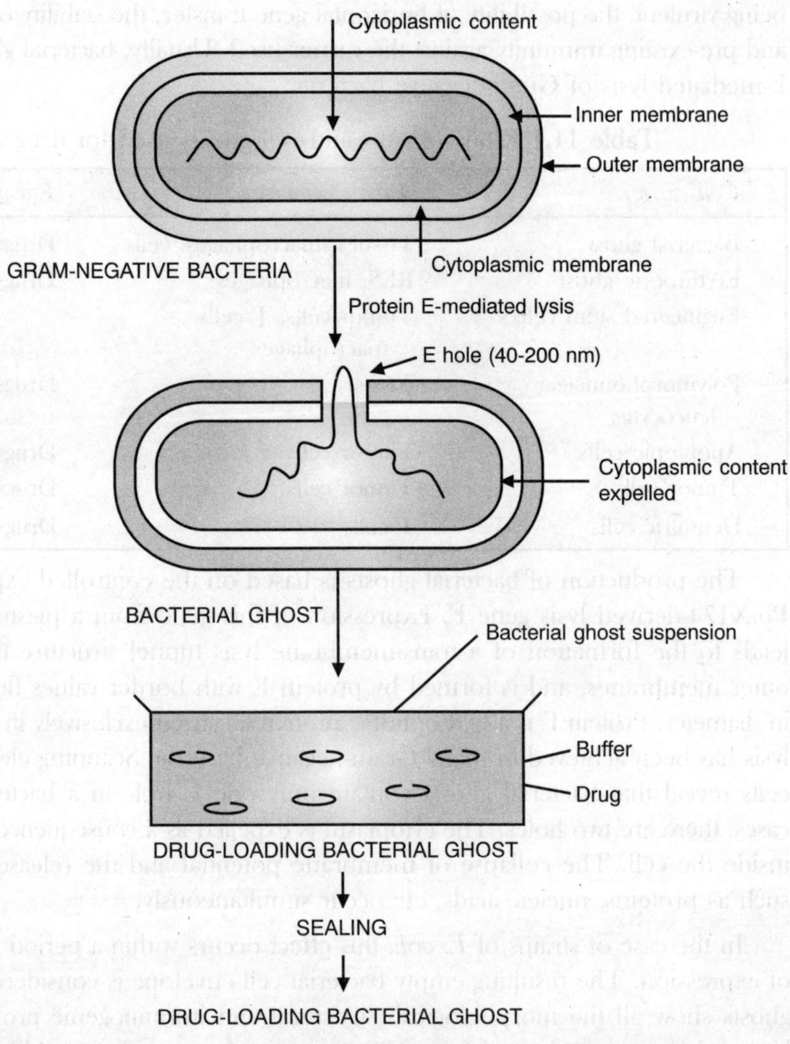

Fig. 11.1. Production and drug loading of bacterial ghosts.

Prokaryotic Delivery System

Bacterial ghosts have important therapeutic applications. They can be loaded with drugs, proteins and other substances, and can be targeted selectively to macrophages, tumors or endothelial cells. Bacterial ghosts have been used as efficient drug delivery systemsw in the field of anticancer drugs. Bacterial ghosts obtained have been used as a delivery system of doxorubicin to human colorectal carcinoma cells. Cytotoxicity assays revealed that doxorubicin-loaded ghosts show better antiproliferative capacity in Caco-2 cells than when free doxorubicin is used at the same concentration.

Experiments have also been carried out using *E. coli* ghosts containing streptavidin, in order to increase the affinity of streptavidin for biotinylated compounds. Streptavidin-loaded ghosts permit specific targeting to mucosal surfaces of the gastrointestinal and respiratory tracts, and also to phagocytic cells. Bacterial ghosts have been used as veterinary vaccines for the immunization of different animal species. *Pasteurella multocida* is a pathogen that causes morbidity and mortality in rabbits and its importance as a human pathogen has also been recognized. *P. multocida* ghosts have been used to immunize rabbits and mice. Similar results have been obtained in the immunization of cattle against pasteurellosis using *Pasteurella haemolytica* ghosts. *Actinobacillus pleuropneumoniae* is a highly contagious microorganism and is the cause of porcine pleuropneumonia, infecting 30-50% of pig populations. However, *Actinobacillus pleuropneumoniae* vaccines provide limited protection, since they decrease mortality but not morbidity in swine.

Comparative studies have been carried out on immunization using a aerosol infection model for pigs vaccinated with loaded-ghosts or formalin inactivated *Actinobacillus pleuropneumoniae* bacterins. The results obtained showed that immunization with bacterial ghosts is more efficient in protecting pigs than bacteria. Bacterial ghosts can also be used as carriers of therapeutic DNA or RNA. The use of nucleic acid vaccines currently offers a technique for the development of prophylactic or therapeutic vaccines, based on the use of DNA plasmids to induce immune responses by direct administration of DNA-encoding antigenic proteins into animals, and this is also suitable for the induction of cytotoxic T cells.

Bacterial ghosts loaded with DNA produce a high level of gene expression. They can be used to enhance the mucosal immune response to target antigens expressed in the bacterial ghost system. They can also be used for the specific targeting of DNA-encoded antibodies to primary antigens located in cells. Ghosts of *Vibrium cholerae* have been tested as antigen carriers of *Chlamidia trachomatis* as potential vaccines for the control of genital infections produced by this bacteria. Recombinant *Vibrium cholerae* ghosts, previously cloned with a major outer membrane protein of *C. trachomatis*, afforded a high level of protective immunity against *Chlamydia* in a murine model. *Mannheimia haemolytica* is a pathogen that causes ovine mastitis. *M. haemolytica* ghosts loaded with plasmid DNA stimulate the elicitation of efficient immune responses in mice, with no symptoms of acute or subacute toxicity during the experiment.

ROLE OF RBCS

Erythrocytes constitute the largest population of blood cells and are produced in the bone marrow. They are mature blood cells that produce haemoglobin and carry out the exchange of oxygen and carbon dioxide between the lungs and the body tissues. The term "*erythrocyte ghost*" attempts to define the resulting cell-like structure when erythrocytes are subjected to a reversible process of osmotic lysis. For more than 30 years, many studies, both *in vivo* and *in vitro*, have been carried out to explore the use of erythrocyte ghosts as delivery systems of drugs and other substances. Erythrocyte ghosts are obtained from fresh erythrocytes coming from human blood or the blood of different animal species such as the rat, mouse, rabbit, etc, and are loaded with different types of substance, mainly drugs, peptides and enzymes, using different encapsulation methods.

The most frequent methods for collecting erythrocyte ghosts are osmosis-based methods such as hypotonic dialysis. Autologous erythrocyte ghosts offer a drug delivery system that can act as a reservoir of the drug or substance encapsulated, providing the sustained release of the drug into the organism together with selective targeting of the drugs to the reticuloendothelial system (RES) of the liver, spleen and bone marrow. The main advantages of carrier erythrocytes as drug delivery systems are their high degree of biocompatibility, the possibility of encapsulating the drug in a small amount of cells, the sustained release of the encapsulated drug or substance into the body, the selective targeting to the RES, and the possibility of encapsulating substances of high molecular weight such as peptides. Among the drawbacks of these systems are the rapid leakage of some drugs out of the loaded erythrocytes and other problems related to their standardized preparation, storage and potential contamination.

Erythrocyte ghosts can be obtained by diverse procedures such as hypotonic dilution, hypotonic pre-swelling, osmotic pulse, hypotonic hemolysis, hypotonic dialysis, electroporation, drug-induced endocytosis and chemical methods. Of the different ways of obtaining carrier erythrocytes, hypotonic dialysis is undoubtedly the most frequently used encapsulation method. The reasons why it is so popular are its simplicity, its ease of application for a large number of drugs, enzymes and other substances, and because it is the method that best conserves the morphological and haematological properties of the erythrocyte ghosts obtained. Hypotonic dialysis is based on the exposure of red cells to the action of a hypo- tonic buffer, inducing cell swelling and the formation of pores that permit the drug to enter erythrocytes by means of a passive mechanism. Morphological inspection of erythrocyte ghosts is usually performed using transmission (TEM) or scanning (SEM) electron microscopy. Some physical parameters of red cell membranes can also be studied from the diffusion of haemoglobin.

The haemolytic methods employed in the production of erythrocyte ghosts normally affect the haemolytic volume, surface area and tension. Haematological parameters determine the effects of the procedure used to collect erythrocyte ghosts on their haematological properties. Among others, parameters such as reduced glutathione (GSH), mean corpuscular volume (MCV) or red cell distribution width (RDW), may be evaluated using a haematology analyzer. Erythrocyte ghosts obtained by hypotonic dialysis show a decrease in the mean corpuscular volume and an increase in size dispersion. Erythrocyte ghosts show a greater degree of haemolysis than normal erythrocytes.

RBCs Delivery System

Erythrocyte ghosts can be used as potential drug delivery systems for enzymes, proteins and peptides, allowing sustained release into the systemic circulation and the delivery of these substances into the RES. *In vitro* release of drugs from loaded erythrocyte ghosts is usually tested using autologous plasma or an isoosmotic buffer at 37°C; alternatively, a dialysis bag may be used. The *in vitro* release of drugs and substances from loaded erythrocytes is usually a first-order process, suggesting that the drug crosses the plasma membrane through a passive diffusion mechanism. However, zero-order release kinetics from loaded erythrocytes has also been described. *In vitro* studies about the release kinetics of different drugs, enzymes and peptides from loaded erythrocytes have shown a slow release of the encapsulated substance. When loaded erythrocyte ghosts are administered *in vivo*, changes in the pharmacokinetics of the encapsulated drugs occur, involving a systemic drug clearance related to the biological half-life of the erythrocytes. Increased serum half-lives and the areas under the curve of drugs encapsulated in loaded erythrocyte ghosts, in comparison with the free drug, have been observed in animals and humans.

At the same time, erythrocyte ghosts show a greater accumulation in tissues such as liver and spleen. Surface treatment of erythrocyte ghosts with substances such as glutaraldehyde, ascorbate, $Fe^{(\pm 2)}$, diamide, band 3-cross-linking reagents, trypsin, phenyl-hydrazine and the N-hydroxy-succinimide ester of biotin (NHS-biotin), enhances the recognition of erythrocyte ghosts by macrophages *in vitro* and liver targeting *in vivo*. Red cells may be used as carriers for some drugs such as antineoplastics, anti- infective agents, antihypertensives, corticosteroids, etc. Thus, carrier erythrocytes have been widely studied as delivery systems of antineoplastic drugs for targeting the RES located in organs such as liver and spleen. Different antineoplastic drugs have been encapsulated in erythrocyte ghosts in both *in vitro* and *in vivo* experiments. Increases have been obtained in average survival times in the treatment of mice bearing hepatomas, using methotrexate-loaded carrier erythrocytes.

Better recognition and capture of erythrocyte ghosts by macrophages have been obtained by using biotinylated erythrocytes containing methotrexate, by alterations to the membrane using band-3 cross-linkers of erythrocyte ghosts containing etoposide, or by treatment of erythrocytes containing doxorubicin with glutaraldehyde. Anti-infective agents such as gentamicin, metronidazole, primaquine or imizol have also been encapsulated in erythrocytes. Human erythrocytes containing gentamicin have proven to act as an efficient slow-release system *in vivo*. Erythrocyte ghosts containing dexamethasone have been used *in vivo* in rabbits and humans. A sustained release of dexamethasone *in vivo* in animals and humans was observed using carrier erythrocytes. An increased anti-inflammatory effect of the drug using carrier erythrocytes was observed in rabbits. Moreover, new prodrugs of anti-opioid drugs such as naltrexone and naloxone have been encapsulated in erythrocytes to solve stability problems of the primary drug within the erythrocyte.

The encapsulated prodrugs are transformed into the active compound, following their release from erythrocyte ghosts. In the fields of biochemistry and enzymatic therapeutics, the encapsulation of enzymes in erythrocytes has been studied in some depth. Enzymatic deficiencies or the treatment of specific illnesses may be approached using carrier erythrocytes loaded with enzymes. The encapsulation of enzymes in erythrocytes solves some of the problems associated with enzyme therapy, such as the short half-life deriving from the action of plasma proteases, intolerant reactions, and the immunological disorders or allergic problems associated with the use of enzymes in therapeutics. *In vitro* or *in vivo* studies with enzyme carrier erythrocytes have been performed using L-asparaginase, hexokinase, alcohol dehydrogenase, aldehyde dehydrogenase, alcohol oxidase, glutamate dehydrogenase, uricase, urokinase, lactate 2-mono oxigenase, arginase, rhodanase, recombinant phosphotriestearase, delta-aminolevulinate dehydratase, urease, pegademase, brinase and alglucerase.

One of the best examples of the use in therapeutics of carrier erythrocytes containing *enzymes*, is that of L-asparaginase encapsulated in human erythrocytes. This has been successfully used in the treatment of acute lymphoblastic leukaemia in paediatrics. Erythrocyte ghosts may act as carrier systems for the delivery of peptides and proteins. One of the main therapeutic applications of carrier erythrocytes in this field is that of anti-HIV peptides. Nucleoside analogues successfully inhibit the replication of immunodeficiency virases. In view of the importance of the monocyte-macrophage system in infection by HIV-1, it would be of maximum therapeutic interest to have available, the specific delivery of these therapeutic peptides into macrophages, which act as an important reservoir for the virus. Carrier erythrocytes containing anti-HIV peptides such as azidothimidine (AZT) and didanosine (DDI), significantly reduced the pro-viral DNA content in comparison with the administration of free peptides in a murine AIDS model.

Similar results have been obtained with 2′,3′-dideoxycytidine 5′-triphosphate (ddCTP), 2′,3′-dideoxycytidine (ddCyd) and AZT prodrugs encapsulated in erythrocytes. Anti-neoplastic peptides such as 2-fluoro-ara-AMP (fludarabine) and 5′-fluoro-2′-deoxyuridine 5′-monophosphate (FdUMP), a pro-drug of 5-fluro-2′-deoxyuridine (FdUrd), have been encapsulated in human carrier erythrocytes, behaving as a slow-release delivery system. Macrophage uptake *in vitro* of antisense oligonucleotides may be increased by using carrier erythrocytes. Other peptides, such as erythropoietin, heparin, dermaseptin S3, interleukin-3 or vaccines, have also been encapsulated in erythrocytes to increase their stability, acting as a slow release system with a prolonged half-life, or for their specific targeting to bacterial membranes. Erythrocyte ghost derivatives can also be used as drug delivery systems.

Nanoerythrosomes are erythrocyte membrane derivatives formed by spheroid vesicles, obtained by consecutive extrusion under nitrogen pressure through a polycarbonate filter membrane of a erythrocyte ghost suspension to produce small vesicles having an average diameter of 100 nm. *In vitro* and *in vivo* studies, carried out with nanoerythrosomes loaded with daunorubicin, have shown that when linked covalently to nanoerythrosomes, the drug produces slow release of daunorubicin to the organism over a prolonged period of time and also that, in comparison with the free drug, cytotoxicity is greater. The advantage of nanoerythrosomes, as compared with erythrocyte ghosts as drug delivery system, is that the former are able to escape from the reticuloendothelial system faster. *In vitro* studies have shown that the nanoerythrosome-daunorubicin complex is rapidly adsorbed and phagocytosed by macrophages.

Liver, spleen and lungs are the organs in which nanoerythrosomes show the greatest capacity of accumulation. Another derivative of erythrocyte ghosts are reverse biomembrane vesicles loaded with drugs. Reverse biomembrane vesicles are produced by spontaneous vesiculation of the ghost erythrocyte membrane by endocytosis, using an appropriate vesiculating medium, producing small vesicles containing the drug within the parent ghost. *In vivo* studies carried out using reverse biomembrane vesicles from erythrocyte ghosts loaded with doxorubicin in rats have revealed increases in the half-life and bioavailability of the drug, the liver and spleen, being the main organs for the clearance of this drug delivery system.

CLINICAL APPROACH

In gene therapy, a therapeutic transgene is introduced into the patient with a view of supplementing the functions of an abnormal gene. To achieve the delivery of genetic material into the target cell, it is necessary to have a suitable carrier. One important aim in the field of gene therapy is the design and development of gene carriers that encapsulate and protect the nucleic acid, and selectively release the vector/nucleic acid complex to the target tissue, so that the genetic material will be released at the cellular level later. In practice, there are several ways to achieve this. The first is through the use of modified viruses containing the genetic material of interest. The use of viruses for gene delivery has some drawbacks since it is limited to specific cells susceptible to being infected by the virus, and also the administration itself of the virus, has some immunological problems among others. The second alternative is to use living cells modified genetically, such as stem cells, to deliver transgenic material into the body. Stem cell therapy is a new form of treatment, in which cells that have died or lost their function are replaced by healthy adult stem cells.

One advantage of this kind of cell is that it is possible to use samples from adult tissues or cells from the actual patient, for culture and subsequent implantation. Within the framework of stem cell research, the use of stem cells as delivery systems is a novel and attractive technique in the field of gene therapy, in which the cells of the patients themselves are genetically engineered, in order to introduce a therapeutic transgene used to deliver the genetic material. A promising therapeutic strategy is the use of stem cells such as lymphocytes or fibroblasts as drug delivery systems. Experimental studies using stem cells as such systems have been tested in different therapeutic applications, especially in the field of cancer therapy. Considering the affinity of stem cells for tumor tissue, engineered stem cells have been successfully used for direct drug delivery to cancer cells.

In vitro cultures have been made of human mesenchymal stem cells from bone marrow that are transduced with an adenovirus vector carrying the human interferon beta-gene, which exerts therapeutic action against cancer. Engineered stem cells administered *in vivo* allow the delivery of the genetic material to cancer cells. This new drug delivery system has proven to be efficient in the treatment of experimental neoplasms, such as lung cancer, in mice. *In vivo* studies have also been carried out with neural stem cells engineered using adenoviral vectors to express interleukin-12, an oncolytic gene, whose efficiency has been demonstrated in the treatment of intracranial malignant gliomas in mice. The used of haematopoietic stem cells has allowed antiviral genes to be introduced in both T cells and macrophages for the treatment of AIDS.

The use of stem cells as vehicles for gene therapy has also been suggested for the treatment of ischaemic heart disease. Stem cells have also been employed in the field of antiepileptic therapy. Glial precursor cells, which release adenosine, have been derived from adenosine kinase embryonic stem cells. In these experiments, the fibroblasts were engineered to release adenosine by inactivating adenosine metabolising enzymes. After encapsulation within polyethersulfone hollow-fibre capsules, and the introduction into the brain ventricles in a rat epilepsy model, the local release of adenosine allows drug-resistant focal epilepsy to be treated. These engineered cells were shown to suppress seizure activity.

LEUCOCYTIC CARRIER

Polymorphonuclear leucocytes (PMN) can be used as carriers of antibiotics in view of their selective targeting to sites of infection. Simply incubating PMN in the presence of high concentrations of antibiotic for 1 hr at 37°C guarantees cell loading with the antibiotic. PMN loaded with the macrolide azithromycin have been found to be efficient in an *in vitro* model that permits the delivery of the antibiotic in a bioactive form to *Chlamydia* inclusions in polarized human endometrial epithelial (HEC-1B) cells infected with *Chlamydia trachomatis*. PMN carriers allow the accumulation of large amounts of antibiotic in endometrial epithelial cells and its retention over long periods of time.

PROGRAMMED CELLULAR DEATH

Programmed cellular death or apoptosis is a process that is controlled genetically in which the cells induce their own death in response to different types of stimulus such as the binding of death-inducing ligands to cell surface receptors. A new strategy for drug delivery, called *apoptopic induced drug delivery* (AIDD), allows drug delivery to tumor cells upon the initiation of apoptosis by using a biological mechanism to achieve drug delivery. This new system is based on the fact that apoptosis produces many changes in cell morphology that can be taken advantage of to achieve drug delivery. Apoptosis is reflected in enhanced membrane permeability, which favors the release of the encapsulated drug from the apoptotic cells to the tissue.

Phagocytosis of drug-loaded apoptotic carrier cells by tumor cells facilitates the localization of the drug within the tumor cell. One advantage of the *apoptotic induced drug delivery system* (AIDD) is that the drug carrier cells may be genetically engineered to modify their properties. *In vitro* studies have been performed using S49 mouse lymphoma cells in which apoptosis is produced by exposure to dexamethasone. The cytotoxicity of RG-2 cells caused by temazolamide-loaded-S49 apoptotic cells was from 4 to 7 times higher than that of control temazolamide-loaded S49 cells.

DRUG LOADED TUMOR CELL SYSTEM

A novel strategy for drug delivery based on the use of cell systems is the *drug-loaded tumor cell system* (DLTC), developed for drug delivery and targeting in lung metastasis. The tumor cells as drug carriers permit drug targeting to the blood-borne cancerous cells and the lungs as potential metastatic organs. In practice, there is affinity between the plasma membrane of malignant tumor cells and the metastatic addressins expressed by the endothelial cells of the targeted organ. *In vivo* studies have been carried out with DLTC based on Doxorubicin-loaded B16-F10 murine melanoma cells. Doxorubicin accumulation in the mouse lung was several times higher than that seen after administering free Doxorubicin.

ANTIGEN-PRESENTING CELLS

Dendritic cells (DC) are antigen-presenting cells. They ingest antigen by phagocytosis, degrade it, and present fragments of the antigen at their surface. Dendritic cells have huge potential for immunization against a broad variety of diseases, because they travel throughout the body in search of pathogens indicative of infection or disease. They are very important for the induction of T cell responses, which result in cell-mediated immunity. Selective targeting of drugs incorporated in dendritic cells to T cells allows the response of these cells to be manipulated *in vivo.* It has been shown that when incorporated into dendritic cells, the drug O-galactosylceramide improves their anti-tumor activity.

12

Systemic Therapies

As the end organ for the treatment of local diseases or as the route of administration for systemic therapies, the lung is a very attractive target for drug delivery. The lung provides direct access to the site of disease for the treatment of respiratory illness, without the inefficiencies and unwanted effects of systemic drug delivery. In addition, it provides an enormous surface area and a relatively low enzymatic environment for the absorption of drugs to treat systemic diseases. Inhaled medications have been available for many years for the treatment of lung diseases. Inhalational delivery has been widely accepted as being the optimal route of administration of first line therapy for asthmatic and chronic obstructive pulmonary diseases. Drug formulation plays an important role in producing an effective inhalable medication. In addition to being pharmacologically active, it is important that a drug be efficiently delivered into the lungs, to the appropriate site of action and remain in the lungs until the desired pharmacological effect occurs.

A drug designed to treat a systemic disease, such as insulin for diabetes, must be deposited in the lung periphery to ensure maximum systemic bioavailability. For gene therapy, anti cancer or anti infective treatment, cellular uptake and prolonged residence in the lungs of the drug may be required to obtain the optimal therapeutic effect. Thus, a formulation that is retained in the lungs for the desired length of time and avoids the clearance mechanisms of the lung may be necessary. The human lung contains airways and approximately 300 million alveoli with a surface area of 140 m^2, equivalent to that of a tennis court. As a major port of entry, the lung has evolved to prevent the invasion of unwanted airborne particles from entering into the body.

Airway geometry, humidity, mucociliary clearance and alveolar macrophages play a vital role in maintaining the sterility of the lung, and consequently, they can be barriers to the therapeutic effectiveness of inhaled medications. The size of the drug particle can play an important role in avoiding the physiological barriers of the lung and targeting to the appropriate lung region. Nanoparticles are solid colloidal particles ranging in size from 10 to 1000 nm.[2] Studies have demonstrated that they are taken up by macrophages, cancer cells, and epithelial cells. Their small size ensures the particles containing the active pharmacological ingredient will reach the alveolar regions. However, the use of an aerosol delivery system that generates nano-sized particles for inhalation, places these particles at risk of being exhaled, leaving very few drug particles to be deposited in the periphery of the lung. Residence time is not long enough for the particles to be deposited by sedimentation or diffusion.

INHALED DRUG SYSTEM

The origin of inhaled therapies can be traced back 4000 years ago to India, where people smoked the leaves of the *Atropa belladonna* plant to suppress cough. In the 19th and early 20th centuries, asthmatics smoked asthma cigarettes that contained stramonium powder mixed with tobacco to treat the symptoms of their disease. Modern inhalation devices can be divided into three different categories, the refinement categories, the refinement of the nebulizer and the of compact portable devices, the pressurized metered dose inhaler (pMDI), and the dry powder inhaler (DPI).

Role of Compressed Gas

Nebulizers have been used for many years to treat asthma and other respiratory diseases. There are 2 basic types of nebulizers, jet and ultrasonic nebulizers. The jet nebulizer functions by the Bernoulli principle by which compressed gas (air or oxygen) passes through a narrow orifice, creating an area of low pressure at the outlet of the adjacent liquid feed tube. This results in the drug solution being drawn up from the fluid reservoir and shatter into droplets in the gas stream. The ultrasonic nebulizer uses a piezoelectric crystal, vibrating at a high frequency (usually 1 to 3 MHz), to generate a fountain of liquid in the nebulizer chamber; the higher the frequency, the smaller the droplets produced.

Nebulizers can aerosolize most drug solutions and provide large doses with very little patient coordination or skill. However, treatments using these nebulizers can be time consuming and inefficient, with large amounts of drug wastage e.g. 50% loss with continuously operated nebulizers. Most of the prescribed drug never reaches the lung with nebulization. The majority of the drug is either retained within the nebulizer (referred to as residual or dead volume) or released into the environment during expiration. On average, only 10% of the dose placed in a continuous output jet nebulizer is actually deposited in the lungs. Advances in technology have led to the development of novel nebulizers that reduce drug wastage and improve delivery efficiency. Breath-enhanced jet nebulizers such as the Pari LC Star, increase aerosol output by directing auxiliary air, entrained during inspiration, through the nebulizer, causing more of the generated aerosol to be swept out of the nebulizer and available for inhalation. Drug wastage during exhalation is reduced to the amount of aerosol produced by the jet airflow rate that exceeds the storage volume of the nebulizer.

Adaptive aerosol delivery monitors a patient's breathing pattern in the first 3 breaths and then targets the aerosol delivery into the first 50% of each inhalation. This ensures that the aerosol is delivered to the patient during inspiration only, thereby eliminating drug loss during expiration that occurs with continuous output nebulizers. A number of metered dose liquid inhalers, including AERx, AeroDose and Respimat, have been developed to produce a fine aerosol in the respirable range by forcing the drug solution through an array of nozzles, using vibrating mesh or electronic micropump platforms with 30 to 75% of the emitted dose being deposited in the lungs.

Role of Pressurized MDI

The pressurized metered-dose inhaler (pMDI) was a revolutionary invention that overcame the problems of the hand-bulb nebulizer, and it is the most widely used aerosol delivery device today. The pMDI emits a drug aerosol driven by propellants, such as chlorofluorocarbons (CFC) and more recently, hydrofluoroalkanes (HFAs) through a nozzle at high velocity (>30 m/sec). pMDIs deliver only a small fraction of the drug dose to the lung. Typically, only 10 to 20% of the emitted dose is deposited in the lung. The high velocity and large particle size of the spray causes approximately 50% to 80% of the drug aerosol to impact in the oropharygeal region. Hand-mouth discoordination is another obstacle in the optimal use of the pMDI.

Crompton and colleagues found 51% of patients experienced problems coordinating the actuation of the device with inhalation, 24% of patients halted inspiration upon firing the aerosol into the mouth, and 12% inspired through the nose instead of the mouth when the aerosol was actuated into the mouth. The delivery efficiency of a pMDI depends on a patient's breathing pattern, inspiratory flow rate and hand-mouth coordination. The studies by Bennett and Dolovich demonstrated that for any particle size between 1 to 5 μm mass median aerodynamic diameter (MMAD), deposition was more dependent on inspiratory flow rate than any other variable. Fast inhalations (>60 L /min) result in a reduced peripheral deposition because the aerosol is more readily deposited by inertial impaction in the conducting airway and oropharyngeal regions. When aerosols are inhaled slowly, deposition by gravitational sedimentation in peripheral lung regions are enhanced.

Peripheral deposition has also been shown to increase with an increase in tidal volume and a decrease in respiratory frequency. As the inhaled volume is increased, aerosols are able to penetrate more distally into the lungs. A period of breath holding on completion of inhalation enhances deposition of particles in the periphery, thus preventing the particles from being exhaled during the expiratory phase. Thus, the optimal conditions for inhaling pMDI aerosols are from a starting volume equivalent to the functional residual capacity, the actuation of the device at the start of inhalation, inspiratory flow rate of <60 L/min, followed by a 10 second breath-hold at the end of inspiration. Spacer tubes, valved holding chambers and mouthpiece extensions have been developed to eliminate coordination requirements and reduce the amount of drug deposited in the oropharynx, by decreasing the particle size distribution and slowing the aerosol's velocity.

Spacer geometry and materials of manufacture influence the quality and quantity of aerosol available. The aerosols from a pMDI and the holding chamber are finer than that with the pMDI alone, with an approximate 25% decrease in the mass median aerodynamic diameter (MMAD), compared with the original aerosol. This finer aerosol is more uniformly distributed in the normal lung, with increased delivery to the peripheral airway. However, in patients with airway obstructions, the addition of a holding chamber to the pMDI may not change the distribution of the aerosol.

Role of Dry Powder Inhalers

Dry powder inhalers (DPIs) were designed to eliminate the coordination difficulties associated with the pMDI. There are a wide range of DPI devices on the market from single-dose devices loaded by the patient to multi unit dose devices provided in a blister pack, multiple unit doses sealed in blisters on a strip that moves through the inhaler or reservoir-type (bulk powder) systems. Lung deposition varies among the different DPIs. Approximately 12% to 40% of the emitted dose is delivered to the lungs with 20 to 25% of the drug being retained within the device.

Poor drug deposition with DPIs can be attributed to inefficient deaggregation of the fine drug particles from coarser carrier lactose particles or drug pellets. Slow inspiratory flow rate, high humidity and rapid, large changes in temperature are known to affect drug deaggregation and hence the efficiency of pulmonary drug delivery with DPIs. With most DPIs, drug delivery to the lungs is augmented by fast inhalation. Borgstrom and colleagues demonstrated that increasing inspiratory flow from 35 L/min to 60 L/min through the Turbuhaler, increased the total lung dose of terbutaline from 14.8% of nominal dose to 27.7%. This is in contrast to the MDI which requires slow inhalation and breath holding to enhance lung deposition of the drug. Each DPI has a different air flow resistance that governs the required inspiratory effort.

The higher the resistance of the device, the more difficult it is to generate an inspiratory flow great enough to achieve the maximum dose from the inhaler. However, deposition in the lung tends to increase when using high resistance inhalers. Active DPIs are being investigated to reduce the importance of a patient's inspiratory effort. By adding either a battery driven propeller that aids in the dispersion of the powder, or using compressed air to aerosolize the powder and converting it into a standing cloud in a holding chamber, the generation of a respirable aerosol becomes independent of a patient's inspiratory effort.

DRUG AEROSOL

Aerosol particle size is one of the most important variables in defining the dose deposited and the distribution of drug aerosol in the lung. Fine aerosols are distributed on peripheral airways, but deposit less drug per unit surface area than larger particle aerosols which deposit more drug per unit surface area, but on the larger, more central airways. Most therapeutic aerosols are nearly always heterodisperse, consisting of a wide range of particle sizes. These aerosols are described by the log-normal distribution, with the log of the particle diameters plotted against particle number, surface area or volume (mass) on a linear or probability scale and expressed as absolute values or cumulative %. Since delivered dose is very important when studying medical aerosols, particle number may be misleading as smaller particles contain less drug than larger ones. Particle size is defined from this distribution by several parameters. Mass median diameter of an aerosol refers to the particle diameter that has 50% of the aerosol mass residing above and 50% of its mass below it.

The aerodynamic diameter relates the particle to the diameter of a sphere of unit density that has the same settling velocity as the particle of interest, regardless of its shape or density. MMAD is read from the cumulative distribution curve at the 50% point. Geometric standard deviation (GSD) is a measure of the variability of the particle diameters within the aerosol, and is calculated from the ratio of the particle diameter at the 84.1% point on the cumulative distribution curve to the MMAD. For a log-normal distribution, the GSD is the same for the number, surface area or mass distributions. A GSD of 1 indicates a monodispersed aerosol, while a GSD of >1.2 indicates a heterodispersed aerosol. Particles can be deposited by inertial impaction, gravitational sedimentation or diffusion (Brownian motion), depending on their *size*. While deposition occurs throughout the airways, inertial impaction usually occurs in the first 10 generations of the lung, where air velocity is high and airflow is turbulent.

Most particles above 10 μm are deposited in the oropharyngeal region with a large amount impacting on the larynx, particularly when the drug is inhaled from devices requiring a high inspiratory flow rate (DPIs) or when the drug is dispensed from a *device* at a high forward velocity (MDIs). The large particles are subsequently swallowed and contributed minimally, if at all, to the therapeutic response. In the tracheobronchial region, inertial impaction also plays a significant role in the deposition of particles, particularly at bends and airway bifurcations. Deposition by gravitational sedimentation predominates in the last 5 to 6 generation of airways (smaller bronchi and bronchioles), where air velocity is low. In the alveolar region, air velocity is negligible and thus the contribution to deposition by inertial impaction is also negligible. Particles in this region have a longer residence time and are deposited by both sedimentation and diffusion. Particles not deposited during inhalation are exhaled.

Deposition due to sedimentation affects particles down to 0.5 μm in diameter, whereas below 0.5 μm, the main mechanism for deposition is by diffusion. Targeting the aerosol to conducting or peripheral airways can be accomplished by altering the particle size of the aerosol. It is difficult to predict the actual site of deposition, since airway calibre and anatomy differ among people. However, in general, aerosols with a MMAD of 5 to 10 μm are mainly deposited in the large conducting airways and the oropharyngeal region. Particles 1 to 5 μm in diameter are deposited in the small airways and alveoli with greater than 50% of the 3 μm diameter particles being deposited in the alveolar region. In the case of pulmonary drug delivery for systemic absorption, aerosols with a small particle size would be required to ensure peripheral penetration of the drug. Particles <3 μm have approximately 80% chance of reaching the lower airways, with 50 to 60% being deposited in the alveoli. Nanoparticles <100 nm are deposited mainly in the alveolar region.

THERAPEUTIC ROLE

The therapeutic effect of aerosolized therapies is dependent on the dose deposited and its distribution within the lung. If a drug aerosol is delivered at a suboptimal dose or to a part of the lung, devoid of the targeted disease or receptors, the effectiveness of therapy may be compromised. For example, the receptors for the β_2 agonist, salbutamol and the muscarinic$_3$ (M_3) agonist, ipratropium bromide, are not uniformly distributed throughout the lung. Autoradiographic studies have shown β_2 adrenergic receptors are present in high density in the airway epithelium from the large bronchi to the terminal bronchioles. Airway smooth muscle has a lower β-receptor density, greater in the bronchioles than bronchi. However, greater than 90% of all β receptors are located in the alveolar wall, a region where no smooth muscle exists and whose functional significance is unknown. Another autoradiographic study has shown a high density of M_3 receptors in submucosal glands and airway ganglia, and a moderate density in smooth muscles throughout the airways, nerves in intrapulmonary bronchi and in alveolar walls.

The location of these receptors in the lung suggests that ipratropium bromide needs to be delivered to the conducting airways, while salbutamol requires a more peripheral delivery to the medium and small airways to produce a therapeutic effect. Since particle size affects the lung deposition of an aerosol, it can also influence the clinical effectiveness of a drug. Rees *et al.* reported the varying clinical effect of 250 µg of aerosolized terbutaline from a pMDI, given in three different particle sizes of <5 µm, 5 to 10 µm, and 10 to 15 µm.

In asthmatics, the greatest increase in forced expiratory volume in one second (FEV_1) was found with the smallest particle size (<5 µm), suggesting that the smaller particle aerosol was considerably more effective than larger particle size aerosols in producing bronchodilation, since it has the best penetration and retention in the lungs in the presence of airway narrowing. Using three monodisperse salbutamol aerosols (MMAD of 1.5 µm, 2.8 µm, 5 µm), Zanen and colleagues demonstrated in patients with mild to moderate asthma that the 2.8 µm particle size aerosol produced a superior bronchodilation, compared with the other two aerosols. In patients with severe airflow obstruction (FEV_1 < 40%), Zanen *et al.* demonstrated that the optimal particle size for β_2 agonist or anticholinergic aerosols is approximately 3 µm. They examined the effect on lung function of equal doses of three different sizes of monodisperse aerosols, 1.5 µm, 2.8 µm and 5 µm, of salbutamol and ipratropium bromide. Their findings suggest that small particles penetrate more deeply into the lung and more effectively dilate the small airways than larger particles, which are filtered out in the upper airways.

The 1.5 µm aerosol induced significantly less bronchodilation than the 2.8 µm aerosol, suggesting that this fine aerosol may be deposited too peripherally to be effective, since smooth muscle is not present in the alveolar region. The optimal site of deposition in the respiratory tract for aerosolized antibiotics depends on the infection being treated. Pneumonias represent a mixture of purulent tracheobronchitis and alveolar infection. Successful therapy would theoretically require the antibiotic to be evenly distributed throughout the lungs. However, those confined to the alveolar region would most likely benefit from a greater peripheral deposition. *Pneumocystis carinii* pneumonia, the most common life-threatening infection among patients infected with HIV, is found predominately within the alveolar spaces, with relapses occurring in the apical region of the lung after treatment with inhaled pentamidine given as a 1µm MMAD aerosol. The mechanism suggested for this atypical relapse is the poorer apical deposition of the aerosol.

Regional changes in intrapleural pressure result in the lower lung regions receiving relatively more of the inspired volume than the upper lung, when sitting in an upright position or standing. This influence on deposition has been shown to occur in an experimental lung model, analyzing sites of aerosol deposition in a normal lung. The experiment showed a 2:1 ratio in the overall deposition for a 4 µm aerodynamic diameter aerosol between the lower and upper lobes when in the upright position. Chronic lung infection with *Pseudomonas aeruginosa,* in patients with cystic fibrosis or non-CF bronchiectasis, resides in the airway lumen with limited invasion of the lung parenchyma. Infection starts in the smaller airways, the bronchioles, and moves into the larger airways. The optimal site of deposition for inhaled antimicrobial therapy would, therefore, be a uniform distribution on the conducting airways.

Mucus plugs in the bronchi and bronchioles may prevent deposition of even small particle aerosols in regions distal to the airway obstruction, possibly the regions of highest infection, and thereby limiting the therapeutic effectiveness of the aerosolized antibiotic. Until recently, aerosol drug delivery has been limited to topical therapy for the lung and nose. The major contributing factor to this restriction was the inefficiencies of available inhalation devices that deposit only 10% to 15% of the emitted dose in the lungs.

While appropriate lung doses of steroids and bronchodilators can be achieved with these devices, for systemic therapies, large amounts of the drug are necessary to achieve therapeutic drug levels systemically. Recent advances in aerosol and formulation technologies have led to the development of delivery systems that are more efficient and that which produce small particle aerosols, allowing higher drug doses to be deposited in the alveolar region of the lungs, where they are available for systemic absorption. Most macromolecules cannot be administered orally because proteins are digested before they are absorbed into the bloodstream. In addition, their large size prevents them from naturally passing through the skin or nasal membrane; therefore, they cannot be administered intranasally or transdermally without the use of penetration enhancers. Thus, the easiest route of administration for proteins has been through intravenous or intramuscular/ subcutaneous injection.

It has been known for many years that proteins can be absorbed from the lung as demonstrated with insulin in 1925. Macromolecules < 40 kiloDaltons (kDa) (<5-6 nm in diameter) appear rapidly in the blood following inhalation into the airways. Insulin which has a molecular weight (mw) of 5.7 kDa and a diameter of 2.2 nm peaks in the blood 15 to 60 min after inhalation. Macromolecules >40 kDa (>5-6 nm in diameter) are slowly absorbed over many hours; inhaled albumin (68 kDa) and alpha$_1$-antitrypsin (45-51 kDa) have a T_{max} of 20 hrs and between 12 to 48 hrs respectively. The lung is the only organ through which the entire cardiac output passes. Before the inhaled drug can be absorbed into the blood from the lung periphery, it has several barriers to overcome such as lung surfactant, surface lining fluid, epithelium, interstitium and basement membrane, and the endothelium.

Drug absorption in the lung periphery is regulated by a thin alveolar-vascular permeable barrier. An enormous alveolar surface area with epithelium, consisting of a thin single cellular layer (0.2 to 0.7 μm thickness), promotes efficient gas exchange through passive transport, but also provides a mechanism for efficient drug delivery into the bloodstream. Although the mechanism of absorption is unknown, it has been hypothesized that macromolecules either pass through the cells via absorptive transcytosis (adsorptive or receptor mediated), paracellular transport between bijunctions or trijunctions or through large transitory pores in the epithelium caused by cell injury or apoptosis. Thus, the high bioavailability of macromolecules deposited in the lung (10 to 200 times greater than nasal and gastrointestinal values) may be due to its enormous surface area, very thin diffusion layer, slow surface clearance and anti-protease defense system.

INFLUENCE OF ENVIRONMENT

Like all major points of contact with the external environment, the lung has evolved to prevent the invasion of unwanted airborne particles from entering into the body. Airway geometry, humidity and clearance mechanisms contribute to this filtration process. The challenge in developing therapeutic aerosols is to produce an aerosol that eludes the lung's various lines of defense.

Aquatic Droplets

Progressive branching and narrowing of the airways encourages impaction of particles. The larger the particle size, the greater the velocity of incoming air, while the greater the bend angle of bifurcations and the smaller the airway radius, the greater the probability of deposition by impaction. Drug particles are known to be hygroscopic and grow in size in high humidity environments, such as the lung which has a relative humidity of approximately 99.5%. The addition and removal of water can significantly affect the particle size and thus deposition of a hygroscopic aerosol. A hygroscopic aerosol that is delivered at relatively low temperature and humidity into one of high humidity and temperature would be expected to increase in size when inhaled into the lung.

The rate of growth is a function of the initial diameter of the particle, with the potential for the diameter of fine particles less than 1 µm to increase 5-fold, compared with 2 to 3-fold for particles greater than 2 µm. The increase in particle size above the initial size should affect the amount of drug deposited, and particularly, the distribution of the aerosolized drug within the lung. Ferron and colleagues have predicted that for initial sizes between 0.7 µm and 10 µm, total deposition of hygroscopic aerosols increases by a factor of 2. For particles with an initial size of 1 µm, Xu and Yu were able to predict changes in the distribution pattern due to particle growth. The calculations showed a shift from deposition due to sedimentation to primarily impaction on more central airways.

Protective Mechanism

Once deposited in the lungs, inhaled drugs are either cleared from the lungs, absorbed into the circulatory or lymphatic systems, or metabolized. Drug particles deposited in the conducting airways are primarily removed through mucociliary clearance, and to a lesser extent, are absorbed through the airway epithelium into the blood or lymphatic system. Ciliated epithelium extends from the trachea to the terminal bronchioles. The airway epithelial goblet cells and submucosal glands secrete mucus forming a two-layer mucus blanket over the ciliated epithelium: a low-viscosity periciliary or sol layer covered by a high-viscosity gel layer. Insoluble particles are trapped in the gel layer and moved towards the pharynx (and ultimately to the gastrointestinal tract) by the upward movement of mucus generated by the metachronous beating of cilia.

In the normal lung, the rate of mucus movement varies with the airway region and is determined by the number of ciliated cells and their beat frequency. Movement is faster in the trachea than in the small airways, and is affected by factors influencing ciliary functioning and the quantity and quality of the mucus. For normal mucociliary clearance to occur, airway epithelial cells must be intact, ciliary structure and activity normal, the depth and chemical composition of the sol layer optimal, and the rheology of the mucus within the physiological range. Mucociliary clearance is impaired in lung diseases such as immotile cilia syndrome, bronchiectasis, cystic fibrosis and asthma. In immotile cilia syndrome and bronchiectasis, the ciliary function can be either impaired or nonexistent.

In cystic fibrosis, the ciliary structure and function are normal, however, the copious amounts of thick, tenacious mucus present in the airways impairs their ability to clear the mucus effectively. In these diseases, clearance of aerosolized drugs deposited in the conducting airways is generally decreased and secretions are cleared from the lung by cough. In addition to mucociliary clearance, soluble particles can also be removed by absorptive mechanisms in the conducting airways. Lipophilic molecules pass easily through the airway epithelium via passive transport. Hydrophilic molecules cross via extracellular pathways such as tight junctions or by active transport via endocytosis and exocytosis. From the submucosal region, particles are absorbed either into systemic circulation, bronchial circulation or lymphatic systems. Drugs deposited in the alveolar region may be phagocytosed and cleared by alveolar macrophages or absorbed into the pulmonary circulation.

Alveolar macrophages are the predominant phagocytic cell for the lung defense against inhaled microorganisms, particles and other toxic agents. There are approximately 5 to 7 alveolar macrophages per alveolus in the lungs of healthy, non-smokers. Macrophages phagocytose insoluble particles deposited in the alveolar region are either cleared by the lymphatic system or moved into the ciliated airways along currents in alveolar fluid and then cleared via the mucociliary escalator. This process can take weeks or months to complete. As discussed above, soluble drug particles deposited in the alveolar region can be absorbed into the systemic circulation. The pulmonary epithelium appears to be more resistant to soluble particle transport than to the endothelium or the interstitium. The lung-blood barrier may behave as a molecular sieve, allowing the passage of small solutes but restricting the passage of macromolecules.

Conhaim and colleagues proposed that the lung barrier was best fitted to a three pore size model, including a small number (2%) of large-sized pores (400 nm pore radius), 30% of medium-sized pores (40 nm radius) and 68% of small-sized pores (1.3 nm). The rate of protein absorption from the alveoli is size dependent. Effros and Mason demonstrated an inverse relationship between alveolar permeability and molecular weight. In rats, after intratracheal instillation of DDAVP (1-desamino-8-D-arginine vasopressin) (mw = 1.1 kDa), peak serum DDAVP levels occurred at 1 hr compared with 16 to 24 hrs after the intratracheal instillation of albumin (mw = 67 kDa). However, some proteins are cleared from the lung more rapidly than expected for their size. After intratracheal instillation or aerosolization of human growth hormone (mw = 22 kDa), peak serum levels were observed between 0.5 to 4 hrs, indicating a rapid, saturable clearance from the lung that is suggestive of receptor-mediated endocytosis.

Vasoactive intestinal polypeptide (VIP) is believed to be completely degraded during the passage across the pulmonary epithelium and into the bloodstream. Nanoparticles can pass rapidly into the systemic circulation. The distribution of radioactivity, after the inhalation of a ^{99m}Technetium (Tc)-labeled ultrafine carbon particles (5 to 10 nm), was detected in the blood one min post-inhalation and peaked between 10 and 20 min. This blood radioactivity level was sustained up to 60 min. 8% of the initial lung radioactivity was measured in the liver 5 min post- administration and remained stable over time. The rapidity of the appearance of radioactivity systemically makes the translocation from the lung unlikely due to phagocytosis, by macrophages or endocytosis by epithelial and endothelial cells, but by passive diffusion.

NANOSIZED AEROSOLS

Delivery of nano-sized aerosols to the lung may result in very little drug being deposited in the lung. The majority of particles <500 nm inhaled will not have enough residence time in the lung to deposit, and therefore will be exhaled. However, if the nanoparticles were delivered in larger carrier particles, they could be sufficiently deposited in the lung. The carrier particle would dissolve after contact with the lung surface fluid, releasing the nanoparticle at the target tissue or cells. Sham and colleagues demonstrated that nanoparticles (173 to 242 nm) could be delivered into the lung in larger respirable lactose carrier particles produced by spray-drying. The dry powder containing the nanoparticles had a MMAD of 3.0 μm.

pMDI formulations are typically micronized drugs in the 2 to 3 μm range suspended in a hydrofluoroalkane (HFA) propellant. Solution pMDI such as QVAR produce smaller drug particles on propellant evaporation, resulting in better deposition and distribution than a micronized formulation. However, for insoluble drug particles in the propellant, the efficiency of pMDI is limited. A study by Dickinson *et al.* proposed the use of nanoparticles suspended in propellant as a method of increasing the delivery efficiency of insoluble drugs in pMDIs. They produced hydrophilic nanoparticles using a *reverse* phase microemulsion technique that captures nanoparticles by snap freezing, followed by freeze-drying. The nanoparticles of pure drug (salbutamol) and the drug in a non-polymer matrix (lecithin-based), with and without lactose, were dispersed in HFA-227 and in aerosol performance assessed by cascade impaction. The size of the salbutamol nanoparticles ranged from 34 to 216 nm. Dispersion of the nanoparticles in a HFA-227:hexane (95:5 v/v) blend resulted in a homogeneous fine suspension that showed no signs of sedimentation or creaming over several months.

Rapid release of salbutamol from the nanoparticle was observed (approximately 4 min) as expected from the large surface area of the particles and the high water solubility of the drug. A high fine particle fraction (ex-device, % < 5.8 μm) of 58.3% to 65.5% and a low MMAD (1.2 to 1.5 μm) were observed with the nanoparticle formulations. This data suggests that a high fraction of the nanoparticles would be distributed in the alveolar region of the lung and represents the best aerosol that can be produced using a pMDI. Budesonide is a potent corticosteroid used as an inhaled anti inflammatory agent to treat asthma. It is available as a dry powder inhaler and as a suspension for inhalation with a nebulizer. A new formulation for nebulization has been developed that contains nanocrystals of budesonide that give the suspension solution-like qualities. The particles are 75 to 300 nm in diameter, compared with 4400 nm for the marketed budesonide suspension.

In a randomized crossover study, 16 healthy volunteers were given the nanocrystal budesonide formulation (0.5 mg and 1.0 mg doses), Pulmicort respules and placebo via nebulization using a Pari LC jet nebulizer. Nebulization times were shorter for the nanocrystal formulation, compared with Pulmicort respules (~7.1 min vs. 8.7 min). Similar AUCs were observed with the formulations, suggesting similar pulmonary absorption. However, a higher C_{max} (1212 pg/mL vs. 662 pg/mL) and shorter T_{max} (8.4 min vs. 14.4 min) for nanocrystal budesonide compared with the same dose of Pulmicort, suggests a more rapid drug delivery or absorption with the nanocrystal formulation.

Role of Radiolabeled Nanoparticles

Radiolabeled nanoparticles have been used for many years in pulmonary ventilation studies. Ultrafine ^{99m}Tc labeled carbon particles (Technegas) is a relatively new advance in ventilation scintigraphy. Technegas consists of nanoparticles of carbon with a diameter of approximately 5 nm, that behaves more like a 0.2 μm m particle. Technegas is generated by the electrostatic heating of a graphite crucible to 2500°C in which a saline solution of ^{99m}Tc-pertechnetate had been placed and dried. The aerosol is dispersed in a lead-lined chamber in an atmosphere of 100% argon gas that is then inhaled by the patient. It is deposited in alveoli by inhalation and distributes similarly as the inert gas radioisotopes. Once they are inhaled, the particles adhere to the alveolar structures without appreciable movement for at least 40 min. Pulmonary delivery of nanoparticles is also being investigated for lymphoscintigraphy to assess the spread of or the staging of lung cancer.

Lung cancer usually exhibits metastasis proliferation, spreading through the lymphatic system and the blood circulation. Lymphatic drainage is responsible for the alveolar clearance of the deposited particulates and drugs up to a certain particle diameter (500 nm). Thus, radiolabeled nanoparticles could be used to visualize the lymph nodes to determine the presence of tumors. The lymphatic uptake of solid lipid nanoparticles has also been studied as an imaging method to stage lung cancer. The lipid nanoparticles were radiolabeled with the lipophilic tracer, D,L-hexamethylpropylene amine oxime (HMPAO), tagged with 99 m-Tc. The lipid nanoparticles were prepared by the melted homogenization method and had a mean diameter of 200 nm.

The radiolabeled nanoparticles were aerosolized using an ultrasonic nebulizer and delivered to rats until 200,000 cpm was achieved over the lung. After inhalation, the total activity in the lung was observed, followed by a fast clearance rate ($t_{1/2}$ = 10 min) that decreases activity in the lung to 25% of the total dose. A significant uptake (16.7%) was detected in the regional lymph nodes during the first 45 to 60 min, suggesting that aerosol delivery to the lungs of solid lipid nanoparticles could be used as an effective colloidal carrier for lymphoscintigraphy. Drainage into the lymph nodes following the lung instillation of nanoparticles of insoluble iodinated CT x-ray contrast agents was studied in beagle dogs. Nanoparticles of the contrast agent were prepared by microfluidization. A particle size of 150 to 200 nm was achieved. The nanoparticles were suspended in 2 different surfactant solutions. 1.5 mL of the suspension was instilled using a fiber optic bronchoscope at specific sites in the small airways and alveoli.

The nanoparticles were transported from the lung to the draining lymph nodes, 6 to 9 days post instillation as visible on the CT radiographs. No adverse clinical signs were observed in the dogs. However, microscopic lung lesions were observed at the instillation sites for both formulations and vehicle. The lesions consisted of inflammatory infiltrates, mainly macrophages, in intra-alveolar, interstitial and perivascular locations. A few small sites had fibrosis and granulomatous nodules with the destruction of the lung parenchyma. The presence of foamy macrophages was observed in the lymph nodes. The microscopic findings suggest that instillation of these nanoparticles of contrast agent may be harmful to the lung. The authors suggested that administering the nanoparticles as an aerosol, rather than by instillation, would prevent high concentrations in focal areas believed to be responsible for these lesions.

Vaccine Administration

Mucosal vaccine administration is an attractive method of inducing an immune response, since many pathogens invade the body through mucosal surfaces in the nose, lung and gut. As it is the first contact point, the mucosa has developed barriers to protect the body. The mucosa associated lymphoid tissue (MALT) is one of these barriers. It contributes 80% of the immunocytes and secretes more immunoglobulins than any other organs in the body. Antigens are delivered locally in the respiratory tract to nasal-associated and bronchus-associated lymphoid tissues (NALT and

BALT, respectively) and a mucosal immunity is induced. Using nanoparticles, systemic immunity may also be induced. Several studies have investigated the use of nanoparticles as carriers for the nasal delivery of vaccines.

Using tetanus toxoid as a model antigen, Vila and colleagues have studied the use of chitosan nanoparticles as well as polyethyleneglycol and polylactic acid (PEG-PLA) nanoparticles as nasal vaccine carriers. They compared PEG-PLA nanoparticles with PLA alone. Tetanus toxoid was entrapped in the hydrophobic PLA core and protected from interacting with enzymes such as lysozymes, by a hydrophilic PEG coating. Upon incubation with lysozymes *in vitro,* PLA particles aggregate and do not reach the epithelium, whereas PEG-PLA nanoparticles remain stable and size unmodified. The nanoparticles were produced using a double emulsion technique. PEG-PLA tetanus toxoid nanoparticles had a similar diameter to the PLA particles (196 nm vs. 188 nm), but had a lower loading efficiency of 33.4% compared with 48.1% with PLA. The IgG antibody response induced by PEG-PLA was superior at weeks 2 to 24, after intranasal instillation of 30 μg of tetanus toxoid (10 μl per nostril) on days 1, 8 and 15 in male BALB/ c mice.

In a similar study, the same group compared radiolabeled PEG-PLA, PEG-PLA with gelatin stabilizer to radiolabeled PLA encapsulated tetanus toxoid. They reported that 1 hr after intranasal administration, PEG-PLA nanoparticles produced a radioactivity level 10-fold higher in the blood than PLA which remained constant for 24 hrs. The radioactivity detected in the lymph nodes, lungs, liver and spleen was 3 to 6 fold higher for PEG-PLA than PLA nanopartides 24 hrs post instillation. The results of this work suggest that the PEG-PLA nanoparticles are partially taken up by the M cells of the NALT, as well as being transported to the submucosa and drained into the lymphatic system and blood stream. Recent work by the same group has investigated the potential use of chitosan nanoparticles for nasal administration of vaccines. Chitosan is a hydrophilic natural polysaccharide that is biodegradable and has mucoadhesive properties. The nanoparticles are formed spontaneously by adding the counter anion sodium TPP into the chitosan solution, without the use of energy sources or organic solvents required for the production of PEG-PLA nanoparticles.

Again, using tetanus toxoid as the model antigen, the investigators studied the effect of chitosan dose (200 μg and 70 *μg*) and molecular weight (23, 38 or 70 kDa) on the efficacy of the nanopartides. The nanoparticles produced were 300 to 350 nm and had a positive surface charge (+40 mV). The loading efficiency of tetanus toxoid was 50 to 60%, irrespective of the molecular weight of chitosan. *In vitro,* the formulations exhibited a rapid release over the initial 2 hrs followed by a slow release for 16 days, with the greater initial release at lower molecular weights of chitosan. 30 or 10 μg of antigen (associated with 200 and 70 μg of chitosan) was given intranasally to BALB/ c mice on days 1, 8 and 15. The IgG levels induced by the nanoparticles were significantly higher than those elicited by free tetanus toxoid. The response lasted for the 24 weeks studied with the IgG titres increasing over time.

Anti-tetanus IgA titers were detected in the saliva, bronchoalveolar and intestinal lavage fluids 24 weeks post administration. The results were independent of the administered dose and were significantly higher for the nanoparticle than the free tetanus toxoid. Jung and colleagues evaluated tetanus toxoid-loaded polymer nanoparticles as potential nasal vaccine carriers in mice. The nanoparticles were produced with various diameters (100 nm, 500 nm) using a novel polyester, sulfobutylated poly (vinyl alcohol)-graft-poly(lactide-co-glycolide), SB(43)-PVAL-g-PLGA. The surface charge was –43 to 59 mV. Mice were immunized with tetanus toxoid nanoparticles or free toxoid in solution at weeks 1, 2 and 3, either by oral, intranasal or intraperitoneal administration. Four weeks after the first intranasal immunization, IgG and IgA titers were significantly higher than baseline.

Oral immunization with the nanoparticles produced a weak IgG antibody response. Only 10% of the oral dose was administered to the nose (2.89 vs. 28.9 μg), however, intranasal immunization appeared to be more effective in inducting an immune response. Particle size had an effect on the titer levels. Particles >1 μm did not induce an immune response, but no difference was observed between the 500 nm and 100 nm nanoparticles which both induced significantly levels of IgG and IgA. These studies suggest that nasal delivery of vaccines using biodegradable nanoparticles are a promising method of inducing mucosal and systemic immunity.

Eradication of Mycobacterium

Intracellular bacterial infections caused by pathogens such as *Mycobacterium tuberculosis* are difficult to eradicate because they are generally inaccessible to free antibiotics. By loading antibiotics into nanoparticles, it is expected that delivery to the infected cells would improve since nanoparticles have been shown to localize preferentially in organs with high phagocytic activity and in circulating macrophages as well. The encapsulation of antibiotics has several advantages: (i) It modifies their pharmacokinetic characteristics by prolonging the antibiotics half-life and increasing the area under the concentration time curve (AUC), while decreasing its apparent volume of distribution. (ii) It improves the targeting of the drug to the phagocytic cells. (iii) It reduces toxicity of the antibiotics, such as the hepatotoxicity of anti tuberculosis drugs and the nephrotoxicity of aminoglycosides.

Antibiotics encapsulated in nanoparticles have been shown to be superior at treating intracellular infections when administered intravenously. However, the pulmonary delivery of these nanoparticles have only been investigated recently. Although effective therapy for tuberculosis is available, treatment failure and drug resistance is typically the result of patient's noncompliance. To improve compliance, investigators have been studying ways to reduce the dosing frequency of the drugs. Poly (lactide-co-glycolide) (PLG) nanoparticles as an aerosolized sustained release formulation for anti tuberculosis drugs, isoniazid, rifampicin and pyrazinamide, has been investigated since pulmonary tuberculosis is the most common form of the infection. The majority of the nanoparticles were 186 to 290 nm in diameter. Drug encapsulation efficiency was 56.9% to 68%. Aerosolized nanoparticles had a MMAD of 1.88 μm, with 96% of the particles in the respirable range (<6 μm). A single nebulization to guinea pigs resulted in sustained plasma drug concentrations for 6 to 8 days and in the lung for 11 days.

The half-life and mean residence time of the drugs was significantly prolonged, compared with the oral free drugs. Nebulizing the nanoparticles every 10 days to guinea pigs infected with *Mycobacterium tuberculosis* resulted in no detectable bacilli in the lung after 5 doses of treatment, compared with 46 daily doses of orally administered drug to achieve the equivalent efficacy. The use of lectin-based PLG nanoparticles as an aerosolized sustained release formulation of isoniazid, rifampicin and pyrazinamide has also been studied in guinea pigs. Mucoadhesive drug delivery systems such as chitosan have been previously investigated as a method of prolonging residence at a site of absorption. The main drawback of mucoadhesive systems is that its residence time is limited by the turnover time of the mucous gel layer, which is only a few hrs. Attaching the polymeric nanoparticles to cytoadhesive ligands such as lectins could prolong the duration of adhesion, thereby prolonging residence time.

Lectins bind to epithelial surfaces via specific receptors. Wheat germ agglutinin (WGA) is the least immunogenic lectin and has known receptors on the alveolar epithelium as well as the intestinal wall. WGA lectin-PLG nanoparticles were prepared by a two-step carbodiimide procedure. Their size ranged from 350 to 400 nm with drug encapsulation efficiency between 54% and 66%. The nanoparticles were delivered via nebulization to guinea pigs. 88% of the aerosol was in the respirable range (<6 μm) with a MMAD of 2.8 μm (GSD of 2.1). Three doses of nanoparticles were administered every 15 days for 45 days. The WGA-PLG nanoparticles resulted in a prolonged T_{max}, increased AUC and mean residence time after inhaled delivery. All three drugs were present in the lungs, liver and spleen at concentrations above the minimum inhibitory concentration 15 days post dosing, compared with orally-administered free drug.

Chemotherapeutic studies in guinea pigs infected with *Mycobacterium tuberculosis* showed that 3 doses administered every 15 days for 45 days yielded undetectable mycobacterial colony forming units, which was only achievable with 45 doses of the oral free drugs. The study results suggest that WGA-based PLG nanoparticles could be potential drug carriers for anti tuberculosis through aerosol delivery, reducing the drug dosing frequency.

Pulmonary Gene Delivery

Pulmonary gene delivery and DNA vaccinations are attractive therapies for a variety of lung diseases such as cystic fibrosis, asthma, chronic obstructive pulmonary disease, lung cancer and infections caused by *Mycobacterium tuberculosis*, influenza or SARS-associated coronavirus. Gene delivery requires carriers to transfer DNA into the nuclei of cells. There are two approaches for delivery: viral and non viral carriers. Viral delivery systems, although very efficient at transfection, are problematic due to their inherent immunogenicity. Non viral are safer but their transfection efficiency is low. Recently, biodegradable polymer-based nanoparticles have been investigated as a non viral pulmonary gene delivery system, taking advantage of their prolonged residence time in the lung and ability to be taken up by macrophages and dendritic cells, and to escape degradation by lysosomes. Asthma is characterized by elevated eosinophilic inflammation in the airway and increased airway hyperresponsiveness.

Chronic inflammation can lead to structural damage and airway remodeling. IFN-γ is a cytokine that promotes T-helper type 1 (Th1) responses which down regulates the Th2 immune responses present in asthma. Recombinant IFN-γ has been shown to reverse inflammation in murine models of asthma. However, its short half-life and severe adverse effects at high doses have prevented its therapeutic use. An intra-nasal IFN-γ gene therapy had been developed as an attempt to circumvent the drawbacks to its use. Kumar and colleagues studied the effects of a chitosan-IFN-γ plasmid DNA nanoparticle in a BALB/c mouse model of allergic asthma (using ovalbumin-sensitization).

Mice treated with the chitosan nanoparticles exhibited a significantly lower airway hyperresponsiveness (to methacholine challenge), reduced number of eosinophils and a significant decrease in epithelial denudation, mucus cell hyperplasia and cellular infiltration. Production of IFN-γ was increased post-treatment while IL-5 and IL-4 and ovalbumin-specific IgE were reduced. Chitosan IFN-γ nanoparticles induced IFN-γ gene expression predominately in epithelial cells and worked within 3 to 6 hrs after intranasal administration. Poly (D,L-lactide-co-glycolide (PLGA)–polyethyleneimine (PEI) nanoparticles are also being investigated for pulmonary gene delivery. PLGA had been extensively evaluated for its sustained-release profile and ability to be taken up by macrophages. PEI is a cationic polymer. Its high positive charge density suggests that it would be a promising candidate as a non viral vector.

PLGA nanoparticles with PEI on their surface had a mean particle diameter between 207 and 231 nm, surface charge > 30 mV and a DNA loading efficiency of >99%. Internalization of the nanoparticles in the human airway submucosal epithelial cell line, Calu-3, was observed and DNA detected 6 hrs after administration. However, *in vivo* efficiency of this system still needs to be studied. Respiratory syncytial virus (RSV) infection is a major cause of respiratory tract infections and is associated with approximately 17000 deaths annually on a worldwide basis, with no anti viral therapy or vaccine available. RSV NS1 protein appears to antagonize the host type 1 interferon-mediated response.

Zhang and colleagues hypothesized that blocking the NS gene expression might inhibit RSV replication and thus provide effective antiviral therapy. Small interfering RNA (siRNA) targeting the NS1 gene (siNS1) were encapsulated in chitosan nanoparticles. BALB/c mice were intranasally treated with siNS1 chitosan-nanoparticles before or after RSV infection. A significant decrease in virus titers in the lung was observed, in addition to a decrease in inflammation and airway hyperresponsiveness, compared with controls. The effect of siNS1 lasted at least 4 days. The data show that siNS1 nanoparticles may be a promising anti viral therapy against RSV infection.

13

Protein Trafficking

The underlying mechanisms of protein and peptide membrane trafficking, absorption, and secretion have traditionally been under debate. In the past, the virtual absence of peptides in portal blood following a meal was sufficient evidence to conclude that luminal protein digestion had to proceed all the way to free amino acids before absorption could occur. Coupled with an apparent high capacity of the small intestine to absorb free amino acids through various mechanisms, researchers largely ignored the possibility of additional absorptive systems, either intestinally or otherwise. Limitations associated with gastrointestinal protein and peptide drug delivery substantiated this hypothesis. However, this hypothesis exhibits a number of flaws in that it does not explain how certain dipeptides, such as carnosine (β-alanylhistidine) and anserine (β-alanyl-1-methylhistidine), could be absorbed intestinally intact. Traditionally, transport of these dipeptides was explained by receptor-mediated endocytotic processes.

Upon further investigation, it has been determined that other active transport mechanisms exist for di- and tripeptide transport, similar to those observed for a majority of nutrients and xenobiotics. In fact, the bulk of nitrogen absorption occurs largely through the epithelial apical absorption of intact di- and tripeptides and to a much lesser extent, free amino acids. The lack of intact di- and tripeptides in portal blood after a meal is dependent on the activity of intracellular, cytosolic peptidases, which exhibit little affinity for derived or nonclassical amino acids, such as β-alanine. The basal secretion of digested substrates in the form of free amino acids into blood is indicative of the overall efficiency of these absorptive mediators.

Still, the need for concentrative transport mechanisms to mediate peptide absorption is not readily apparent given that peptide concentrations in the intestinal lumen are most certainly higher than plasma levels. However, considering potential physicochemical barriers to peptide permeation and the significance of peptide absorption to cellular energy and overall nutrient requirements, it seems logical that these important substrates should be actively absorbed. The luminal morphology of the gastrointestinal tract varies significantly from relatively no folding in the esophagus to folds of Kerckring, villi, and microvilli in the small intestine. Polarized columnar epithelial cells of the small intestinal villi are the primary mediators for gastrointestinal absorption of orally administered drugs and nutrients. The presence of tight junctional complexes and high nutrient selectivity of these cells impart further barrier functionality to the small intestine, excluding many compounds that may be harmful to the organism or detrimental to cellular function. Based on these barriers, the physicochemical nature of a compound dictates the route and extent of its intestinal absorption. Compounds traverse biological barriers through either paracellular or transcellular routes. Paracellular absorption occurs by diffusion of dissolved solute between cells through the tight junctional complex, or zonula occludens, and the tortuous intercellular pathways.

The paracellular pathway is quite restrictive depending on the pore size and charge of the tight junctions as well as the cell barrier's porosity. In contrast, the transcellular route comprises several potentially parallel pathways for drug permeation, including passive transcellular diffusion, ion channels, facilitated diffusion, active transport, and endocytosis. The perceived mechanism of barrier permeation is highly dependent on a number of physicochemical properties of the drug, or xenobiotic, such as its net overall charge, hydrophilicity, shape/conformation, size, and molecular weight. Transcellular transport of a compound is also highly dependent on a number of physiological factors, some of which are discussed below. Although the passive paracellular diffusion of hydrophilic amino acids and peptides is possible, size restrictions, secondary structural considerations, and charge-charge interactions through this pathway largely preclude it from serving as a primary mechanism of permeation of many small peptides, oligopeptides, and protein-based compounds.

Due to these physiological restrictions, a majority of amino acids and peptides require that absorption occur via the transcellular route. However, the physicochemical properties of amino acids and peptides, including hydrophilicity and charge, generally favor a hydrophilic permeation route (i.e., paracellular diffusion) and prevent their passive transcellular diffusion through the hydrophobic cellular membrane. Given these restrictions as well as those observations concerning the intestinal absorption of undigested dipeptides, such as carnosine and anserine, researchers have realized that the bulk of di- and tripeptide absorption occurs through the function of active, concentrative transporter proteins. Numerous classes of transporter proteins have been identified to date, each with different and sometimes overlapping substrate specificities, capacities and affinities as well as specific tissue, cellular, and temporal expression patterns. Not surprisingly, the physicochemical properties of a compound dictate its interactions with transporter proteins. Considering that transport is a multifaceted process, variability due to overlap of the substrate selectivity of transporters in passive diffusion due to the lipophilic character and potential solubility differences may result in fluctuations in the observed net membrane transport for a substrate.

Notwithstanding other potential sources of variability, the net observable transport of amino acids and di- and tripeptides is mediated by a number of different transporter families. Di- and tripeptide transport is, however, generally believed to be attributable to the activity of the proton-coupled oligopeptide transporter [POT; SLC15 (solute carrier)] superfamily of transporters. Several comprehensive reviews can be found describing the common characteristics of the oligopeptide transporter proteins. To date, four mammalian members of the POT superfamily have been identified and described functionally. The currently known peptide transporters include peptide transporters 1 and 2, PepT1 (SLC15A1) and PepT2 (SLC15A2), and peptide/histidine transporters

1 and 2, PHT1 (SLC15A4) and PHT2 (SLC15A3). Interestingly, the human intestinal peptide transporter, HPT1 (CDH17), is a member of the cadherin family, which to our knowledge has been identified as the only active di- and tripeptide transporter that is not a member of the POT superfamily. Human PepT1, the most widely studied of the POT family members, was first cloned from a rabbit intestinal cDNA library. Later studies determined that PepT1 was a low-affinity high-capacity transport system for di- and tripeptides, with little to no affinity for amino acids. Subsequent identification and cloning of the high-affinity low-capacity PepT2 transporter from a human kidney cDNA library suggested that these two transporter systems act concertedly to facilitate the absorption and conservation of di- and tripeptides. Interestingly, this hypothesis does not account for the relative activities of either PHT1 or PHT2, which have only recently been described.

Discerning the importance of a single transporter from the functional effects of other transporters illustrates one of the many problems inherent in the study of drug transport (i.e., overlapping specificity of these systems may confound the applicability of transport data to actual in vivo physiological systems). Also, current methodology limits the feasibility of studying multiple transporter systems simultaneously, further hindering our efforts to understand the relevance of a particular transporter to overall transcellular flux. A fundamental understanding of a drug transporter's net effect on absorption is essential when analyzing a drug's physiological behavior and response [pharmacokinetics/pharmacodynamics (PK/PD)] after peroral administration. Difficulties in characterizing the intestinal transepithelial transport of drugs underscores the need for a complete understanding of the biophysical and biochemical barriers that could potentially alter a drug's PK activity (i.e., absorption, distribution, metabolism, and excretion). Members of the POT superfamily of transporter proteins facilitate the absorption and secretion of a wide range of di- and tripeptides and of peptidomimetic pharmaceutical agents.

CYTOLOGICAL FEATURES

Members of the POT superfamily share a common topological map consisting of 12 putative α-helical transmembrane domains with intracellularly localized N- and C-termini. Two characteristic protein signatures of the POT family members have been identified, known as the PTR2 family signatures: (1) [GA]-[GAS]-[LIVMFYWA]-[LIVM]-[GAS]-D-x-[LIVMFYWT]-[LIVMFYW]-G-x(3)-[TAV]-[IV]-x(3)-[GSTAV]-x-[LIVMF]-x(3)-[GA], and (2) [FYT]-x(2)-[LMFY]-[FYV]-[LIVMFYWA]-x-[IVG]-N-[LIVMAG]-G-[GSA]-[LIMF]. A third consensus sequence has been proposed by Fei et al., (GTGGIKPXV). Based on their phylogenetic analysis, Saier et al. have also proposed three different signature sequences associated with the POT superfamily. Despite these identified consensus sequences, there is little identity between the PepT and PHT members of this superfamily. For instance, while the human PepT1 (hPepT1) shares 83.5% identity with its rat homolog, its identity with the human PHT1 (hPHT1) is just 20.4%. Interestingly, hPepT1 shares just 26.0% identity with the intracellularly localized hPHT2. As mentioned previously, PepT1 is arguably the most widely studied member of the POT superfamily of transporters. However, relatively little information is known concerning its tissue and cellular localizations as well as species variations. The cloned human PepT1 cDNA sequence encodes a 708-amino acid protein with an estimated molecular weight of 79 kDa and an isoelectrical point of 8.6. PepT1 expression has been demonstrated in several animal species, with each isoform exhibiting high homology with other species. Although hPepT1 shows high homology with its various orthologs, studies have demonstrated different tissue expressions of PepT1 among different species.

These expressional differences have not been limited to PepT1, as rPHT1 was not expressed in the gastrointestinal tract, in contrast to its human homolog. Furthermore, correlation of the expression data from one physiological system may not correlate with another. Consequently, great care should be taken in discerning too much from the data and extrapolating to other physiological models. PepT1 is expressed primarily in small intestinal epithelial cells, the S1 (pars convoluta) segment of renal proximal tubules, hepatic bile duct epithelial cells, and the pancreas. Localization studies have also demonstrated that PepT1 is expressed to a lesser extent in the kidney, placenta, and prostate, while in the small intestine, expression is maximal in the duodenum. Ogihara et al. further demonstrated PepT1 localization to the apical plasma membrane of enterocytes in rats. Interestingly, PepT1 expression decreases along the descending small intestinal segments, with ileal expression appearing to be lower than jejunal.

Recently, Terada et al. demonstrated that PepT1 is also relatively highly expressed in the stomach of cancer patients, although interindividual differences were apparent. Interestingly, histological results suggest that gastric PepT1 may originate from intestinal metaplasia, which is characterized by the transdifferentiation of gastric epithelial cells to an intestinal phenotype. The functional and subsequent regulatory significance of this expression remains to be elucidated. Additional studies have demonstrated the intracellular localization of PepT1 isoforms to lysosomal compartments of renal tubular cells. The functional significance of this finding remains to be fully elucidated, although it is surmised that PepT1 may prevent accumulation of di- and tripeptides in renal lysosomes by transporting substrates to the cytosolic compartment for subsequent hydrolysis. Of note is the fact that the apical expression of PepT1 has been well established in both prenatal and mature animals, although the cellular localization does vary and is highly regulated. Interestingly, PepT2 is more widely expressed than PepT1, with primary expression in the S2 and S3 segments of the apical membranes in renal proximal tubular cells (pars recta), brain astrocytes, epithelial cells of choroids plexus, retina, mammary gland, and bronchial epithelial cells.

The cloned human PepT2 cDNA is 2190 bp long, encoding a predicted protein of 729 amino acids that shares approximately 50% identity and 70% similarity with its corresponding PepT1 ortholog. The gene encoding PepT2 has been mapped to chromosome 3q13.3-q21. In vitro translation of rabbit PepT2 cRNA resulted in an unglyosylated 83-kDa product and a core-glycosylated 107-kDa product. PepT2 mRNA expression has been identified in human, rat, mouse, and rabbit. PepT2 was first identified due to the functional assessment that the renal peptide transport system was similar but not identical to its intestinal counterpart (PepT1). In fact, Liu et al. identified PepT2 by screening a human kidney cDNA library with a probe derived from the rabbit intestinal PepT1 cDNA. Since its first identification and characterization, considerable research has been focused on elucidating the functional expression and activity of PepT2 in the kidney. Moreover, recent studies have hypothesized that its primary role is the active reabsorption (secretion) of renally eliminated small oligopeptides. While the expression of both PepT1 and PepT2 have been demonstrated in the proximal tubules of microdissected rat nephrons, all other renal sections were negative for their expression.

The positive expression of PepT1 and PepT2 in proximal tubules correlated with in vivo uptake of a fluorescent dipeptide derivative probe (D-Ala-Lys-AMCA) in the inner cortex and outer stripe in rat kidney. Interestingly, the uptake of the substrate was competitively inhibited by cefadroxil and Gly-Gln, indicative of PepT2-mediated absorption. However, PepT2 is much more widely distributed than just the kidney and as such must play a more important role in peptide absorption and accumulation in other tissues as well. In particular, PepT2 transcripts, protein, and functional activity have been shown in choroid plexus and brain, as nicely reviewed by Smith et al. As such, this transporter is believed to play a significant role in neuropeptide homeostasis and the efflux of peptides and peptidomimetic xenobiotics from cerebrospinal fluid.

Immunoblotting analysis determined the neural PepT2 tissue localization, with protein expression apparent in cerebral cortex, olfactory bulb, basal ganglia, cerebellum, and hindbrain sections of adult brain. The strongest signals were found in the cerebral cortex, while PepT1 protein expression was not found in brain. Interestingly, expression levels were maximal in the fetus and declined with age, indicative of a complex regulatory mechanism to be discussed later. Also, the protein was expressed exclusively on the apical membrane (CSF-facing), indicative of its role in peptide homeostasis and transport. Shen et al. demonstrated that PepT2 is the primary POT member responsible for Gly-Sar uptake in the choroid plexus of PepT2 knockout mice. However, the lack of an observable pathological phenotype indicates that other systems are able to compensate for the loss of activity. Furthermore, while Gly-Sar is a model substrate for PepT transporters, recent studies in our laboratory have demonstrated that it is not a substrate for the human PHT1 isoform in transiently transfected COS-7 cells.

The rat PHT1 isoform was first suggested to be expressed in the apical membranes of the rat choroid plexus, although our results may not reflect species differences in the function of the PHT1 isoforms. Taken together, these data suggest that redundancies may exist for the transport of di- and tripeptides at the choroid plexus and presumably, other biological barriers. Interestingly, although many studies have demonstrated that PepT2 is not expressed in the gastrointestinal tract, R¨uhl et al. investigated peptide transport activity in the neuromuscular layers of whole-mount preparations from mouse, rat, and guinea pig stomach and small and large intestines. Surprisingly, D-Ala-Lys-AMCA specifically accumulated in both ganglionic layers of the enteric nervous system, and its accumulation was inhibited by Gly-Sar, D-Phe-Ala, Gly-Gln, and cefadroxil, but not free histidine or benzylpenicillin (a PepT2 substrate).

Immunohistochemical analyses demonstrated that the dipeptide uptake was localized to enteric glial cells [as demonstrated by 100% uptake in of glial fibrillary acidic protein (GFAP$^+$) and S100$^+$ cells] and periganglionic tissue-resident macrophages. The researchers surmised that PepT2-mediated dipeptide transport in enteric glia could contribute to the clearance of neuropeptides in the enteric nervous system. The findings described above underscore the importance of POT members with respect to peptide homeostasis, uptake, and accumulation. Recent evidence indicates that enteric glia play a significant role in modulation of gastrointestinal functions and may be involved in signaling processes of the enteric nervous system. Loss of enteric glial cells in genetically modified animals resulted in neuronal degeneration or changes in the neurochemical coding of enteric neurons, emphasizing the importance of enteric glia for proper neuronal maintenance in the enteric nervous system, as stressed by R¨uhl et al. The researchers also surmised that PepT2 expression in the enteric glia is unrelated to nutritive peptide absorption, due to its spatial localization and regional distribution throughout the intestine. However, the functional significance of PepT2 expression and function in the enteric neuronal system remains to be elucidated. Recently, two putative human PHT (hPHT1 and hPHT2) transporters have been identified with expression observed in several human tissues. The hPHT1 mRNA sequence is approximately 2.7 kb long, encoding a translated 577-amino acid protein with an estimated molecular weight of 62 kDa and a predicted pI value of 9.2. Four N-linked glycosylation sites were predicted along with several protein phosphorylation sites. Interestingly, though, PHT1 was first cloned in the rat and was the first POT transporter identified in brain.

The rat cDNA was 2751 bp, with an open reading frame of 1719 bp that encoded a protein of 572 amino acids. The rPHT1 exhibits approximately 86.5% identity to its human ortholog and was found to be expressed primarily throughout the whole brain, in hippocampus, choroid plexus, cerebellum, and pontine nucleus. The rat PHT1 isoform was also found in both neuronal and small nonneuronal cells by in situ hybridization, while the protein was expressed abundantly in brain and retina, with lower expression in lung and spleen, as determined by Northern blotting. Rat PHT1 expression was not observed in the pancreas, kidney, intestine, liver, heart, and skeletal muscle. In contrast to its rat ortholog, hPHT1 was found to be expressed at low levels in the human gastrointestinal tract. Interestingly, mRNA expression was also demonstrated in brain, colon, heart, kidney, leukocytes, liver, lung, ovary, pancreas, placenta, prostate, skeletal muscle, small intestine, spleen, testis, and thymus, which were confirmed by Southern blotting. Furthermore, mRNA expression was also demonstrated in retinal pigment epithelium.

The functional significance of hPHT1 expression in these tissues remains to be elucidated. However, our laboratory has recently demonstrated hPHT1 expression in the enterocytes of intestinal tissue segments, as well as hPHT1 functional activity in transiently transfected COS-7 cells. The contribution of hPHT1 to overall peptide transport remains to be elucidated. Similar to hPHT1, hPHT2 has not been widely studied and little is known concerning its physiological significance. Formerly named PTR3, PHT2 was first isolated from the human placenta and has an open reading frame of 1.7 kb, encoding a protein with 581 amino acids with an estimated molecular mass of 64.6 kDa. The rat PHT2 (rPHT2) ortholog has been partially evaluated, having been cloned from a rat brain cDNA. The encoding cDNA was 1979 bp long with an open reading frame of 1748 bp (including the termination codon) and encoded a protein of 582 amino acids that exhibited 49% identity to PHT1 and 80% homology to the human orthologs.

Three N-linked glycosylation sites on the rPHT2 protein are predicted and protein phosphorylation sites (PKA and PKC) were identified. Human PHT2 was found to be widely expressed in various tissues, with mRNA expression demonstrated throughout the gastrointestinal tract, with increased expression in the colon. Furthermore, hPHT2 mRNA expression was shown in the brain, colon, heart, kidney, leukocytes, liver, lung, ovary, pancreas, placenta, skeletal muscle, small intestine, spleen, testis, and thymus, which was all confirmed by Southern blot analysis. Interestingly, Sakata et al. demonstrated rPHT2 mRNA expression in lung, spleen, and thymus, with lower expression in brain, liver, adrenal gland, and heart by RT-PCR. Strong expression was also determined by in situ hybridization in immunocytes, specifically eosinophils, macrophages, and other phagocytes. Further confounding the elucidation of PHT2's function were observations by Sakata et al. suggesting an intracellular localization of PHT2 in the Golgi, lysosomes, autophagosomes, and vacuoles of HEK-293T and baby hamster kidney (BHK) cells. Interestingly, some cells had apparent PHT2 localization on the outer nuclear membrane. Unfortunately, one cannot ascertain the functional significance of these transporters by investigating their tissue and cellular localizations.

Further, peptide transport may also be potentially mediated by a number of other transporter proteins that exhibit overlapping specificity for POT member substrates or play a role in mediating the energy forces driving transport. The primary role of PepT1, PepT2, and PHT1 transporters seems to be the active absorption and accumulation of peptide nutrients. The role of PepT2 may be extended to the clearance of neuropeptides, while the role of PHT2 remains to be elucidated. Further complicating matters is as yet unidentified but functionally characterized basolateral transporters that facilitate the active transcellular flux of these substrates. Interestingly, similar to differences in PepT1 and PepT2 tissue localizations, the intestinal and renal basolateral transporters are also different. In addition, non-POT members may also serve as a confounding factor in understanding peptide transport. First described in 1994 by Dantzig et al., hPT1 shares only 16% identity and 41% similarity with the hPepT1 amino acid sequence.

The PT1 cDNA coding region is approximately 2.5 kb long and encodes a 120-kDa protein consisting of 832 amino acids. HPT1 was first identified as a result of a monoclonal antibody that inhibited the transport of β-lactam antibiotics in Caco-2 cells. Chinese hamster ovary (CHO) cells subsequently transfected with HPT1 demonstrated dipeptide and β-lactam transport that was not competitively inhibited by amino acids. Interestingly, hPT1 function has not been fully characterized, although a rat isoform has been shown to be regulated by dietary protein content. Although the molecular, structural, and functional characterization of PT1 has yet to be completed (presumably due to patent issues), the concept of a cadherin family member mediating peptide transport is an important finding and is indicative of the potential overlap in substrate affinities/selectivity between families. There are other reports of nonclassical peptide transporter behavior, but those transporters will not be elaborated on here. In short, considerably more research is required to fully elucidate all of the functional peptide transporter systems.

PHYSIOLOGICAL FUNCTIONS

Basis of Transport

POT transporters use a proton gradient and the membrane potential as the main driving forces, as reviewed previously, to mediate the uptake of peptides or peptide- based drugs. In bacteria, yeast, and plant cells, the proton gradient driving force is supplied primarily by membrane ATPases, while in mammalian cells it is generally provided by electroneutral proton–cation exchangers (e.g., Na^+/H^+ antiporters). For example, PepT1 -mediated intestinal absorption of peptide-based substrates occurs in an asymmetrical manner, with the proton gradient generally considered to be the predominant force influencing transport in the upper small intestine. Further studies have demonstrated that charged substrates present different binding affinities based on their degree of ionization. Di- and tripeptides, and peptidomimetic compounds with no net charge at the site of absorption are best transferred across the cell membranes by PepT-like transporters. PepT substrates share the same substrate-binding site regardless of substrate charge.

Proton coupling occurs in the H^+-binding site of PepT1, where a H^+ is bound prior to anionic or neutral substrate uptake but is not required for cationic substrates. Irie et al. have developed a computational model to illustrate the H^+-coupled substrate transport of neutral and charged molecules, establishing a PepT1 mechanistic model based on two assumptions: (i) H^+ binds not only to the H^+-binding site but also to the substrate-binding site, and (ii) H^+ at the substrate-binding site inhibits the interaction of neutral and cationic substrates but is necessary for that of anionic substrates.

Identification of Carrier

Some molecular requirements on the structure of PepT-like transporters have been recognized as essential for substrate recognition and transport. Site-directed mutagenesis analyses of single amino acids located within different transmembrane domains (TMD2, TMD4, and TMD5) have shown that Y56, Y64, Y167, N171, and S174 residues modify or inactivate PepT-like transport activity and/or substrate binding completely. Other studies have shown that mutations on W294 and E595 reduced Gly-Sar uptake significantly. Sequence alignments of PepTs have shown the presence of conserved histidyl residues (H57, H121, and H260). Experimental evidence suggests that the histidyl residue H57 is involved in H^+-binding and is essential for peptide transport activity. H121 is involved in substrate recognition, while the role of H260 remains to be elucidated. In hPepT2, H87 has been shown to be absolutely essential to maintain transport activity. Doring et al. demonstrated that the N-terminal region of the protein (TMD1 to TMD9) confers all the phenotypic characteristics, while studies utilizing PepT1/PepT2 chimeras suggest that the first 400 residues (TMD1 to TMD6) contain the substrate-binding pocket and the region that determines pH dependence.

Supporting these findings, Doring et al. further demonstrated that a section between TMD2 and TMD3 (amino acid residues 60 to 91) plays a significant role in the pH-dependent transport process. Furthermore, Fei et al. demonstrated that TMD7 to TMD9 play a role in defining substrate affinity. Doring et al. also determined that the first 59 amino acid residues of both PepT1 and PepT2 contribute significantly to substrate affinity. The functional significance of the C-terminal region of oligopeptide transporters is unknown at this time; however, portions of the C-terminus may aid in transporter trafficking, membrane insertion, and/or transporter regulation, as evidenced by its inclusion in the functionally active rat pineal gland PepT1 (pgPepT1) splice variant. Studies using the substituted cysteine accessibility method (SCAM) have provided more detail concerning PepT structure. TMD5, the most conserved region across species, displays solvent accessibility along its entire length, lining the putative aqueous channel akin to related antiporters. Similarly, TMD7 also exhibits solvent accessibility.

Preliminary molecular modeling, along with some experimental evidence, has suggested that the extracellular end of TMD7 may shift following substrate binding, providing the basis for channel opening and substrate translocation. However, additional studies are required to clearly elucidate the molecular PepT structure. Taken together, these findings suggest that the N-terminal TMDs form a porelike structure, while TMD7 to TMD9 compose the substrate-binding pocket.

Basic Specifications

The three-dimensional structure of any of the POT proteins has not been elucidated. As such, the structural requirements for molecular recognition have been based on uptake or transport experiments, as well as on computational modeling. Several studies have summarized the primary chemical moieties that a peptidomimetic molecule should contain to function as a PepT substrate or inhibitor. For years it was believed that a peptide bond was essential for a molecule to have affinity for PepT, however, nowadays it has been well established that this is not the case. Furthermore, an entire range of molecules, with varying functional moieties, are also transported by PepT1 with affinity constants similar to those of some dipeptides. Distinctively, substrates or inhibitors with high PepT1 affinity (<0.5 mM) have been recognized as molecules containing the following characteristics: (i) L-amino acids, (ii) an acidic or hydrophobic moiety at the C-terminus, (iii) a weakly basic group in α-position at the N-terminus, (iv) a ketomethylene or acid amide bond, and in the case of a molecule with a peptide bond, (v) to have a trans conformation.

Bailey et al. have corroborated experimentally experimentally some of the structural characteristics already proposed and identified some structural requirements necessary to increase PepT1 affinity for a D-enantiomer over its corresponding L-enantiomer. Furthermore, analysis of the binding and transport characteristics of PepT1 have led to the development of several molecular models attempting to establish a PepT–substrate template. Comparatively, PepT1 is considered alow-affinity (K_m of 200 pM to 10 mM), high-capacity transport system, while PepT2 has a higher substrate affinity (K_m of 5 to 500 μM) with a lower transport capacity. Interestingly, PepT2 affinity has been attributed to the requirement of an α- or β-aminocarbonyl moiety for substrate recognition. Moreover, the general structural requirements for PepT2 substrate or inhibitor recognition correspond with those described for PepT1.

However, Those et al. also reported that the spatial localization of a side-chain amino-protecting group in a dipeptide containing a diaminocarbonic acid, and its intramolecular distance from the alpha C-atom, are integral features for transforming a PepT2 substrate into a PepT2 inhibitor. The molecular and structural characteristics for PHT-like transporters have not yet been determined. Given their recent identification and cloning, it is not surprising that only a few studies have been published regarding their functional properties. In fact, additional studies are required to fully characterize the functional aspects of PHT-mediated transport. As such, additional studies should be designed systematically to provide the required information to analyze the essential structural elements for substrate or inhibitor recognition.

Physiological Substrates

The established physiological substrates of PepT-like transporters are di- and tripeptides, which potentially include different combinations of the 20 L-α-amino acids. These combinations correspond to approximately 400 dipeptides and 8000 tripeptides, not including respective combinations of nonclassical or derived amino acids, such as hydroxyproline, trimethyllysine, and ornithine, homocysteine, methylhistidine, and hydroxylysine. These numbers also do not include the presence of D-enantiomers of the 20 classical amino acids, whose incorporation at the di- and tripeptide N-terminal results in substrates with good affinities and high rates of transport. These observations have resulted in the conclusion that PepT-like proteins are capable of mediating the transport of a huge variety of molecules, with a large diversity of physicochemical characteristics. Several peptidomimetic drugs, prodrugs, and nonpeptidic compounds have been established as exogenous substrates of the PepT-like transporters.

Pharmacologically active compounds that are demonstrated substrates of PepT-like transporters include β-lactam antibiotics; ACE inhibitors such as captopril, enalapril, and fosinopril; bestatin; prodrugs of L-dopa; prodrugs of various nucleosides, such as valacyclovir and valganciclovir; prodrugs of some bisphosphonates, such as L-Pro-L-Phe-alendonate and L-Pro-L-Phe-pamidronate; and antibacterial peptide analogs such as alafosfalin. The magnitude of the interaction of β-lactam antibiotics with PepT1 and PepT2 has been analyzed recently in the context of old and new cephalosporins and penicillins, establishing a ranking of affinity constants that may help establishing structure–function studies to further analyze structural substrate/inhibitor requirements. The peptide/histidine transporters PHT1 and PHT2 have demonstrated dipeptide transport activity, both with high affinity for histidine, and PHT1 for carnosine as well.

Rat PHT1 when expressed in *Xenopus* oocytes exhibited a high-affinity proton-dependent histidine uptake that was inhibited by several di- and tripeptides but not by other amino acids. Studies in our laboratory suggest that hPHT1 mediates the transport not only of carnosine and L-histidine, but also valacyclovir, in a proton-dependent sodium-independent manner. Interestingly, these studies also concluded that unlike PepT1 and PepT2, the dipeptide glycylsarcosine (Gly-Sar) is not a substrate for hPHT1. Taken together, these data are indicative of both the overlapping specificity, in terms of valacyclovir, as well as the functionally distinct specificity of these transporters. Nevertheless, there is still much to elucidate concerning the affinity and substrate specificity of these peptide transporters.

CONTROL

Oligopeptide transporter research has predominantly been focused on delineating the functional characteristics of each transporter isoform. Although considerable research is still needed in this area, particularly with undercharacterized isoforms such as PHT1, our focus must now shift toward developing a greater appreciation of their molecular characteristics. In particular, understanding the underlying mechanisms of oligopeptide transporter regulation by various physiological and exogenous stimuli is integral to comprehending their role as mediators of nutritional and pharmacological substrate absorption. Clearly, alterations in oligopeptide transporter expression could potentially result in significant changes that affect the PK/PD phenomena of peptide- based drug substrates. PepT1 has been the most widely studied isoform and has been shown to be regulated via a number of different mechanisms under varying conditions. Studies on the regulation of other POT members (PepT2, PHT1, and PHT2) and PT1 are scarce at best.

Factors such as diet, development phase, hormonal influences, pathological conditions (diabetes, intestinal disorders), and various pharmacological agents have all been demonstrated to regulate PepT1 expression and are also discussed in various reviews. Some of these factors are discussed in greater detail below. However, it is evident that additional studies are required to better elucidate the pre- and post-transcriptional changes that the oligopeptide transporters undergo following various stimuli.

Influence of Diet

Changes in PepT1 expression due to dietary influences are relatively well established. Erickson et al. demonstrated that a high-casein diet induced a parallel 1.5- to twofold increase in both the PepT1 and rat PT1 mRNA expressions in the middle and distal regions of the rat small intestine when contrasted with low-casein diet control. No apparent change was observed in the proximal region of the rat small intestine, leading to the suggestion that the middle and distal regions may be the most sensitive to dietary fluctuations. Shiraga et al. further investigated the molecular mechanisms of PepT1 expressional changes in the Caco-2 cell model. Incubation of cells with dipeptides (Gly-Sar, Gly-Phe, Phe-Val, Lys-Phe, and Asp-Lys) and individual amino acids (phenylalanine, arginine, and lysine) resulted in activation of the PepT-1 gene promoter.

Subsequent induction of the PepT1 mRNA was caused by transcriptional activation through an AP-1 binding site and an amino acid–response element present at the –295 and –277 nucleotides relative to the transcription start site in this region. Taken together, these expressional changes represent an important evolutionary adaptation to maximize nutrient absorption in response to substrate availability. Thamotharan et al. also utilized Caco-2 cells to demonstrate that pretreatment with Gly-Sar and cefadroxil for 24 hours increased the V_{max} value of glycylglutamine (Gly-Gln) transport twofold without any significant change in its K_m. Subsequent Western blot analysis revealed a greater than twofold increase in the protein expression of PepT1 in the apical membrane, and a threefold increase in PepT1 mRNA was also observed. Interestingly, an inhibitor of the trans-Golgi network, brefeldin, significantly increased PepT1 mRNA with no demonstrable effect on Gly-Gln transport, suggesting that dipeptides induce PepT1 transcription.

Moreover, Walker et al. demonstrated that previous exposure with Gly-Gln (4 mM) for 3 to 4 days in Caco-2 cells resulted in a twofold increase in the V_{max} of Gly-Sar transport, without any change in its K_m. Results suggest that Gly-Gln treatment increased the cellular mRNA level and subsequent PepT1 membrane expression by twofold. Thamotharan et al. demonstrated that a brief fast or starvation can also up-regulate PepT1 expression on both the mRNA and protein levels. Brush border membrane vesicles prepared from jejunum of fed and 1-day-fasted state rats exhibited a twofold increase of Gly-Gln (V_{max}) uptake upon fasting without a subsequent change observed in the K_m. In addition, PepT1 protein and mRNA increased threefold in both the intestinal brush border membrane and intestinal mucosa, respectively, suggesting a transcriptional regulatory mechanism.

In addition, Ogihara et al. demonstrated that PepT1 expression at the microvillous plasma membrane is regulated by diet, whereas the transporter protein gradient along the crypt–villus axis was maintained between the fasting and fed states. Another study comparing rats starved for 4 days and semistarved (50% of freely fed control) for 10 days resulted in a 179% and a 161% increase in jejunal PepT1 mRNA expression of the starved and semistarved groups, respectively, compared to the freely fed control group. These results suggest that PepT1 gene expression can also be modulated by malnutrition. For better insight into this regulatory mechanism, Naruhashi et al. correlated the functional transport of a model PepT1 substrate, cefadroxil, in both starved and fed rats with PepT1 mRNA expression. Briefly, the transport of cefadroxil was quantified across the small intestines of both groups via an Ussing chamber, and PepT1 mRNA expression was determined using quantitative reverse transcriptase polymerase chain reaction.

A good correlation was observed between mRNA expression levels and the permeability coefficients ($r^2 = 0.859$), suggesting that cefadroxil transport is directly proportional to PepT1 expression. Although higher in the proximal regions, PepT1 mRNA expression in the starved rats was increased in all small intestinal segments. Similarly, cefadroxil transport was also higher in the proximal small intestinal regions of starved rats in contrast to the fed rats, indicative of its correlation to PepT1 mRNA expression. Furthermore, Howard et al. investigated the effect of luminal nutrients on the expression of PepT1 in starved and fed rats. Removal of luminal nutrients increased ileal PepT1 mRNA expression compared with orally fed rats. These results allude to the importance of a continuous amino acid supply for the maintenance of proper gut mucosal function. Collectively, these studies suggest that up regulation of PepT1 expression occurs whether fasting is brief or prolonged and may be indicative of a nutrient-sensing feedback mechanism.

Ontogenic Control

Studies have also demonstrated developmental stage-dependent differences in oligopeptide transporter expression and dipeptide absorption during peri- and postnatal periods in rabbits and guinea pigs. More recently, Miyamoto et al. demonstrated a dramatically higher intestinal PepT1 mRNA expression in 10-day-old rats, which rapidly decreased and then leveled off reaching a plateau at adult expression levels by day 28 after birth. Further investigation revealed that rat PepT1 mRNA and protein levels in the duodenum, jejunum, and ileum attained maximal levels by days 3 to 5 after birth. Interestingly, these mRNA and protein levels proceeded to fall rapidly to 11 to 13% of maximal expression by day 14, but then rose again to 23 to 58% of maximal expression by day 24. Higher expression of PepT1 at weaning, especially in the ileum, may be related to an adaptive response to a reduced protein diet or nourishment at this time period.

The age-dependent PepT1 and PepT2 expression patterns observed in the developing rat kidney were similar to that observed for PepT1 in the small intestine and colon. However, the expression of PepT1 and PepT2 was found to be more consistent in rat kidney than in the intestine. The results suggest that nutritional state may play a less prominent role in regulating the renal expression of these transporters or that an autocrine effect from PepT1 substrates is involved in its regulatory control. In a recent immunocytochemical study, PepT1 expression was measured using an antibody specific to the C-terminal sequence of PepT1 in rat duodenum at different stages of growth. PepT1 was found to be distributed exclusively in the apical brush border membrane of enterocytes from both prenatal and mature animals. Interestingly, PepT1 expression extends to the subapical cytoplasm, basal cytoplasm, and basolateral membrane of enterocytes immediately after birth.

Since the nature of oligopeptide transporter expression at the basolateral membrane has not been defined clearly, atypical PepT1 distribution at birth could be due to enteroendocrine mechanisms following first exposure to enteral (i.e., luminal) nutrition or humoral signals generated during parturition.

Photoperiodic Effect

Recently, Pan et al. demonstrated the effect of diurnal rhythm on PepT1 expression utilizing an in situ intestinal loop model and everted rat intestines, maintained in a 12-hour photoperiod with free access to food. Kinetic studies demonstrated enhanced Gly-Sar uptake and PepT1 protein and mRNA expression in dark phase (8:00 PM; beginning of the dark phase)-treated rats comparative to light phase (8:00 AM; beginning of the light phase)-treated rats in a 24-hour study cycle. A subsequent study determined that fasting conditions may disrupt the effect of diurnal rhythm on intestinal PepT1 expression and transport activity. In this PK study of ceftibuten oral absorption, higher plasma concentrations were observed at 8:00 P.M. (dark phase) in contrast to 8:00 a.m. (light phase) in fed rats, but no differences in 4-day-fasted rats. Recently, a PepT1 gene product has been identified in the rat pineal gland (pgPepT1), which encodes 150-amino acid protein encompassing the three C-terminal membrane-spanning domains of intestinal PepT1 and three additional N-terminal residues. PgPepT1 forms a functional PepT1-like transporter through oligomerization. Interestingly, pgPepT1 mRNA and protein (approximately 16 kDa) levels were found to be >100-fold higher at night, suggestive of a diurnal regulatory mechanism. The neural pathway that controls pineal melatonin production has been suggested to play an important role in regulating the expression of pgPepT1. Thus, a peptide transporter in the pineal gland could play an important role in circadian pineal physiology and contribute to clearance of active or degraded neuropeptides.

Clinical Study

Many disease state-dependent mechanisms that may alter oligopeptide transporter expression and function could be conceived. Inflammatory disorders, particularly those associated with the release of proinflammatory peptides, are examples of such disease state-dependent variability. A recent study using colonic mucosal samples from diseased (chronic ulcerative colitis, microscopic colitis) and control patients (normal colon) revealed aberrant PepT1 expression under inflammatory conditions. Further studies utilizing coloniclike cells (HT29-Cl.19A) stably transfected with hPepT1 tagged on the N-terminus with green fluorescence protein, with and without inflammatory peptides, indicated that in some states of chronic inflammation, hPepT1 may be expressed anomalously. Recently, Vavricka et al. investigated the effects of tumor necrosis factor alpha (TNFα), interferon-γ (IFN-γ), and inflammatory cytokines and mediators [interleukins (IL)-1β, IL-2, IL-8, IL-10] on PepT1 regulation in Caco-2/b be monolayers and in mice.

From these studies it appeared that TNFα and IFN-γ play the most prominent roles in activating PepT1 protein expression in the colon. These findings were supported further by the observation of higher Gly-Sar uptake in TNFα- and IFN-γ-treated cells when contrasted to controls. Interestingly, Barbot et al. revealed that transcriptional up-regulation of PepT1 occurred in the brush border membrane of enterocytes in *Cryptosporidium parvum* (an amino acid malnutrition disease)-infected rats from day 4 to day 50. Intestinal transplantation is a surgical procedure performed in patients suffering from short bowel syndrome with complications of parenteral nutrition and irreversible intestinal failure. Syngeneic small intestinal transplantation in rats resulted in increased PepT1 protein expression levels, indicating that transporter expression may be useful to evaluate intestinal graft function in addition to the current histopathological examination of the mucosa. Studies conducted in nephrectomized rats also revealed increased renal high-affinity peptide transport activity and PepT2 expression, while PepT1 levels were undetectable.

Two weeks after surgery, Gly-Sar uptake rate (V_{max}) in renal brush border membrane vesicles was significantly increased compared to sham-operated controls. Utilizing Western and Northern blotting techniques, Gangopadhyay et al. further illustrated that an increase in PepT1 protein and mRNA expression occurred in streptozotocin-induced diabetic rats. Diabetes also increased PepT1 activity in the brush border membrane of the intestinal mucosa, as demonstrated by an increased V_{max} for Gly-Gln. Furthermore, both PepT1 protein and mRNA expression were up-regulated in the brush border membrane of renal tubules. Interestingly, Watanabe et al. compared the intestinal absorption and pharmacokinetics of cephalexin in type 1 diabetes (streptozotocin-induced) and hyperinsulinemic type 2 diabetic GK and Zucker-fa/fa (Zucker) rats.

Pharmacokinetic parameters k_a, C_{max}, and $AUC^{0\rightarrow\infty}$ were significantly higher in type 2 diabetic GK and Zucker rats compared to control rats, whereas the type 1 diabetic streptozotocin-induced rats demonstrated no significant differences from the control. Expression studies supported the PK findings demonstrating increased protein expression in both type 2 diabetic rat models. In contrast, the intestinal PepT1 mRNA levels were not significantly different between controls and any of the study groups. These findings suggest that the PepT1-mediated intestinal absorption of drugs may be enhanced in hyperinsulinemic type 2 diabetic rats. When these results are compared to those determined by Gangopadhyay et al., the streptozotocin-induced diabetic results are contradictory. This contradiction may be attributed to differences in the experimental conditions, potentially including dietary differences.

Endocrine Control

The interplay between insulin and leptin regulation of glucose and lipid homeostasis is well known and forms the basis for the complexities of the metabolic syndrome- associated phenotypes. As discussed above, recent findings in both diabetic and Zucker rats, may indicate that these hormones may also have a concerted effect on protein/peptide homeostasis in part through the modulation of PepT1 expression. In terms of a classical biochemistry view, it would appear obvious and essential that the three major sources of nutrients (lipid, carbohydrates, and protein) would be subject to coordinate regulation. Toward this aim, recent studies performed in Caco-2 cells demonstrated that within 30 to 60 minutes of adding physiologically relevant insulin concentrations (5 mM), enhanced Gly-Gln uptake was observed and attributed to increased PepT1 expression.

In support of this conclusion, kinetic studies revealed a twofold increase in V_{max} with no significant change in K_m, which was reduced by the tyrosine kinase inhibitor genistein. Additionally, the stimulatory effect of insulin was observed even after treatment with brefeldin, which disrupts the Golgi apparatus, thereby modifying the cells' ability to process newly synthesized PepT1 protein for insertion into the plasma membrane. Furthermore, when colchicine (disrupts the translocation of proteins targeted for membrane insertion) was added 20 minutes before the addition of insulin, it abolished the insulin stimulatory effect completely. Moreover, increased PepT1 protein expression in the apical membrane was demonstrated; however, no corresponding changes in PepT1 gene expression were observed. Nielsen et al. further demonstrated in Caco-2 cell monolayers that short-term insulin pretreatment resulted in a Gly-Sar uptake rate (V_{max}) that was significantly higher than that of the controls, and K_m remained constant. Moreover, the substantial increase in Gly-Sar uptake was not accompanied by changes in hPepT1 mRNA or by measurable changes in the cytosolic pH. Watanabe et al. further investigated the effects of insulin on cephalexin transport and PepT1 mRNA and protein expression in Caco-2 cells.

Consistent with other studies, a 10-minute insulin pretreatment significantly increased the cephalexin uptake and PepT1 protein expression on the apical membrane. Again, no changes were observed in the PepT1 mRNA expression levels. These findings suggest that insulin binds to its receptor on the membranes, promoting PepT1 translocation from the intracellular pool to the apical membrane surface. Collectively, these results indicate that insulin signals an increased PepT1 translocation from a preformed cytoplasmic pool to the apical membrane as opposed to de novo synthesis, which is similar to events observed with the unfolded protein response. These results are acute, as opposed to the studies in diabetic rat models that are reflective of a chronic response, further suggesting an alternative pathway by which insulin might increase PepT1 transcription and translation. In addition to insulin, other hormonal PepT1 regulators have been identified and studied, such as epidermal growth factor (EGF) and leptin. In contrast to insulin, EGF was found to inhibit PepT1 by decreasing the number of transporter molecules on the apical membrane.

Kinetic studies demonstrated a significant decrease in V_{max} on the apical uptake of Gly-Sar, whereas K_m remained constant. In contrast, basolateral Gly-Sar uptake demonstrated no changes in either V_{max} or K_m after EGF treatment. Inhibition of transcellular Gly-Sar flux was observed after 5 days of EGF treatment and reached a maximum after 15 days, indicating that a decrease in peptide transport occurs after long-term EGF treatment. Interestingly, brief exposure of EGF has been demonstrated to enhance the functional activity of PepT1. A dose-dependent increase in the apical Gly-Sar uptake with no change in K_m was observed in Caco-2 cells after 5 minutes of EGF exposure. Furthermore, short-term EGF treatment did not alter PepT1 mRNA expression or the intracellular pH. Unfortunately, the actual mechanism by which short-term EGF treatment increases PepT1 functional expression remains unclear. Down-regulation of PepT2 mRNA was also observed after long-term treatment with EGF in the rat proximal tubule cell line SKPT0 193 cl.2 (SKPT), implicating either a transcriptional regulatory mechanism through an EGF-responsive element or regulation of PepT2 mRNA stability.

However, further studies are required to discern if the EGF regulatory mechanisms for PepT1 and PepT2 share common features. Buyse et al. explored hormonal PepT1 regulation further by demonstrating that leptin secreted by the stomach may reach the small intestinal lumen to mediate transporter expression. The addition of leptin to intestinal perfusate increased the absorption of Gly-Sar across the rat jejunum in in vivo perfusion studies. In addition, uptake studies conducted utilizing Caco-2 cells demonstrated that leptin significantly increased the V_{max} of Gly-Sar and cephalexin without changes in K_m values. Further, leptin addition to the apical side of Caco-2 cells resulted in a 60% increase in membrane PepT1 protein, with a corresponding decrease in intracellular PepT1. Moreover, enhanced Gly-Sar uptake in response to leptin was abolished with colchicine pretreatment but unaffected by brefeldin. Taken together, these results suggest increased trafficking of PepT1 from the intracellular pool to the apical membrane as a possible mechanism to enhance the transcellular flux of substrates across Caco-2 cells. Interestingly, gastrointestinal hormones and/or insulin elicit their regulatory effects through specific binding to their respective surface receptors, while thyroid hormones may regulate the expression of oligopeptide transporters by entering the cells directly.

Significant inhibition of Gly-Sar uptake in Caco-2 cells was observed after pretreatment with 3,5,3′-L-triiodothyronine (T3; 100 nM) for 4 days. Kinetic analysis revealed a twofold decrease in the V_{max} values with no apparent change witnessed in the K_m. Conversely, Western blot analysis illustrated a 70% reduction in the apical membrane expression of PepT1 protein versus untreated cells. Since decreased gene expression was also observed, it was suggested that T3 inhibition of dipeptide transport is due to decreases in the transcription and/or stability of PepT1 mRNA. However, there is no evidence suggestive of a thyroid hormone-responsive element in the promoter region of the PepT1 gene.

Alternatively, it is possible that after binding to its nuclear receptor, T3 could affect transcription through an indirect mechanism(s). Lu and Klaassen recently demonstrated increased PepT1 and PepT2 mRNA expressions in male rat kidneys after undergoing thyroidectomies. These changes are suggestive of nutrient conservation during malnutrition, as evidenced by increased renal reabsorption of di- and tripeptides due to up-regulation of these two transporters.

Influence of Pharmaceutical Agents

Various pharmaceutical agents, such as clonidine, pentazocine, and cyclophosphamide, have also been demonstrated to affect the activity of oligopeptide transporters. Berlioz et al. studied the receptors responsible for the stimulatory effect of clonidine utilizing Caco-2 cells that were engineered to stably express α_{2A}-adrenergic receptors (Caco-2 3B) at a density similar to that found in the mucosa. Interestingly, higher cephalexin transport was observed in Caco-2 3B cells pretreated with clonidine than in wild-type (expressing PepT1 but not having substantial α_2-adrenergic receptors) and HT29 19A clones (a clone that expresses α_2-adrenergic receptors but does not express detectable levels of PepT1). Kinetic analysis revealed an increase V_{max} without altering K_m in these cells and that the stimulatory effect of clonidine was blocked completely by Gly-Sar and α_{2A}-adrenergic receptor antagonists.

Furthermore, disrupting the microtubule integrity by colchicine abolished the clonidine effect, whereas no effect was observed by Na^+/H^+ exchanger blocker (amiloride), suggesting the role of microtubular integrity on the effect of α_2-agonists. Collectively, these results suggest that clonidine can directly activate α_2-adrenergic receptors located on epithelial cells and thus modulate PepT1 expression by increasing translocation to the apical membrane. Pentazocine pretreatment for 24 hours resulted in increased Gly-Sar uptake in Caco-2 cells in a concentration- and time-dependent manner. Kinetic analyses have indicated that pentazocine enhanced the V_{max} for Gly-Sar uptake without significantly altering K_m in Caco-2 cells. Increased PepT1 mRNA expression was observed, suggesting that pentazocine induces PepT1 transcription in Caco-2 cells, leading to increased PepT1 protein expression at the cell membrane.

In contrast, treatment with the immuno-suppressive agents tacrolimus and cyclosporine A decreased PepT1 activity through an unknown mechanism, as demonstrated by reduction in V_{max} of Gly-Sar uptake without any discernable effect on K_m. Similarly, pretreatment with steroid hormones (progesterone and norethisterone at 3, 10, and 30 mM) for 24 hours resulted in a significant decrease in the apical-to-basal transport of cephalexin in Caco-2 cell monolayers.

SNP of Transporter Genes

Single-nucleotide polymorphisms (SNPs) of transporter genes have also been investigated as possible factors influencing interindividual variations in PK and PD profiles. Coding region SNPs (cSNPs) of transporter genes may induce amino acid mutations, resulting in changes in transporter intrinsic activity by altering the protein's affinity to substrates (K_m) and/or translocation ability or capacity (V_{max}). Despite increasing efforts to understand the molecular and functional characteristics of oligopeptide transporters, little information is available on their respective genetic polymorphisms. Zhang et al. reported nine nonsynonymous and four synonymous cSNPs in the PepT1 gene in a population of 44 persons of different ethnic backgrounds. Plasmids were constructed by site directed mutagenesis for each of the nine nonsynonymous variants and transiently transfected into HeLa cells for functional characterization.

Transport kinetics were studied utilizing Gly-Sar as a model substrate; and PepT1 expression was determined by Western blot and immunocytochemical analyses. Results demonstrated that among the nine nonsynonymous SNPs, P586L resulted in significantly reduced transport capacity (V_{max}) and decreased PepT1 protein expression levels. Following its initial cloning from a human cDNA kidney library, a single PepT2 transcript was identified and accepted as a consensus sequence. Currently, five nonsynonymous cSNPs (R57H, L350F, P409S, R509K, and M704L) have been identified. However, the functional activities of only two of these variants, R57H and P409S, have been characterized functionally in transiently transfected HEK293 cells. Uptake studies performed using these transients and contrasted with wild-types revealed elimination of Gly-Sar uptake in the presence of the R57H variant. Interestingly, the P409S SNP had little effect on function since comparable activity was observed in the transient comparative to the wild-type hPepT2. Western blot analysis and immunofluorescence in*Xenopus* oocytes expressing the PepT2 R57H variant verified protein expression at the membrane surface, indicating that the loss of transport activity was not due to changes in expression, but rather, in it functional characteristics.

Mechanism of Splicing

The importance of splice variants in general cannot be underestimated, as they increase protein diversity by allowing multiple, sometimes functionally distinct proteins to be encoded by the same gene. However, elucidating the function of each splice variant has been a tremendous task, especially given the focus on the parent isoform. Inui and colleagues first identified a novel pH-sensing regulatory protein cofactor which modulates the transport activity of hPepT1. The cDNA was cloned from human duodenum having 1704 bp that encoded a predicted protein of 208 amino acid residues.

Functional studies were performed in *Xenopus* oocytes injected with water, hPepT1 cRNA, the 208 hPepT1 splice variant cRNA, and hPepT1 cRNA combined with the hPepT1 splice variant cRNA. Interestingly, results demonstrated a lack of transport activity for Gly-Sar in hPepT1-RF-injected oocytes; however, cotransfection of the hPepT1 splice variant with hPepT1 affected the pH sensitivity of peptide uptake in the double transfect, suggesting a pH-regulatory function for the splice variant and leading to its hPepT1-RF name. Studies by Sadee and co-workers reported multiple splice variants in the hPHT1 gene. In several regions of the hPHT1 sequence deduced, multiple expressed sequence tag (EST) overlap was observed, suggesting possible sequence variations in the human population. Expression analysis in several tissues and by EST alignments suggests the presence of possible hPHT1 splice variants, as indicated by an insert or gap in two regions of the *hPHT1* coding sequence. Indeed, our laboratory identified a previously unreported hPHT1 splice variant from Caco-2 cells utilizing the primers designed for the cloning of the hPHT1 full-length sequence.

This PHT1 cDNA contained a gap of 58 bp between bases 842 and 900 from the start codon. This gap induced a shift in the coding frame, resulting in a predicted protein sequence of 295 amino acids. Interestingly, this hPHT1 splice variant comprised the first six conserved hPHT1 TMDs, similar to the hPepT1-RF, raising potential questions concerning modularity in the hPHT1 gene coding information and its functional relevancy. Recently, Pinsonneault et al. reported haplotype analysis of the PepT2 gene to determine the effect of variants on PepT2 expression and function. Among these variants, PepT2*1 and *2 were shown to be significantly present in 247 human genomic DNA samples from different ethnic groups. A haplotype is the combination of several sequence variants on a single chromosome at a specific locus and may more accurately reflect genetic diversity in the human population. Wild-type hPepT2 and its variants were transfected into CHO cells and their function was investigated after 24 hours. Results indicated significantly different K_m values for Gly-Sar uptake for each variant. Furthermore, varying levels of hPepT2*1 and *2 mRNA were observed in nine heterozygous kidney tissue samples, suggesting the presence of *cis*-acting polymorphisms affecting transcription or mRNA processing.

SCREENING OF DRUGS

Permeation of peptide-based compounds across cellular barriers can occur by several routes. One needs to be mindful that each of these routes of permeation can apply to different peptide-based compounds; however, the physicochemical properties of the compound and the physiological significance of each route will determine the relative importance in controlling the compound's net absorption. In screening compound permeation across cellular barriers, delineation of the key rate-determining steps for controlling absorption needs to be identified. Methods to delineate the functional relevancy of each route are now being enhanced further by focusing the kinetics on potential paracellular and transcellular events. Adson et al. demonstrated that the delineation of the paracellular and transcellular mass transfer resistances from observed cell-based assay permeability studies was possible. Several studies conducted by Burton and colleagues further demonstrated that hydrogen-bonding potential may provide a better means for predicting the potential passive transcellular diffusion of peptides.

Based on these observations, several studies conducted with larger oligopeptides demonstrated that conformation and lipophilicity may dictate the extent of passive paracellular and transcellular peptide transport across Caco-2 and bovine brain microvessel endothelial cell (BBMEC) barriers. These studies focused on passive diffusion; however, permeation of smaller di- and tripeptides and peptide- based analogs of similar size may also have affinity for oligopeptides transporters that can be rate controlling for their absorption. Although it is clearly evident that oligopeptide transporters play a major physiological role in the absorption of various nutrients and peptidelike xenobiotics, the pharmaceutical industry has yet to fully realize and exploit the potential advantages associated with targeted delivery to these transporters. Probably the most readily apparent application of oligopeptide transporter research is the marketability of Valtrex (valacyclovir) comparative to Zovirax (acyclovir). Comparing clinical trial data (below), the addition of a valyl moiety to the ether end of acyclovir increased the bioavailability from approximately 15% with Zovirax to 54% with Valtrex. Although this increase in acyclovir bioavailability is significant, this case is the exception and not the rule, as it is arguably difficult to directly synthesize compounds with high affinity for a particular transporter and acceptable biological activity.

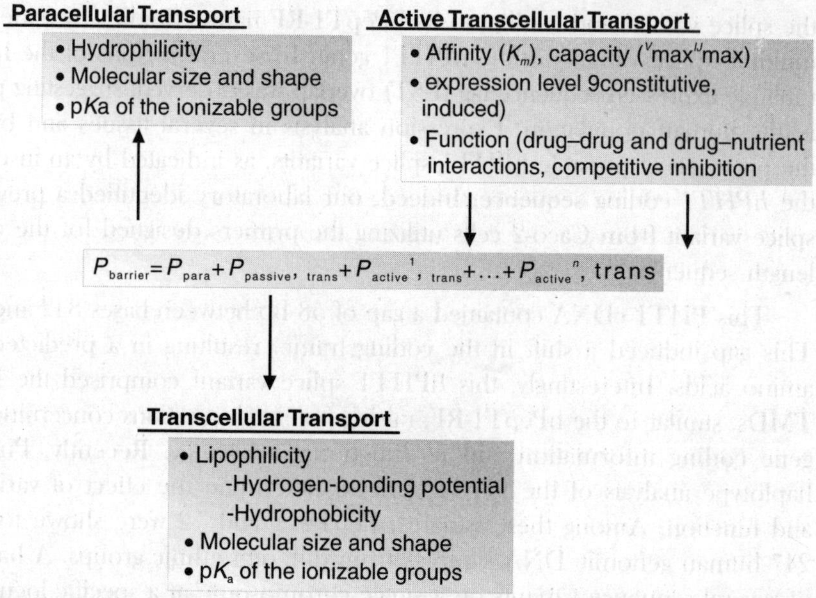

Fig. 13.1. Some of the rate-limiting mass transfer resistances/characteristics of the parrallel paths that oligopeptide-based compounds typically encounter as they transverse cell barriers.

As such, the focus of industry has been to screen compounds for their potential transporter affinities or effective permeability across cellular monolayers and choose lead compounds that are balanced in their physicochemical properties and potential transporter interactions. However, given the relative youth of the transporter field, many scientists have had problems interpreting when a transporter may be important. Although this is largely dependent on the compound and whether or not it is a substrate for a transporter, decision making in this area apparently remains vague. Some have suggested altering the Biopharmaceutics Classification System (BCS) to include elimination criteria to provide criteria to help us understand when transporters will significantly alter the pharmacokinetic profile of a new drug. However, one still needs to know if the compound of interest is a substrate for a particular transporter. This is normally done by permeability screening assays.

Traditionally, the Caco-2 cell line, which is derived from a colonic adenocarcinoma, is used to measure relative gastrointestinal permeability of compounds. However, given varying growth conditions, selective pressures, and intrinsic cellular heterogeneity, it may be difficult to reproduce permeability data accurately between laboratories and individuals. For instance, even increasing the number of passages within the same laboratory changes the expression of numerous transporters in Caco2 cells. Given the widespread use of these screening models, these differences in transporter expression may not properly translate to the actual physiological condition resulting in selection of lead candidates based on erroneous data. These obvious concerns have led a number of laboratories to look at the contribution of individual transporters through either transient or stable transfection of the transporter of interest into different cell lines and then measuring transport kinetics. Our laboratory developed stably transfected Madin-Darby canine kidney (MDCK)/hPepT1-V5&His clonal cell lines expressing varying levels of epitopetagged hPepT1 protein to quantify the relationship between hPepT1 expression and its functional kinetics.

The utility of this cell line is that it may serve as a useful in vitro model for screening and quantifying peptide and peptide-like drug transport as a function of hPepT1 expression in drug discovery. Unfortunately, though, even these altered cell lines are limited in that they can model physiology only under extremely controlled conditions. As mentioned above, oligopeptide transporters alone are regulated by a whole host of different mechanisms. Of particular importance, dietary protein increases the gut intralumenal amino acid and oligopeptide concentrations, ultimately resulting in increased POT expression patterns. Given that POT up-regulation is virtually immediate, we (and others) surmise that the increased membrane expression is due to mobilization from intracellular pools of unfolded protein (unfolded protein response).

Although the initial thought is that these findings can be utilized to increase the bioavailability of orally administered peptidomimetic drugs through concomitant prescription of a high-protein diet, this presents other potential issues, especially with respect to food–drug interactions. Studies suggest that dipeptides may regulate their own transport by modifying PepT1 gene expression via two possible mechanisms: (i) increasing the stabilization of mRNA encoding the gene, and (ii) increasing the gene transcription. In fact, the effect of dietary protein intake on oligopeptide transporter regulation is probably influenced by a combination of these mechanisms. In short, dietary protein increases the gut intralumenal amino acid and oligopeptide concentrations, ultimately resulting in altered oligopeptide transporter expression patterns.

With respect to a compound that exhibits a narrow therapeutic index that happens to be a good substrate for a POT transporter, the concerted up-regulation of the protein in response to food could dramatically increase drug absorption and subsequent blood concentrations. Such an incidence could result in toxic drug effects. Although some may think that such an occurrence could be averted through simply assessing PK parameters in response to food, subtle hormonal regulatory effects, such as up-regulation due to insulin, could present adverse clinical events. Unfortunately, many people do not consider that a small, sugary snack could seriously affect their drug therapy. However, if such a drug were a POT substrate, the resulting insulin secretion in response to the snack could result in serious adverse effects. Although we are not aware of any compounds or situations where these problems have been manifested, we feel that screening tools should be improved to better assess the regulatory mechanisms associated with POT member up/down-regulation to address such potential issues.

Examples

Valacyclovir is an l-valine ester prodrug of acyclovir that is used for the treatment of herpes, varicella zoster (VZV), and cytomegaloviruses (CMV). Valacyclovir was developed to increase the oral absorption and plasma levels of acyclovir. Increased plasma concentrations of acyclovir are important in maintaining antiviral activity, especially in immunocompromised patients and in treatment of less sensitive viruses, such as VZV and CMV. Suboptimal exposures can lead to more resistant viral strains. To achieve high enough exposures, acyclovir needs to be dosed intravenously or in multiple high peroral doses. In the design of valacyclovir, the following criteria were met: It was as safe as acyclovir, efficiently converted, and gave exposures after oral administration that were comparable to plasma levels of intravenously dosed acyclovir. Several reviews describe the development, pharmacokinetics, and efficacy of valacyclovir. Initially, two prodrugs of acyclovir with modifications on the purine ring were synthesized. The development of these prodrugs was discontinued due to incomplete conversion and to chronic toxicity that was worse than that of acyclovir.

The toxicity was suspected to be due to phosphorylation of the prodrug. Subsequently, 18 amino acid esters of acyclovir were synthesized. The bioavailability was measured following oral dosing to rats. Of the 18 prodrugs, 10 exhibited greater recovery of acyclovir in the urine than acyclovir itself when dosed orally. Of the 10, the L-valyl ester, valacyclovir, performed the best, with 63% excreted as acyclovir in the urine. Valacyclovir was also not phosphorylated, reducing the risk of the potential toxicity observed with the initial prodrugs utilized. Further studies were established to determine if the high absorption of valacyclovir was due to carrier-mediated transport. The absorption of acyclovir and valacyclovir was studied in cynomologus monkey intestinal brush border membrane vesicles, where the influx of valacyclovir into the vesicles was six- to 10-fold higher than the influx of acyclovir.

Additional studies in Caco-2 cells showed that transport of valacylovir was seven times higher than acyclovir transport. In rats, l-amino acid ester analogs of acyclovir show better absorption than D or D-L analogs, indicating that the transport was also stereoselective. Several studies in Caco-2 and transfected cell lines have shown that valacylovir is transported by rat and human PepT1, even though the prodrug does not have a peptide bond. Recent work has, however, shown that several additional transporters may be involved in valacyclovir transport in humans, and that PepT1 may not be the predominant transporter of this prodrug in humans. Most recently it has been suggested that valacyclovir transport by PHT1 and hPT1 may also be contributing absorption pathways. The absolute bioavailability of acyclovir when 100mg of oral valacyclovir prodrug is dosed to healthy human subjects is 54%, compared to only 15 to 20% (200- to 600- mg doses) after acyclovir was dosed orally.

Larger variability was observed after dosing of acyclovir than following valacyclovir dosing. The bioavailability of acyclovir from orally dosed valacylovir is similar in rats and cynomolgus monkeys. Over 99% of valacyclovir that is not absorbed is converted to acyclovir. When valacyclovir is administered orally, there is a slightly less than dose proportional increase in acyclovir exposure and an increase in T_{max} with increasing doses. The slightly reduced absorption of valacyclovir, with increasing dose is not likely, due to saturable conversion to acyclovir, because of low urinary recovery of valacyclovir, and because the valacyclovir/acyclovir ratio remains the same with increasing dose. The reduced absorption may be due to saturation of absorption sites along the gastrointestinal tract. Although valacyclovir is more soluble than acyclovir (174 mg/mL versus 1.3 mg/ mL), solubility is unlikely to limit absorption of either compound. This is supported by the fact that many of the other amino acid ester analogs that have been studied also exhibit improved solubility over acyclovir. However, these analogs have very diverse bioavailability, probably due to differences in their carrier-mediated transport.

CONCLUDING REMARKS

It is clear that the passive routes of permeation are very restrictive and limited in the capacity for proper significant absorption of peptide-based drugs, particularly across the gastrointestinal epithelium. Based on this, increasing attention has been focused on developing agents that may be able to traverse these barriers through increasing their affinity and capacity for peptide transporters. Until recently, most of the attention for facilitating the active oral absorption of peptide-based drugs has been focused on the PepT1 isoform. Recent evidence has suggested that the presence of other peptide transporters, PHT isoforms and PT1, may require greater consideration. Assessing and delineating the functional relevancy and overlap of these isoforms are needed to properly understand their impact on peptide uptake and absorption. Although attention to proper assessment of the functional relevancy of other oligopeptides transporters is merited, there are several universal considerations that will also hold true in all cases.

The most important consideration is the impact of diet on the function and expression of the transporters, which has been studied with PepT1 and to a lesser extent with PepT2. Hormonal regulation is another key element that may be altered in certain disease states, in particular metabolic syndrome, which could significantly alter the membrane expression and the clinical impact of oligopeptides transporters. Further studies on the functional relevancy of the oligopeptide transporter splice variants and SNPs are also warranted, especially when one considers the two PepT1 splice variants that do convey functional relevancy. Many other factors do exist, including the fact that we are still discovering new oligopeptides transporter isoforms which will require significant efforts for their proper elucidation. In short, it is apparent that considerably more research is required to characterize all of the functional peptide transporter systems to recognize their full clinical impact.

14

Amphipathic Detergents as Carrier

Bile acids are amphipathic physiological detergents that play essential roles in promoting absorption, excretion, and transport of cholesterol, lipids, lipophilic nutrients, and other hydrophobic compounds in the liver and the intestine. The two primary bile acids in humans are *cholic acid* (CA) and *chenodeoxycholic acid* (CDCA). In the intestinal bacterial flora these can be converted to secondary bile acids, *deoxycholic acid* (DCA) and *lithocholic acid* (LCA), respectively. Bile acids are synthesized from cholesterol in the liver and stored in the gallbladder, from which they are postprandially released into the small intestine. From the small intestine the bile acids are recycled back to the liver via portal blood.

A bile acid molecule may shuttle up to a dozen times between the liver and the intestine during one day, each time crossing the membrane domains of hepatocytes and enterocytes. This enterohepatic circulation of bile acids is highly efficient in healthy human individuals: Less than 10% of the total bile acid pool escapes ileal reabsorption, and thus enters the colon. In the colon, secondary bile acids are formed which may be either passively absorbed into colonocytes or lost through fecal excretion. This small loss into feces is compensated for by hepatic de novo synthesis of bile acids from cholesterol, catalyzed by cascades of cytochrome P450 (CYP) enzymes located in endoplasmic reticulum, in mitochondria, or in cytosol. Thus, conversion of cholesterol into bile acids also provides an important route for elimination of cholesterol from the human body. Disruption of the biliary secretion of bile acids may result in cholestasis, in which intrahepatic and systemic accumulation of bile acids causes cytotoxicity. In the liver this may eventually progress to fibrosis and cirrhosis.

BILE IN LIVER AND GUT

Enterohepatic circulation of bile acids is mediated by specific transporters, most of which are integral plasma membrane proteins, expressed in hepatocytes and enterocytes in a polarized manner. The chief transporters involved in bile acid cycling and bile formation discussed in more detail below. Following their synthesis, bile acids are excreted from hepatocytes into bile canaliculi. This represents a major driving force for generation of bile flow and is a rate-limiting step in overall bile acid transport. The efflux of monovalent bile acids from hepatocytes occurs mainly via the canalicular bile salt export pump (BSEP, *ABCB11*), which belongs to the superfamily of ATP-binding cassette (ABC) transporters. The ABC transporters couple the energy released from ATP hydrolysis to their transport activity.

Most eukaryotic ABC proteins share several evolutionarily conserved domains and typically contain two hydrophobic transmembrane domains, each of which spans the membrane six times. While BSEP is responsible for the efflux of monovalent bile acids from hepatocytes into bile, the multidrug resistance-associated protein 2 (MRP2, *ABCC2*) exports divalent and sulfated and/or glucuronidated bile acids as well as other conjugated anions, such as chemotherapeutic agents and antibiotics, into bile. MRP2 is also a member of the ABC transporter family which is located at the canalicular membrane of hepatocytes. In addition to BSEP and MRP2, other transporters localized at the canalicular membrane of hepatocytes are involved in bile formation. Bile acids are the most prominent (60 to 70%) solid component of bile, but hepatic bile also contains phospholipids, notably phosphatidylcholine, and cholesterol.

Phospholipids are translocated from the inner to the outer leaflet of the canalicular membranes of hepatocytes by yet another ABC transporter, the multidrug resistance protein 3 (MDR3, *ABCB4*). The transport system for cholesterol at the canalicular membrane of hepatocytes is the protein heterodimer ABCG5/ABCG8. The efflux of lipids mediated by MDR3 and ABCG5/ABCG8 promotes the formation of mixed micelles containing bile acids, cholesterol, and phospholipids. This process not only solubilizes cholesterol but also helps to protect cholangiocytes against the harmful detergent effects of bile acids in the biliary tree. After their passage to the intestinal lumen, bile acids are efficiently taken up into ileocytes via the apical sodium-dependent bile acid transporter (ASBT, *SLC10A2*).

Characteristic of the SLC10 (solute carrier) family of transporters, ASBT contains seven transmembrane domains, and its bile acid uptake activity is electrogenically coupled with cotransport of sodium. It has been proposed that transcellular shuttling of bile acids from the apical membrane domain of enterocytes to the basolateral membrane may be facilitated by the small cytosolic protein called the ileal bile acid-binding protein (I-BABP). It has been suggested that I-BABP interacts physically with ASBT, although the functional significance of this interaction remains unclear. At the basolateral membrane domain of ileal enterocytes, bile acids are extruded into portal blood by the heterodimeric organic solute transporter OSTα/OSTβ. The OSTα protein is predicted to contain seven membrane-spanning domains, whereas the smaller subunit OSTβ has a single transmembrane domain.

Coexpression of OSTα and OSTβ polypeptides is required for correct localization of the heterodimer at the cell membrane and for transport activity. Consistent with its role as the intestinal bile acid efflux transporter, distribution of the OSTα/OSTβ heterodimer along the intestine closely mirrors that of the bile acid uptake system ASBT. Finally, to complete the enterohepatic circulation, bile acids are extracted from the portal blood circulation by the liver. The sodium-taurocholate cotransporting polypeptide (NTCP, *SLC10A1*) at the sinusoidal membrane of hepatocytes is the chief uptake system for bile acids from portal blood into the parenchymal cells of the liver. 14

NTCP belongs to the same sodium-dependent SLC10 transporter family as ASBT, and is similarly likely to contain seven transmembrane domains and a cytoplasmic carboxy terminus. The amino acid identity between the human NTCP and ASBT is approximately 35%. While NTCP is responsible for the sodium-dependent uptake of bile acids into hepatocytes, certain members of the organic anion transporter (OATP, *SLCO*) family, notably OATP1B1 (gene symbol *SLCO1B1*; previously known as OATP-C/OATP2/*SLC21A6*), may also contribute to bile acid extraction from the portal blood at the basolateral membrane of hepatocytes in a sodium-independent manner. Under normal physiological conditions, only negligible amounts of bile acids are effluxed back to portal blood at the basolateral hepatocyte membrane. However, in cholestatic states, the expression of the two basolateral bile acid overflow systems of the ABC transporter family, multidrug resistance-associated proteins 3 (MRP3, *ABCC3*) and 4 (MRP4, *ABCC4*), is increased.

Thus, in cholestasis, MRP3 and MRP4 may transport substantial amounts of bile acids from hepatocytes back into the systemic circulation for subsequent renal excretion. In addition to MRP3 and MRP4, OSTα and OSTβ are also expressed at the basolateral membranes of hepatocytes in humans. It is thus conceivable that OSTα/OSTβ may also contribute to alternative bile acid efflux during cholestasis. Similar to I-BABP in enterocytes, proteins that are involved putatively in intracellular trafficking of bile acids from the basolateral to the canalicular membrane have been identified in hepatocytes. One such intracellular protein capable of binding bile acids with a high affinity in the human liver is the hepatic bile acid-binding protein (HBAB), which may thus assist in the rapid transcellular vectorial transport of bile acids in hepatocytes.

BILE IN HEPATIC DISEASES

Chronic cholestatic liver diseases such as primary biliary cirrhosis and primary sclerosing cholangitis are characterized by an impairment of bile formation or of bile flow. Altered expression or function of bile acid transporters can be either a cause or a consequence of cholestasis, thus leading to hepatotoxicity due to accumulation of bile acids and cholephilic toxins in hepatocytes. Among the genes encoding transporters that are involved in bile acid transport or bile formation are several that have been identified or proposed as disease genes in the pathogenesis of cholestasis. Progressive familial intrahepatic cholestasis type 2 (PFIC2) is caused by mutations in the *ABCB11* gene, which encodes BSEP. These mutations in the*ABCB11* gene lead to a rapidly progressive hepatic dysfunction in early infancy. In such patients the biliary bile salt levels can be reduced to less than 1% that of normal subjects.

In a recent case report, specific *ABCB11* mutations identified in an adolescent cholestatic patient correlated with reduced BSEP protein expression in vivo and decreased bile acid transport activity in vitro. Another case report suggested that heterozygous BSEP deficiency may predispose to transient neonatal cholestasis. Furthermore, defective or altered function or expression of BSEP may contribute to certain types of drug-induced cholestasis and may be associated with intrahepatic cholestasis of pregnancy. Defective MDR3 expression has also been associated with the inherited liver disease PFIC, type 3. PFIC3 is characterized by high bile acid concentrations and elevated γ-glutamyl transpeptidase activity in serum. Several PFIC3-associated mutations in the *ABCB4* gene may lead to either absent or severely decreased MDR3 expression at the canalicular membrane of hepatocytes. Similar to BSEP, there is increasing evidence suggesting that deficiency of impaired activity of MDR3 may be involved in cholestasis induced by drugs such as oral contraceptives. Inherited mutations in the *ABCC2* gene encoding the canalicular transporter MRP2 are linked to the Dubin-Johnson syndrome, characterized by reduced efflux of conjugated bilirubin into bile. Some of these mutations have been reported to result in an absence of the MRP2 protein from the canalicular membrane of hepatocytes. In contrast to the PFIC syndromes, hepatic function is preserved in the Dubin-Johnson syndrome.

Mutations in the *SLC10A2* gene encoding ASBT have been identified that can cause primary bile acid malabsorption, a rare disorder of the intestine characterized by congenital diarrhoea, steatorrhea and reduced plasma cholesterol levels. The ASBT variants carrying these mutations exhibit severely reduced bile acid transport activity in vitro. No mutations in the *SLC10A1* gene encoding NTCP leading to clinically manifest defects in hepatic bile acid uptake have been characterized thus far. However, a recent study identified ethnicity-dependent single-nucleotide polymorphisms in the *SLC10A1* gene that were associated with a considerable decrease in transport function in vitro. Thus, genetic heterogeneity in the *SLC10A1* gene may play a role in the etiology of hypercholanemia. Furthermore, certain human diseases, such as advanced stage primary biliary cirrhosis and cholestatic alcoholic hepatitis, are associated with reduced NTCP expression. However, this change in NTCP expression may be a consequence of cholestatic liver injury rather than a cause of it.

PHYSIOLOGICAL ROLE OF BILE

In addition to their role as physiological detergents, bile acids possess crucial regulatory properties which allow them to control their own transport and metabolism within the enterohepatic circulation through multiple feedforward and feedback mechanisms. Hepatocytes and enterocytes possess numerous signaling pathways that are activated or modulated by bile acids, and ultimately serve to maintain intracellular concentrations of potentially toxic bile acids at a constant level. An important mechanism toward controlling bile acid levels within cells is to adjust the cellular uptake or efflux of bile acids by regulating the expression and/or activity of uptake and efflux proteins, as discussed in detail below. It should be noted, however, that additional mechanisms are also operational in preventing intracellular bile acid concentrations from reaching toxic levels. One such mechanism is to regulate the de novo synthesis of bile acids according to the existing intracellular bile acid content.

To reduce bile acid synthesis, the expression levels of the key CYP enzymes involved in de novo bile acid synthesis (i.e., CYP7A1, CYP8B1, and CYP27A1) are suppressed. Furthermore, expression levels of several phase II enzymes that in addition to their role in drug detoxification may convert bile acids into less toxic and more hydrophilic derivatives are induced in response to elevated levels of bile acids. These metabolizing enzymes include uridine 5′-diphosphate-glucuronosyltransferase 2B4 (UGT2B4) and dehydroepiandrosterone sulfotransferase (SULT2A1). In this review we focus on the mechanisms that regulate the expression of bile acid transporters at the transcriptional level. However, it should be remembered that the activity of bile acid transporters is also known to be regulated at other levels, particularly through posttranslational protein modification and protein–protein interactions. The relative importance of transcriptional and posttranslational events in controlling bile acid transport activity in either normal physiology or pathophysiology remains largely unelucidated. Both are likely to be highly important. It seems likely that the mechanisms involving modification at the protein level could elicit the most rapid changes in transporter activity, whereas transcriptional changes may be responsible for more intermediate- and longer-term regulation of transport.

MOLECULAR ROLE OF BILE

Nuclear and steroid receptors form a large family of transcriptional regulators, with over 100 members in all metazoan organisms and almost 50 members in humans. Most nuclear/steroid receptors share a conserved overall structural design: a ligand-independent activation function at the amino terminus, a central conserved DNA-binding domain, and a carboxy-terminal region containing regions mediating ligand binding, dimerization, and ligand-dependent transactivation. Most nuclear/steroid receptors bind to their DNA response elements as either hetero- or homodimers, which is reflected in their preferred DNA-binding motifs typically containing two hexameric half sites. These hexamers, the general consensus sequence for which is AGGTCA, can be arranged as direct (DR), inverted (IR), or everted (ER) repeats, separated by a variable and receptor-specific number of base pairs. The full transcriptional activity of most, but not all, nuclear/steroid receptors depends on a physical interaction by an agonist with their ligand-binding pocket.

These ligands are typically small lipophilic molecules such as hormones, fatty acids, oxysterols, or bile acids. Their binding induces a conformational shift in the carboxy termini of the receptors, allowing their interaction with transcriptional coactivators. These coactivators may act by modifying histones or other promoter-associated proteins or by altering local chromatin structure in a way that increases the rate of transcriptional initiation. Conversely, in the absence of an agonistic ligand, or when bound to an antagonistic ligand, the carboxy termini of nuclear and steroid receptors associate with transcriptional corepressors that render the proximal promoters less permissive for transcription. The dependence of the transcriptional activity of most nuclear and steroid receptors on specific ligands allows them to monitor intracellular environment and to elicit rapid transcriptional responses to changes in the concentrations of specific compounds.

Control

The chief sensor of intracellular bile acid levels and the main executor of bile acid- induced transcriptional programs is the nuclear receptor farnesoid X receptor (FXR). Bile acids interact directly with the ligand-binding domain of FXR. In transactivation and coactivator recruitment assays, CDCA is the most efficient FXR activator, followed by DCA and CA. LCA alone can weakly activate FXR; however, it strongly antagonizes CDCA-mediated stimulation of FXR. This apparent antagonism of FXR function may contribute to LCA-induced cholestasis. Bile acids are not the only ligands that interact with FXR directly. Recently, it has been suggested that the oxysterol 22(*R*)-hydroxycholesterol, an intermediate in the synthesis of bile acids and steroid hormones, can also interact directly with the ligand-binding pocket of FXR and mediate gene activation.

Traditionally, oxysterols have been considered to be agonistic ligands for another member of the nuclear receptor family, the liver X receptor (LXR), which functions as a chief regulator of cholesterol homeostasis. Furthermore, androsterone, a testosterone metabolite, can interact directly with the FXR ligand-binding domain and enhance transcriptional activity of FXR through coactivator recruitment. It is possible that different ligands induce different conformational states of FXR, leading to distinct patterns of gene regulation. In this review we only discuss the significance of bile acids as FXR ligands and activators of FXR-induced transcriptional programs. In accordance with its function as a bile acid receptor, FXR is expressed abundantly in the tissues exposed to bile acids: liver, intestine, and kidneys.

The consensus DNA-binding motif for FXR is in the format of an inverted repeat-1 (IR-1, inverted hexameric repeat separated by one base pair), to which it binds as a heterodimer with another nuclear receptor, the retinoid X receptor (RXR). However, other configurations, such as IR-0 and ER-8, may allow binding of FXR in the context of specific promoters. Supporting its physiological role in controlling bile acid homeostasis, FXR-null mice exhibit a phenotype similar to that of persons suffering from Byler disease, an inherited cholestatic liver disorder. Upon feeding with cholic acid, these mice lacking FXR exhibit severely increased hepatotoxicity compared to the wild-type counterparts. In further support of the importance of FXR in human bile acid homeostasis, certain hereditary forms of cholestasis are associated with decreased FXR activity. Encouraging reports showing hepatoprotective effects by synthetic FXR agonists in rodent models of cholestasis suggest that specific FXR ligands could also prove to be useful in the treatment of cholestatic liver diseases in humans.

One major concern regarding FXR-based therapy is that this nuclear receptor is involved in other metabolic processes in addition to bile acid homeostasis. Thus, its activation even with specific ligands may affect these unrelated processes in an undesired manner. For example, negative feedback suppression of bile acid–synthesizing enzymes by bile acids is mediated by a complex cascade involving FXR, leading to reduced conversion of cholesterol to bile acids. In addition to controlling bile acid transporter and synthesis genes, FXR is crucially involved in glucose homeostasis via regulation of the genes encoding gluconeogenic enzymes. In addition, FXR activated by bile acids has recently been shown to be an important mediator of liver regeneration. It was hypothesized that FXR activation by increased levels of bile acids would be a signal of reduced hepatic capacity and function. Thus, until the liver reaches the sufficient number of new hepatocytes to cope with the bile acid levels, FXR continues to trigger their proliferation.

Sensors

FXR is not the only nuclear receptor that can use bile acids as ligands that modulate transcriptional activity. In addition, the pregnane X receptor (PXR) and the vitamin D receptor (VDR) can mediate transcriptional responses to certain bile acids. Together with its xenosensor partner, constitutive androstane receptor (CAR), PXR is a nuclear receptor that typically utilizes drugs and xenobiotics as its ligands. In response to these ligands, PXR induces the expression of genes encoding proteins involved in drug detoxification and elimination pathways. In addition to xenobiotics, certain bile acids, such as the highly toxic LCA, can serve as agonistic ligands for PXR. Indeed, activation of PXR can protect mouse livers against LCA-mediated injury. Double-knockout mice lacking both FXR and PXR exhibit more severe disturbances of bile acid metabolism than mice lacking only one of the nuclear receptors, demonstrating that both contribute to bile acid homeostasis. PXR is a master regulator of the gene encoding the CYP3A4 enzyme, which, in addition to its role in detoxifying xenobiotics, also metabolizes bile acids to less toxic and more easily excreted derivatives. Thus, by being both activators of the *CYP3A4* gene and substrates of the CYP3A4 enzyme, bile acids can initiate a hepatoprotective feedforward loop via PXR. Similar to PXR, VDR can utilize the toxic secondary bile acid LCA as an agonistic ligand. Whereas VDR is expressed only weakly in hepatocytes, it is present abundantly in enterocytes. It is thus likely to play a more prominent role in the intestine in response to LCA.

In addition to PXR, VDR can also transactivate the *CYP3A4* gene. Thus, VDR activated by elevated levels of LCA in the intestine may, in a feedforward manner, contribute to enhanced expression of the CYP3A4 enzyme, which is capable of detoxifying LCA. Common to all three nuclear receptors that can utilize bile acids as ligands (i.e., FXR, PXR, VDR) is that they all bind to their respective DNA response elements as heterodimers with the nuclear receptor RXR, with very few exceptions. Positive or negative effects on the transcriptional activity of RXR-containing heterodimers by the RXR ligand 9-*cis*-retinoic acid appear to depend on the exact promoter context.

Transcriptional Control

FXR can also negatively regulate the rate of transcription from specific promoters. In rare cases, FXR is known to repress its target genes, such as those encoding the human apolipoprotein A-I and human apolipoprotein C-III, through direct binding to the promoter. However, FXR more commonly mediates negative transcriptional responses to elevated levels of bile acids indirectly, through inducing expression of its target gene encoding the transcriptional repressor called *small heterodimer partner*, SHP. SHP is an atypical member of the nuclear receptor family; it does not contain a DNA-binding domain and does not depend on a ligand for its activity.

Instead, it interacts directly with a variety of DNA-bound transcriptional activators, interfering with their transcriptional activity. Although SHP can suppress the activity of transcription factors from certain other families, it most commonly targets other nuclear receptors and steroid receptors. Via the SHP pathway, bile acids can influence the activities of a wider range of transcription factors than only those nuclear receptors that are modulated by them directly. Several mechanisms have been suggested for SHP-mediated suppression of trans-activator proteins. SHP may compete with transcriptional coactivators over the same or overlapping interaction surface on transactivators that are bound to the promoter elements.

Alternatively, SHP may interfere with the binding of trans-activators to their DNA response elements. Additionally, SHP may recruit transcriptional corepressors to its target promoters, thus contributing to the reduced transcriptional rate of a given promoter. The importance of SHP in the control of bile acid homeostasis is indicated by the fact that SHP-null mice exhibit an imbalance in bile acid metabolism and abnormal responses when challenged with diets rich in bile acids.

GENE IN BILE TRANSPORT

We discuss next the FXR-dependent effects of bile acids on the expression of transporter genes in hepatocytes and enterocytes. It should be emphasized that although we focus on FXR-dependent effects by bile acids in this review, bile acids also elicit signaling pathways that lead to changes in specific gene expression which are independent of FXR or other nuclear receptors. These alternative bile acid-stimulated pathways include signaling through mitogen-activated protein kinases and through the G-protein-coupled receptor TGR5. Parallel bile acid-stimulated signaling pathways may ensure that the desired transcriptional responses are achieved.

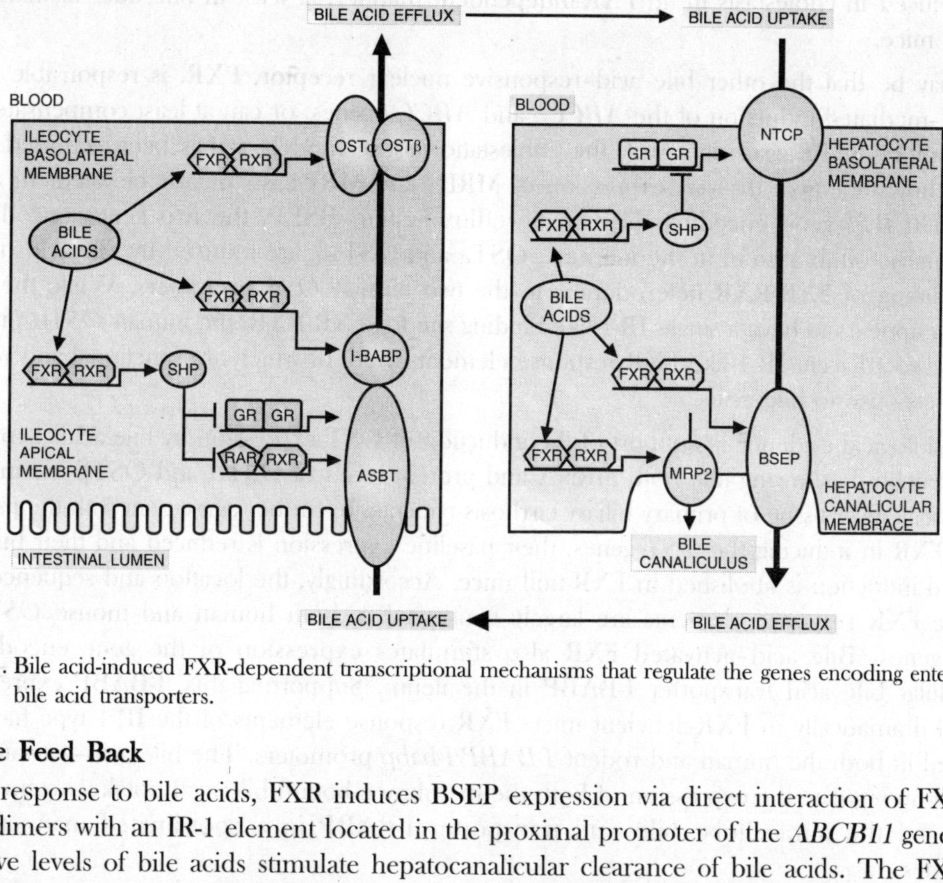

Fig. 14.1. Bile acid-induced FXR-dependent transcriptional mechanisms that regulate the genes encoding enterohepatic bile acid transporters.

Positive Feed Back

In response to bile acids, FXR induces BSEP expression via direct interaction of FXR-RXR heterodimers with an IR-1 element located in the proximal promoter of the *ABCB11* gene. Thus, excessive levels of bile acids stimulate hepatocanalicular clearance of bile acids. The FXR-RXR binding element is conserved between the human and rodent*ABCB11/Abcb11* promoters, supporting its functional importance. In agreement with FXR being a crucial activator of the *ABCB11* gene, the BSEP expression is reduced in mice lacking the *FXR* gene. The *ABCC2* gene encoding the MRP2 transporter provides an illustrative example of the complex interactions between the metabolic nuclear receptors. Both the human and rodent *ABCC2/Abcc2* promoters can be activated by either FXR, PXR, or CAR in the presence of their respective ligands or activators. Interestingly, each of these three nuclear receptors can interact with the same atypical ER-8 element present in the regulatory region of the *ABCC2* gene.

The relative significance or the degree of redundancy between the three transcription factors in the regulation of MRP2 expression is not yet clear. FXR also transactivates the *ABCB4* gene encoding MDR3, the phospholipid transporter at the canalicular membrane of hepatocytes. Thus, bile acids may, in a coordinated manner and via activation of FXR, induce the excretion of both bile acids (BSEP, MRP2) and phospholipids (MDR3) into bile. In agreement with this, FXR-null mice are prone to developing cholesterol gallstones, caused by the deficiency in the excretion of bile acids and phospholipids. Expression of the genes encoding the alternative bile acid export pumps at the basolateral hepatocyte membranes, MRP3 (*ABCC3*) and MRP4 (*ABCC4*), appears to be induced in cholestasis in an FXR-independent manner, at least in bile duct-ligated or bile acid-fed mice.

It may be that the other bile acid-responsive nuclear receptor, PXR, is responsible for the bile acid-mediated induction of the *ABCC3* and *ABCC4* genes, or can at least compensate in the absence of FXR. In agreement with the cholestatic mouse models, it has been reported that in human cholestatic liver disease, expression of MRP3 and MRP4 is similarly elevated. In analogy to the *ABCB11* gene encoding the hepatic efflux system BSEP, the two genes encoding the heterodimeric efflux system in the intestine, OSTa and OSTI3, are induced by bile acids through direct binding of FXR-RXR heterodimers to the two human *OST* promoters. While the *OST*β promoter appears to have a single IR-1-like binding site for FXR-RXR, the human *OST*α promoter contains two adjacent IR-1-like FXR response elements, both of which are functional and required for full response to bile acids.

Physiological evidence in support of the induction of OST expression by bile acids is provided by a recent study showing that both mRNA and protein levels of OSTα and OSTβ are increased in cholestatic liver tissue of primary biliary cirrhosis patients. In further agreement with the proposed role of FXR in inducing the *OST* genes, their baseline expression is reduced and their bile acid-mediated induction is abolished in FXR-null mice. Accordingly, the location and sequence of the IR-1-like FXR response elements are largely conserved in both human and mouse *OST*α and *OST*β genes. Bile acid-activated FXR also stimulates expression of the gene encoding the intracellular bile acid transporter I-BABP in the ileum. Supporting this, I-BABP expression is reduced dramatically in FXR-deficient mice. FXR response elements of the IR-1 type have been identified in both the human and rodent *I-BABP/I-babp* promoters. The bile acid-activated FXR can thus up-regulate the expression of both the membrane-bound bile acid uptake system OSTα/ OSTβ and the intracellular bile acid transporter I-BABP in a coordinated manner within enterocytes.

Negative Feedback

In rodent models of cholestasis, such as bile duct ligation or bile acid feeding, expression of the hepatic bile acid uptake system Ntcp is suppressed at both the protein and mRNA levels. Furthermore, certain human cholestatic states, such as advanced- stage primary biliary cirrhosis and cholestatic alcoholic hepatitis, are similarly associated with reduced NTCP expression. It has recently been suggested that the mechanism of down-regulation of human NTCP expression in response to bile acids involves the FXR-SHP cascade. FXR-induced SHP targets the transactivator glucocorticoid receptor (GR), which interacts with its response element just upstream of the transcription initiation site on the human *SLC10A1* promoter. FXR also suppresses transcription of the *SLCO1B1* gene encoding the sodium- independent bile acid transporter OATP1B1 at the basolateral membrane of human hepatocytes. FXR-mediated repression of the *SLCO1B1* promoter takes place through a multistep regulatory cascade which involves bile acid-mediated repression of the gene encoding the liver-enriched homeodomain factor hepatocyte nuclear factor 1α (HNF1α). HNF1α, in turn, is a strong DNA-binding transactivator of the *OATP1B1* promoter.

The regulatory region of the *HNF1α* gene contains a consensus DNA-binding response element for the nuclear receptor HNF4α, the transcriptional activity of which can be targeted by negative interference by FXR-induced SHP. Similarly to NTCP, treatment of cultured cells with bile acids suppresses expression of the intestinal bile acid uptake transporter ASBT, another member of the SLC10 transporter family in humans. In accordance with this, *SLC10A2* gene expression is reduced in patients with obstructive cholestasis compared to healthy controls. Two transactivators of the human *SLC10A2* promoter, the GR and the nuclear receptor heterodimer retinoic acid receptor (RAR)-RXR, have been suggested as targets for negative interference by the bile acid-induced transcriptional repressor SHP. It remains to be determined whether these two transactivators are targeted simultaneously by SHP to ensure efficient down-regulation of the *SLC10A2* promoter by bile acids, or whether the two pathways are targeted under different circumstances. Interestingly, while both the *SLC10A1* (NTCP) and *SLC10A2* (ASBT) genes are transactivated by GR, the relative locations and configurations of the GR response elements within the two promoters are not conserved.

INTERACTIVE ROLE OF BILE

Transport and metabolism of exogenous compounds, such as drugs, nutrients, and environmental xenobiotics, bear many similarities to those of endogenous bile acids. Indeed, drugs may undergo enterohepatic circulation similarly to bile acids. Several transporters recognize both bile acids and drugs as substrates. For example, although their significance for overall bile acid transport in normal physiology remains uncertain, the two transporters of the human OATP1B subfamily, OATP1B1 and OATP1B3, can transport bile acids in addition to drugs such as methotrexate and rifampicin. The transcriptional regulatory circuits controlling the expression of drug transporter genes often contain feedforward and feedback loops reminiscent of the mechanisms outlined for regulation of bile acid transport above.

It is becoming increasingly apparent that in addition to reprogramming of genes involved in bile acid homeostasis, changes in intracellular concentrations of bile acids also influence the expression levels of hepatic and intestinal drug transporters. Thus, changes in intracellular bile acid levels may affect the efficiency of drug extraction and elimination. HNF4α is a liver-enriched nuclear receptor with a critical role in maintaining the hepatic pattern of gene expression. It binds to the DNA response elements on its target promoters, preferably of the DR-1 or DR-2 configuration, as homodimers. The ligand-binding domain of HNF4α has been suggested to be bound constitutively by endogenous fatty acids; thus, its activity may not be modulated readily by exogenous ligands. In addition to its previous roles in regulation of glucose and cholesterol metabolism, HNF4α has recently emerged as a regulator of hepatic drug transport in humans.

The genes encoding two major hepatic basolateral drug transporters in humans, organic anion transporter 2 (OAT2; *SLC22A7*) and organic cation transporter 1 (OCT1; *SLC22A1*), are transactivated by HNF4α. OAT2 mediates uptake of drugs such as salicylates and cephalosporins from sinusoidal blood, whereas OCT1 transports dopamine, metformin, and verapamil, among other substrates. In the regulatory regions of the human *SLC22A7* gene, HNF4α transactivation is mediated by a single DR-1 element. The human *SLC22A1* promoter, in turn, contains two adjacent lower-affinity DR-2 elements, both of which are required for maximal transactivation by HNF4α. In both promoter contexts, the bile acid-induced transcriptional repressor SHP interferes with transactivation by HNF4α. It thus appears that in conditions of elevated intracellular bile acid concentrations, the expression of two major drug uptake systems at the basolateral hepatic membrane is reduced. This could limit the amount of xenobiotics that enters the hepatocyte for subsequent metabolism when intracellular levels of toxic bile acids are already elevated. Also, the possibility of decreased hepatic extraction of drugs that are substrates of OAT2 and OCT1 should be taken into account when such drugs are administered to patients suffering from cholestasis.

CONCLUDING REMARKS

Enterohepatic bile acid transporters are crucial in maintaining bile acid homeostasis in healthy persons. Their importance for hepatic and intestinal function is emphasized by numerous recent demonstrations that either hereditary or acquired disturbances in the activity and/or expression of bile acid transporters are associated with cholestatic disease conditions. Bile acids regulate the expression of transporters that mediate the enterohepatic circulation of both bile acids and drugs via complex feedforward and feedback mechanisms. The main orchestrator of these transcriptional circuits is FXR, which, in response to bile acids, controls the expression levels of bile acid transporters at all membrane domains of enterohepatic circulation. Increased knowledge of the transcriptional mechanisms governing changes in bile acid transporter gene expression has improved our understanding of the pathophysiology of cholestatic liver diseases. Additionally, this information is likely to provide us with novel tools for designing therapeutic strategies to combat diseases of the liver and the intestine.

15

Drug Resistance

Metallothioneins (MTs) are a family of proteins that bind some, but not all, heavy metal ions essential for eukaryotic cell function (for example, zinc and copper), and some that are both toxic and not required for cell function (for example, cadmium and mercury). A role for MTs in metabolism and detoxication of heavy metals is strongly suggested by the sensitivity of many MT genes to induction by heavy metals and the ability of MT proteins to bind to many inducing metal ions. However, MT genes are also induced by nonmetal toxins and the expression of MTs varies during normal physiological events (proliferation, differentiation, and cell cycle), suggesting a role or roles not directly related to heavy metal stress. One such role may be the homeostatic regulation of zinc availability.

Assessment of function of cells with abrogated MT expression (antisense downregulation of MT and MT gene knockout), increased MT expression by virtue of transient or stable transfection of heterologous MT expression vectors, and in vitro observation of direct and indirect interaction of MT protein with cellular zinc-requiring enzymes and transcription factors has implicated MTs in events modulating resistance to anticancer drug therapy, including zinc-dependent monocyte/macrophage activation, hormone responsiveness, and transcription factor activity.

Evidence exists to suggest that MT regulates immune cell functions by mediating the activity of signal transduction proteins and transcription factors involved in monocyte activation; participates in resisting the effects of damage induced by toxins by regulating the function of the zinc-sensitive transcription factor metal transcription factor 1, the antiapoptotic protein nuclear factor-κB, and the tumor suppressor protein p53, and signaling through the glucocorticoid hormone receptor and other possibly other hormone receptors; and mediates these events in whole or in part by regulating zinc. The importance of inflammation, hormone response, antiapoptotic and zinc-dependent transcription factor function, and zinc regulation in cellular resistance to toxins, coupled with understanding of how MT influences them, sets the stage for rational therapeutic targeting of MT to enhance cancer treatment while sparing normal tissues.

METALLOTHIONEIN PROTEINS

Metallothioneins (MTs) are a family of small (less than 10 kDa), metal-inducible, metal-binding proteins found in a wide variety of organisms including bacteria, fungi, and eukaryotic plants and animals. They are typically composed of 61 or 62 amino acids (depending on the organism and MT isotype) organized into two globular domains (α and β). Both domains contain a high proportion of cysteine residues–eleven in the α-domain and nine in the β-domain–that are responsible for noncovalent, high-affinity metal ion binding by digonal, trigonal, and tetrahedral associations. In mammals, zinc (and, to a minor degree, copper) are the predominant metals bound to MT in the absence of appreciable levels of other metal ions with higher affinity (for example, silver, mercury, copper, and cadmium). Notably, MTs are the single most abundant group of intracellular zinc-binding proteins in eukaryotic cells. Five to 10% of zinc in human hepatocytes is bound to MTs.

An essential biological function for MTs has not been identified, although the remarkable homology among MT protein isotypes and across species implies that such a function or functions exists. Although MTs have long been associated with resistance to toxicity resulting from exposure to toxic metals and generators of reactive oxygen, their capacity to interact directly with metal ions and oxygen radicals has been taken as evidence that they protect by acting simply as "*sacrificial scavengers*" to intercept and directly inactivate toxic molecules.

More recently, it has been suggested that MTs play an indirect role by controlling zinc bioavailability to zinc-requiring proteins that act in a broad range of physiological events, including proteins that themselves directly mediate resistance to toxic events and those that act indirectly by receiving and transducing extracellular signals that alter cellular resistance to toxicity (transcription factors, hormone receptors, metalloproteinases, superoxide dismutase and catalase, among others). In this chapter, we review the evidence surrounding this concept. Understanding the role(s) of MTs in events mediated by these molecules can, potentially, lead to targeting MT expression and/ function as a therapeutic anticancer drug therapy.

Genes for Metallothionein Proteins

Humans possess 16 MT genes clustered on the q13 region of chromosome 16, including at least 11 that encode MT-1 (*MT-1A, -1B, -1E, -1F, -1G, -1H, -1I, -1J, -1K, -1L,* and *-1X*, some of which appear to be incompetent in directing production of MT protein), and single copies of MT-2 (also referred to as MT-2A), MT-3, and MT-4. Related genes (metallothionein-like 5 [*MTL-5*], encoding tesmin protein; and *MT-M* and *MT-E*) have also been reported. Rodents have only four MT genes, clustered on chromosome 8 (*MT-1, MT-2, MT-3,* and *MT-4*), although additional MT-related mouse genes (mouse *MTL-5* and "MT-1 activator") have been reported. MT-1 and MT-2 are expressed at basal and inducible levels in virtually all tissues, with highest capacity for expression in liver.

MT-3 and MT-4 are restricted to certain tissues, with MT-3 produced predominantly in brain in glutaminergic neurons, and at very low levels in pancreas and intestine. MT-4 is limited to squamous epithelial cells of the skin and tongue. A role for MTs in detoxication of metals is supported by the ability of MT to bind to, and be induced by, heavy metals. However, MT-1 and

MT-2 are expressed in virtually all tissues and are correlated with a wide variety of physiological events not directly associated with toxic metal insult. Increased expression of both are associated with proliferation, without metal induction, in a human prostate stem cell line, in rat kidney undergoing compensatory hypertrophy, in proliferating human cancer cells, and in human monocytes undergoing respiratory burst.

The level and intracellular location of MT-1 and MT-2 is developmentally regulated in some cells and tissues. MTs may protect against carcinogenesis, and nuclear MT has been proposed to regulate and/or protect against nuclear oxidant events during cell cycle progression, including oxidant-induced nuclear zinc release—MTs are capable of binding and inactivating reactive oxygen intermediates and may also protect against these toxic species. MTs are, therefore, associated with proliferation, hypertrophy, differentiation, immune cell activation, and resistance to toxins of multiple types in addition to toxic heavy metals.

BIOCHEMICAL SIGNIFICANCE

The multiple circumstances under which MTs are expressed has fuelled speculation about their function(s). Notably, MTs are bound primarily to zinc in mammalian cells, suggesting a role in homeostasis of this multifunctional metal ion. A broad range of proteins important in the processes described above require zinc for activity and zinc- associated MTs could supply zinc to (or, depending on relative affinity, sequester zinc away from) mammalian metalloproteins, including transcription factors and hormone receptors. Unlike strongly redox-active ionic copper and iron, the participation of zinc ions in oxidation-reduction reactions does not readily lead to the formation of damaging free radicals as side products. In addition, zinc can act as an antioxidant to inhibit oxidation by other redox-active metal ions.

Zinc competes with redox-active metal ions (including those of iron and copper) for binding to cellular macromolecules, decreasing their capacity to generate toxic hydroxyl radicals (HO•) through Fenton-like reactions, and association of zinc with sulfhydryl groups also protects proteins (including δ-aminolevulinate dehydratase, alanyl tRNA synthetase, farnesyltransferase, tubulin, dihydroorotase, and zinc finger proteins) against oxidative denaturation. Zinc, therefore, contributes to the structure and activity of enzymes and DNA-binding proteins with less danger of oxidative damage to DNA and other molecules required for cell viability and normal function: a quality that may explain the fact that zinc is an essential cofactor in over 3000 signal transduction proteins and transcription factors, and more than 300 enzymes, including copper/zinc superoxide dismutase with activity in preventing superoxide radical damage, and DNA repair proteins responsible for maintaining the restoring DNA integrity after damage. Zinc is a dietary micronutrient that is absorbed by the intestine, associated primarily with albumin in blood, and transported into and within cells by the ZRT-, IRT-like protein, and cation diffusion facilitator families of zinc transporters.

Albumin has a relatively low affinity for zinc and zinc-albumin constitutes a small, readily available zinc pool that can be rapidly exhausted to produce mild to severe deficiency. Zinc deficiency is, in fact, a reality in a significant portion of the North American population, with 10% of individuals consuming less than half the recommended level. Deficiency is associated with a broad range of clinical disorders (including sickle cell anemia, renal disease, chronic gastrointestinal disorders and acrodermatitis enteropoathica, and HIV infection) and is common in children with diarrhea and among the elderly. Zinc status is tightly linked to antibody and cell-mediated immune integrity in humans and animals. Furthermore, zinc deficient cell culture and rodent models have increased susceptibility to oxidative damage to DNA and proteins, and zinc supplementation offers protection against progression of pathologies associated with free radical-induced damage, including age-related macular degeneration in humans, chemically induced diabetes in mice, and UV-induced DNA damage and death in human fibroblasts in culture.

The antioxidant capacity of zinc, coupled with the requirement of multiple proteins regulating cellular responses to redox damage for zinc, suggest that zinc deficiency can contribute to the development and progression of diseases (including cancer) by enhancing damage to genes and proteins, and by altering or impairing zinc-dependent responses to that damage. The urgent mammalian requirement for zinc, coupled with the ease with which it is depleted, suggest that multiple mechanisms exist to maintain and regulate the cellular supply of zinc ions. Evidence exists to support the concept that MTs are homeostatic molecules, and alteration of MT levels in cells will affect the activities of zinc-requiring proteins that mediate cell functions including signal reception and transduction, gene transcription, and immune cell activation.

OTHER FUNCTIONS

DNA metabolizing and signal transduction proteins require zinc. For example, *glucocorticoid receptor* (GR) and estrogen receptor, among other members of a multigene hormone receptor family, require association with zinc for stability and activity. Removal of zinc from GR by chelation produces an apoprotein that does not bind to glucocorticoid response elements in DNA. DNA binding is restored by zinc addition. Although in vivo requirement of MTs for hormone receptor function has not been demonstrated, MT genes are classically inducible by glucocorticoid hormones and hormone analogues, the cellular level of zinc-associated MT is positively correlated with glucocorticoid receptor activity in rodent cells, and MTs have been demonstrated in vitro to donate or remove zinc from hormone receptors (reversible zinc exchange in vitro between MT-2 and the estrogen receptor zinc finger has been reported in vitro).

Although a role for MT in regulating the zinc status of hormone receptors is an intriguing possibility, there is no direct evidence MTs suppress or enhance receptor activity by modulating zinc availability–under normal conditions, zinc might be directly available to zinc-requiring proteins, without association with "*zinc gatekeepers*" or "*zinc chaperones.*" A critical question arises: Is zinc association with zinc-requiring proteins regulated to control the activity of those proteins? For this to be possible, the availability of zinc within cells must necessarily be limited. A readily available pool of zinc for all proteins requiring them would frustrate strategies to control access to zinc ions. However, virtually all cellular zinc is associated with macromolecules and "free" zinc is extremely limited.

Storage of zinc ions by association with high-affinity molecules, coupled with direct or indirect transfer from that binding species to zinc-requiring (and possibly zinc-regulated) molecules would constitute a point of potential control, especially where zinc levels are limited. Furthermore, chaperone-mediated regulation of essential metals has been observed and described. Copper, for example, is regulated by specific transporters and intracellular chaperones elucidated in yeast, but which imply analogous metal ion regulation in higher eukaryotes, including humans. Evidence now suggests that MTs may act as intracellular zinc chaperones, regulating zinc availability to proteins that mediate resistance to damage and cellular responses to damage events.

EXPRESSION OF MT GENE

MT-1 and *MT-2* genes are widely expressed in many tissues and cell types. Expression is maintained at basal levels in the absence of administration of exogenous agents, and in response to an exceptionally broad range of signaling events mediated by multiple transcription factors including AP1, AP2, SP1, steroid hormone receptors, cytokines, and others. Transcription factors that mediate both basal and induced MT gene expression interact with multiple promoter/enhancer elements, often involving transcription factor interactions that are poorly understood. However, it is clear that metal response elements ([MREs] present in multiple copies in *MT-1* and *MT-2* genes), antioxidant response elements (overlapping with an upstream stimulatory factor-binding site in *MT-1* genes), and glucocorticoid response elements, and signal transducer and activator of transcription response elements are of high importance, in addition to others.

Variations in chromatin structure, DNA methylation, and poorly understood po sttranscriptional events also appear to modulate MT expression. Stress from heat, cold, physical trauma and other conditions–including oxidative stress induced by administrative of exogenous chemicals that generate hydroxyl radicals (hydrogen peroxide) and superoxide radicals (*tert*-butyl hydroquinone and menadione), and mitochondria-specific reactive oxygen generators (antimycin A and 2,4-dinitrophenol)–induce MTs in many organs in whole animals. MTs are classically induced by soluble metal ions to which they bind (zinc, cadmium, mercury, copper, and Bi), and by some metals that do not bind to MTs (nickel and cobalt). These metal ions enter cells by multiple processes including specific eukaryotic metal ion transporters. MREs embedded in the promoters of MT and other genes bind the transcription factor metal transcription factor (MTF)-1, which is essential for basal, metal-induced, and oxidant stress-induced MT induction.

ROLE OF METAL IN TRANSCRIPTION

MTF-1 is an MRE-binding protein that is ubiquitously expressed in mammalian cells. Homozygous knockout of MTF-1 genes in mice is lethal to embryonic mice because of the essential nature of the protein for liver development and stress responses. On the other hand, MTF-1 can be eliminated in adult mice without compromising viability, although MTF-1 conditional knockout mice have increased susceptibility to heavy metal stress. MTF-1 appears to play dual roles–one in liver development that is essential for viability, and the other in response to cell stress that is required only under those stress conditions. MTF-1 contains, in the N-terminal region, six Cys2-His2 "*zinc fingers*" that mediate noncovalent interaction of the protein with DNA.

MTF-1 protein is maintained by constitutive MTF-1 gene transcription and mRNA translation in human and rodent cell lines, although the MTF-1 protein is prevented from mediating high basal MT gene activity, limiting its access into the nucleus. Stress or metal ion exposure induces MTF-1 translocation from the cytoplasm to the nucleus where MTF-1 binds to MREs in the promoter region of *MT-1* and *MT-2* genes, including direct or indirect interactions with components of the RNA polymerase II transcriptional complex. Although MTF-1 binding to DNA is reversibly activated by a variety of metal ions that induce MTs, zinc (and only zinc) directly binds to MTF-1. Of the six zinc fingers, four or five are normally bound to zinc under physiological conditions. On the other hand, one (and possibly two) is a "*zinc-sensing*" finger that is not occupied by metal in the absence of excess added zinc, even when zinc levels are adequate to maintain viability.

MTF-1 is, therefore, both zinc dependent (i.e., requiring zinc for activity) and zinc sensitive (i.e., has variable activity within the range of physiological variation in intracellular zinc concentrations), and variations in zinc availability regulate MTF-1 activity. However, the mechanism(s) by which MTF-1 is regulated appear to be more complex than simple zinc association. MTF-1 can be phosphorylated on metal induction through activation of a kinase signaling cascade including protein kinase C (PKC), phosphoinositiol-3 kinase, c-jun N-terminal kinase, and a tyrosine-specific kinase, suggesting that metal ions can activate MTF-1 by stimulating kinase activity.

Release of MTF-1 from inhibitory regulatory molecules has also been proposed as process to regulate activity. Although it is clear that zinc is a critical component of MTF-1 activation, its association with MTF-1 does not appear to be the sole activating event. MTF-1 activates, not only MT genes, but also other genes important in response to stress and resistance to toxicity. γ-Glutamyl-cysteine synthetase heavy chain (γ-GSChc) is a key enzyme for synthesis of the radical scavenger glutathione (GSH), and has been suggested to be MTF-1-regulated. GSH has been implicated in resistance to several chemotherapeutic agents, including platinum-containing compounds, alkylating agents (for example, melphalan), anthracyclines (including doxorubicin), and arsenic.

A requirement for GSH in multidrug resistance protein-mediated anthracycline resistance (and possibly arsenic resistance) through efflux of these toxic agents has suggested by demonstration of impaired multidrug resistance protein activity following GSH depletion. On the other hand, GSH levels in embryonic livers in MTF-1 knockout mice have been observed to be at least as high as in wild-type littermates. This suggests that MTF-1 direction of γ-GCShc expression may not be critical, at least under these developmental circumstances. In addition to γ-GCShc, the genes encoding the transcription factor C/EBP-α, embryonic α-fetoprotein, the zinc transporter ZnT-1, and placental growth factor are also likely targets of MTF-1. C/EBP-α participates in maintaining the differentiated, nonproliferative state in hepatocytes, and is induced during acute phase response, indicating a role in response to stress. α-Fetoprotein, in addition to other functions, is a scavenger of heavy metal ions and reactive oxygen. ZnT-1 plays an important role in regulating intracellular zinc levels and metabolism, and placental growth factor is a member of the vascular endothelial growth factor family of angiogenic factors and is induced in an MTF-1-dependent manner in fibroblasts. Furthermore, MTF-1 appears to be directly involved in tumor development.

Loss of MTF-1 results in increased activity of transforming growth factor β1 and tissue transglutaminase (both of which are involved in extracellular matrix production), and delayed tumor growth and reduced vasculature density in *ras*-transformed mouse embryonic fibroblasts. Overall, MTF-1 appears to be a critical transcription factor, not only in mediating cellular responses to stresses induced by a variety of toxins including chemotherapeutic agents, but also in physiological events contributing to tumor growth and malignancy. The activity of MTF-1 and other zinc-binding transcription factors depends on a metalloregulatory protein or proteins that do not, themselves, have DNA-binding or transactivation capacity. Such a metalloregulatory protein would respond to an array of disparate signals by adding or removing zinc, and only zinc, from zinc-requiring transcription factors.

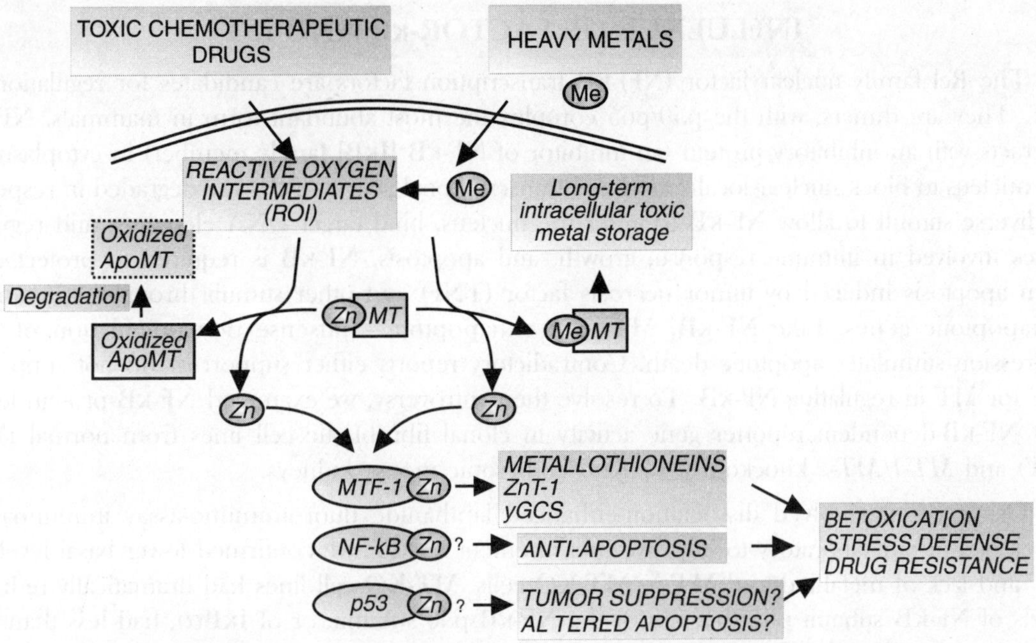

Fig. 15.1. Model of proposed "metal sensor" and "zinc regulatory" roles for metallothionein.

Metallothioneins are candidates for this role. They bind different metals with different avidity—zinc associates with MT with a stability constant (~1011) that is ~100-fold less than that of copper, mercury, silver, or cadmium. Zinc displaced by binding to higher-avidity metals or by redox conditions present during stress would be available to zinc-dependent proteins. Such a high affinity, metal- and stress-responsive zinc carrier, capable of supplying zinc to both high avidity proteins (by direct transfer) or to low-avidity proteins (by release of zinc after reactive oxygen interaction with MTs) would conceivably be important in maintaining an intracellular store of zinc when dietary sources are limited. In support of this, *MT-1*/*MT-2* knockout mice have heightened sensitivity to zinc deficiency leading to retarded kidney development and *in utero* mortality.

In addition, in vitro transcriptional activation of MRE-driven genes by MTF-1 requires zinc-saturated MT—free zinc alone, or zinc-deprived MT, are not sufficient. MT appears, therefore, to be a necessary factor in regulating zinc availability to zinc-dependent MTF-1. Similarly, in rodent cells transfected with a reporter gene driven by an MTF-1-responsive promoter, increased MT-1 levels enhanced zinc-induced gene expression, an observation that correlates MT with MTF-1-mediated transcription in a concentration-dependent manner. Furthermore, mouse fibroblasts adapted to growth in severely zinc-depleted medium have amplified MT genes and elevated MT gene expression, suggesting that MT confers an advantage in coping with scarce zinc, possibly by acting as a high affinity storage depot to conserve zinc for selected physiological processes necessary for survival under conditions when dietary zinc is not readily available, is in high demand because of stress, or both. Thus, MTs can mediate zinc-responsive gene expression and may regulate zinc availability, especially under low environmental zinc conditions where increased zinc transport would have diminished capacity to increase intracellular zinc levels.

INFLUENCE OF FACTOR-κB AND p53

The Rel family nuclear factor (NF)-κB transcription factors are candidates for regulation by MT. They are dimers, with the p50/p65 complex the most abundant form in mammals. NF-κB interacts with an inhibitory protein (an inhibitor of NF-κB [IκB] family member) in cytoplasm or the nucleus to block nuclear localization and transcriptional activity. IκBs are degraded in response to diverse stimuli to allow NF-κB to enter the nucleus, bind target DNA elements, and regulate genes involved in immune response, growth, and apoptosis. NF-κB is required to protect cells from apoptosis induced by tumor necrosis factor (TNF) and other stimuli through activation of antiapoptotic genes. Like NF-κB, MTs are antiapoptotic–antisense down-regulation of MT expression stimulates apoptotic death. Contradictory reports either support or do not support a role for MT in regulating NF-κB. To resolve the controversy, we examined NF-κB protein levels and NF-κB-dependent reporter gene activity in clonal fibroblastic cell lines from normal (MT-WT) and *MT-1/MT-2* knockout (MT-KO) embryonic mouse kidneys.

Using a time-resolved dissociation-enhanced lanthanide fluoroimmuno-assay immunoassay developed in our laboratory to improve measurement of MT, we confirmed lower basal levels of MT and lack of metal-induced MT in MT-KO cells. MT-KO cell lines had dramatically reduced levels of NF-κB subunit p65 (but not of the NF-κB p50 subunit, or of IκBα), had less than half the expression of a transfected NF-κB-dependent reporter, and were more sensitive to apoptosis induced by *tert*-butylhydroperoxide. Decreased nuclear localization of p65 in MT-KO clones was not responsible for differences in NF-κB activity. In fact, MT-KO cells had higher nuclear levels of p65 than MT-WT cells in spite of lower overall cellular levels, suggesting that MT regulates the specific activity of NF-κB.

Reconstitution of MT by stable incorporation of mouse MT-1 expression in multiple clonal populations of MT-KO cells (MT-KO_2^{+mt}) increased NF-κB p65 (but not IκBα or NF-κB p50), and increased NF-κB-dependent reporter activity and resistance to apoptosis. These observations strongly support a role for MT in regulating NF-κB subunit levels and NF-κB function. The mechanism by which MTs mediate NF-κB is not known, but there is a wealth of evidence to suggest that they regulate zinc. However, in vitro evidence reveals that MTs can both donate and sequester zinc depending on relative affinity, overall zinc status, and cellular location of MTs and zinc-dependent proteins. MT-1 and MT-2 bind 7 zinc atoms, but only one or two are readily available for transfer. The remaining zincs require disruption of zinc-thiol coordinate covalent bonds for release.

It is possible that MTs with an incomplete complement of zinc (apo-MT) could sequester zinc when it is scarce, lowering zinc availability to other proteins. For example, moderate overexpression of MT-2 in a human breast cancer-derived cell line (MCF-7) enhanced the transcriptional activity of the zinc-dependent transcription factor and tumor suppressor protein p53, but high-level overexpression suppressed p53 activity. Zinc- depleted MT has also been reported to selectively activate enzymes in which zinc is bound at an inhibitory site, without removing zinc from the catalytic site of metalloenzymes–an effect suggested by the authors to be because of selective sequestration of zinc by MT. Finally, cells selected for survival under low zinc conditions had high MT expression, but transcription of a transfected MTF-1-responsive reporter construct was suppressed, consistent with the interpretation that, in low zinc, MT sequestered scarce zinc ions away from zinc-sensing MTF-1.

These observations are consistent with the hypothesis that, in low zinc, zinc-dependent transcription factor activity (including transcription factors such as p53 and NF-κB that have important roles in modulating apoptosis and sensitivity to toxins) may be (i) more dependent on MT than in the presence of sufficient zinc (if MT has a full complement of seven zinc ions, with one or two available for transfer to zinc-requiring proteins), or (ii) inversely dependent on MT (if MT binds less than seven zinc ions, bound in the molecule's interior and unavailable for ready transfer). If either of these hypotheses is correct, then the case for zinc regulation by MT to mediate NF-κB and/or will be strengthened. If not, then the alternative hypothesis that MTs regulate NF-κB activity independent of zinc will grow in importance. These possibilities have implications for NF-κB function in the face of variable zinc levels in humans. Investigation of the role of MT in regulating NF-κB activity through zinc regulation, by assessing NF-κB function in cells with altered MT and depleted or supplemented zinc levels, is an important area for further investigation.

ROLE OF STEROIDAL HORMONES

A mouse mammary tumor cell line (2305) with a stable, mouse mammary tumor virus- driven vector containing a chloramphenicol acetyltransferase (*CAT*) reporter gene has been used to test the role of MT in GR activity. This vector responds to the artificial glucocorticoid dexamethasone (DEX), which stimulates GR translocation to the nucleus and GR-mediated transcription. An antisense MT-1 expression vector (capable of down-regulating both MT-1 and MT-2) reduced MT-1 mRNA to only 40% of the amount in control cells. This resulted in a small (approximately 10%) but significant decrease in DEX-induced CAT activity. Considering that zinc in fetal bovine serum medium (~4 μM) could be an alternative source and diminish the putative role of MT as a zinc regulator, the serum was chelated to reduce zinc to less than 0.3 μM. Under these conditions, DEX responsiveness was reduced by 25% in MT-expressing cells–a nearly threefold enhancement of suppression. Furthermore, zinc, mercury, and heat shock treatment of 2305 cells (but not copper or cadmium induction) enhanced response of a DEX-inducible CAT reporter construct. The significant factor that correlated with enhancement of hormone response was not the absolute MT protein level, but the level of zinc-associated MT. For example, treatment with cadmium or copper increased MT, but the resulting cadmium- and copper-bound MT had no effect on DEX-induced CAT expression. Heat shock did not increase MT protein, but increased both the amount of MT associated with zinc and DEX-responsiveness. Collectively, these data strengthen the concept that MT mediates GR activity by regulating zinc.

The importance of MT in GR activity is accentuated under conditions of low zinc availability, consistent with the susceptibility of MT knockout mice to adverse effects associated with zinc deprivation. The multiple functions of GR in events related to cancer (growth, development, and death), and the potential of the stress- and metal-inducible protein MT to mediate GR activity, suggests that modulating MT expression may alter tumor cell growth and development, potentially for therapeutic benefit.

INFLAMMATION

Cell-mediated immune responses, including infiltration of macrophages into tumors during inflammation, have profound impact on tumor progression. The implications of inflammation for tumor behavior are complex and can be neutral, lead to tumor rejection, or lead to enhanced tumor growth depending on tumor characteristics and the nature of the inflammatory response. However, suppression of cell-mediated immune responses is associated with many tumors, including melanoma, colorectal, and prostate cancer. Modulation of inflammatory responses, particularly suppression, may be of value in inhibiting human tumor growth. Zinc is essential for proteins mediating activation of monocyte/macrophages. Activation proceeds through a G-protein-linked pathway, including phosphorylation of protein kinases leading to a cascade of events resulting in cytokine production and arachidonate release. PKC and inositol 1,4,5-triphosphate are activated, and active PKC releases NF-κB from IκBα. Released NF-κB is translocated to the nucleus to control transcription of multiple genes mediating activation, including IL-1α, IL-1β, and TNF-α.

There is circumstantial evidence that MTs may mediate innate immunity (which includes monocyte/macrophage action among the action of other cell types)—MT-KO mice are protected against TNF-induced systemic inflammatory response syndrome, and MT-1-overexpres sing transgenic mice are more sensitized to the lethal effects of TNF, compared to wild-type mice, consistent with the hypothesis that MT influences the activity of zinc-requiring monocyte activation proteins. MT expression is elevated in primary human monocytes, and a human monocytic cell line (THP-1), after activation. Although it was initially hypothesized that increased MT was not involved in the activation process and only provided protection against damaging reactive oxygen generated during activated monocytes undergoing respiratory burst, antisense RNA to suppress basal, cadmium, and lipopolysaccharide (LPS)- and cadmium-induced MT in human THP-1 monocytes completely abolished subsequent LPS and granulocyte-colony stimulating factor activation, without reducing viability. Therefore, MT induction concomitant with activation did not appear to mediate resistance to resistance to reactive oxygen.

Loss of activation potential as a result of MT downregulation indicated a role for MT in monocyte activation. MT-KO mice provide a model to study the effect of MT loss on primary monocyte/ macrophages. These animals are phenotypically normal but sensitive to cadmium and oxidant activity, and have reduced zinc uptake suggesting that zinc homeostasis in these animals is disturbed. Innate immune function (monocyte/macrophage activation) is compromised in MT-KO mouse monocytes, consistent with reduced sensitivity to systemic inflammatory response syndrome in MT-KO mice. Characterization of the effects of MT loss and gain on monocyte/macrophage action, and assessment of the putative zinc homeostatic role of MT as the mechanism by which it mediates monocyte/macrophage function, will be important future areas of investigation with implications for strategies for cancer treatment through modulation of inflammation.

Regulatory Activities

Metal salts can inhibit cell activity through direct toxicity to critical cellular molecules and structures. On the other hand, they can also change cell behavior by inducing specific genes (including MT). Therefore, metals may affect cell function by directly damaging cell components, or by transmitting or influencing signals controlling gene expression. To explore this, the ability of low-level metal salt pretreatment to alter monocyte behavior without decreasing viability has been measured. Treatment of THP-1 cells with low levels of zinc, cadmium, or mercury had no effect on viability or proliferation, but significantly induced MT mRNA and dramatically reduced LPS activation potential. Zinc chloride pretreatment of primary human peripheral blood monocytes similarly inhibited activation potential. Interestingly, LPS induction after zinc pretreatment dramatically decreased MT mRNA and protein expression (opposite to the effect of LPS activation of cells not pretreated with metal salts), suggesting a novel zinc-induced alteration in MT gene response to LPS in monocytes.

Zinc levels used were similar to those found in normal human serum, and had no toxic effect as assessed by dye exclusion and radiolabel incorporation during run-on transcription in isolated nuclei. These data indicate that nontoxic pretreatment of monocytes with metal salts induces MT and diminishes human monocyte activation potential. Given the apparent importance of MT in mediating activation revealed by antisense MT and in MT-KO mice, this was unexpected, but of potential therapeutic interest (particularly with respect to zinc) to treat pathological conditions involving chronic upregulation of innate immune functions (such as rheumatoid arthritis, infectious disease, and inflammatory bowel disease and cancer). There are at least three possible explanations.

First, MT levels might need to be in a relatively narrow range to mediate induction of activation: either increasing or decreasing those levels would have a negative effect. Second, metal treatment might affect activity of signaling proteins mediating activation, without decreasing cell viability and without involvement of MT. Third, metal treatment can affect expression of other genes in addition to those encoding MT. Non-MT factors induced by metals might act independently, or in conjunction with MT, to inhibit activation potential. Exploration of these possibilities using cells with genetically altered MT expression, and monocytes isolated from transgenic MT-overexpressing and MT-KO mice with elevated and abolished MT expression (respectively) will be important. The activation potential of these cells, with or without metal pretreatment, will assess the involvement of MT in monocyte activation, and in metal-induced suppression of activation.

CO-RELATIVE STUDY

There is an important connection between immune function, NF-κB and GR activity, and the potential influence of MT. Both NF-κB and GR control the expression of many inflammatory and immune genes, including those regulating monocyte/macrophage activity (both IL-1β and TNF-α activate/are activated by NF-κB in a proinflammatory regulatory loop, and glucocorticoids induce anti-inflammatory genes and inhibit pro- inflammatory gene expression). NF-κB engages in crosstalk with other transcription factors important in innate immune function, including GR. GR transrepresses NF-κB (and vice versa) through a direct protein:protein interaction with the p65 subunit of NF-κB, the component revealed to be regulated by MT in studies of MT-KO and

MT-WT cells. Close examination of the role of MT in regulating these factors and functions separately, and their influence on each other, has important implications for enhancing therapy (combining drugs to target both transcription factors and signaling pathways, and MTs) of pathological conditions involving innate immune action, including cancer. Metallothioneins can act as sacrificial scavengers to directly intercept toxins (including metal ions and reactive oxygen species) that can damage cells and/or transmit signals regulating multiple physiological events, including resistance and sensitivity to toxic agents.

In addition, a new role for metallothioneins as regulators of zinc-requiring and zinc-sensing transcription factors is emerging. Those transcription factors, including NF-κB, p53, GR, and MTF-1 have important roles in regulating cellular events important in inflammation and apoptosis, proliferation, and tumor development and progression. Critical questions that require further investigation include evaluation of the putative role of MT in regulating the zinc status of transcription factors to modulate their activity, and assessment of the potential to alter MT in tumors or normal cells to enhance the effectiveness of chemotherapeutic drugs designed to induce apoptosis, or to alter signaling events in human tumors for therapeutic benefit.

INFLUENCE OF MICROENVIRONMENT

Much of the research into cancer drug resistance has focused on the cancer cells, but it is becoming increasingly clear that host factors can significantly affect the success of chemotherapy. For example, immune modulation, the pharmacological clearance of drug, and poor tolerance to the side effects can all affect the outcome of therapy. Normal cells require cell contacts and growth factors for their survival. Moreover, it is now accepted that the tumor microenvironment, including the contacts between the tumor cells, can also influence the survival of cancer cells during treatment. This effect is clearly related to the ability of the environment to affect signaling pathways important for cell survival, cell cycle checkpoints, and other processes relevant to the response of cells to cytotoxic drugs. The apoptotic response of cancer cells in response to chemotherapeutic agents can therefore be affected by pathways controlled, at least in part, by the microenvironment. The importance of these noncell autonomous mechanisms of drug resistance has attracted an increasing amount of attention and will be discussed in this chapter

Abnormal Environment

The cancer cell occupies an abnormal environment with altered extracellular matrix (ECM), increased amount of proteases, growth factors, as well as abnormal hypoxic conditions and altered stroma. The first clue that the environment could affect tumor cell survival in the presence of cytotoxic agents was noted over 30 yr ago, when it was observed that multicellular spheroids of tumor cells were more resistant to anticancer agents than the corresponding monolayer cultures, and that these differences were not because of a compromised ability of the drug to penetrate the spheroids.

In addition to direct cell-cell contacts, the interactions of cancer cells with various soluble factors and with components of the ECM can drastically affect the apoptotic sensitivity of these cells and their response to chemotherapeutic drugs. The multidrug resistance phenotype that results from direct cell contact with the ECM or other cells has been coined "*cell-adhesion-mediated drug resistance*," or CAM-DR, and appears to be related to an adhesion-dependent suppression of apoptosis.

RESISTANCE AGAINST DRUGS

As mentioned above, the interactions between tumor cells and the environment can be divided into two main categories: the cell-adhesion-type interactions and the soluble factor interactions. The cell adhesion interactions can further be divided into direct cell-cell contact interactions and cell–ECM interactions. Drug resistance that stems from these last two types of interactions has been termed CAM-DR.

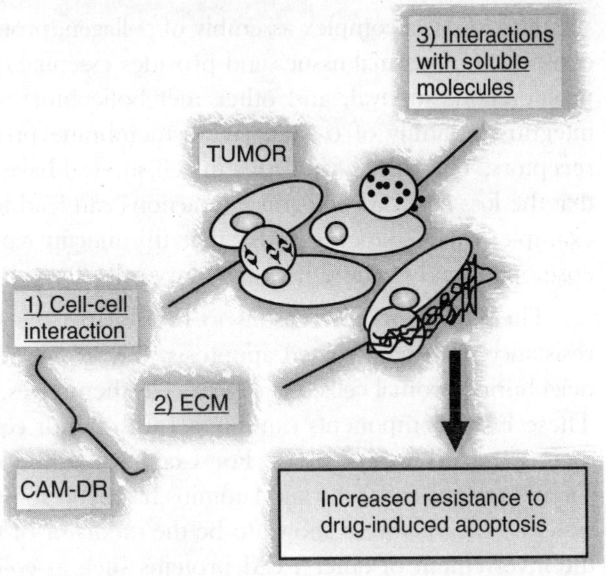

Fig. 15.2. Mechanisms of environment-mediated drug resistance.

Multicellular Resistance

The first indications that cell-cell interactions may influence survival to cytotoxic drugs came to light when it was observed that multicellular spheroids were more resistant to radiation exposure than cells grown in monolayers. This particular type of drug resistance, observed when cells are grown as three-dimensional (3-D) masses, as been coined *multicellular resistance* (MR) and can be considered a subtype of CAM-DR. Indeed, it appears that cells grown as 3-D cultures more closely recapitulate the drug resistance properties of in vivo cells compared to the more typical 2-D (monolayers) cultures used in most laboratories. MR has been shown to encompass a wide variety of cytotoxic agents, including alkylating agents, taxol, and multidrug resistance. In addition, since the initial study, many reports have confirmed that MR can protect cells from ionizing radiation.

Similar findings of MR have been reported in many different experimental systems. Interestingly, it has also been shown that cell-cell interaction disruption (using an anticadherin antibody) can sensitize cancer cells to a host immune response. Although the exact mechanisms of MR are still being elucidated, cadherin molecules have been shown to play an important role. E-cadherin, an epithelial homophilic cell-cell adhesion molecule, was shown to play role in the MR exhibited by human Lovo and MCF-7 cancer cells. Indeed, the presence of a neutralizing E-cadherin antibody could decreases p27 levels and drug resistance in 3-D cultures, but did not affect drug resistance in 2-D cultures of cells that did not express E-cadherin. Similar findings were reported in HT-29 cells where treatment of cells with neutralizing E-cadherin antibodies was found to decrease resistance to a variety of agents such as 5-fluorouracil, paclitaxel, vinblastine, and etoposide but not cisplatin.

This is consistent with findings that p27 plays a major role in chemoresistance and that E-cadherin can upregulate p27. Interestingly, it was shown that forced overexpression of p27 in cells expressing E-cadherin, abolished the effects of neutralizing E-cadherin antibodies on the proliferation of these cells, clearly implicating p27 downstream of E-cadherin. Consistent with this hypothesis downregulation of p27 has been associated with drug resistance in many experimental systems. In addition, it was recently reported that the *PMS2* gene and the overall mismatch repair gene activity was down-regulated in spheroids of EMT-6 cells compared to monolayer cultures, suggesting a new mechanisms as a determinant of MR.

Influence of extracellular matrix

The ECM, a complex assembly of collagen, proteoglycans and other molecules, is an important constituent of normal tissues and provides essential cues for cell development, migration, adhesion, proliferation, survival, and other metabolic functions. ECM components typically interact with integrins, a family of α- and β-transmembrane proteins, which associate to form heterodimeric receptors. The roles of integrins in cell survival have been studied extensively, and it is now known that the loss of ECM–integrin interactions can lead to anoikis, a specialized form of apoptosis. For example, early studies showed that fibronectin can promote anchorage-independent growth of eosinophils and Chinese hamster ovary cells through mechanisms that involve inhibition of anoikis.

There are certainly reasons to believe that the same types of mechanisms may also promote resistance to drug-mediated apoptosis. Indeed, tumor cells are surrounded by ECM produced by neighboring stromal cells, the cancer cells themselves, or other cells in the tumor microenvironment. These ECM components can interact with tumor cell integrins and affect their behavior, including their sensitivity to apoptosis. For example, adhesion to fibronectin or other ECM components such as various collagens and laminins has been shown to prevent drug-induced apoptosis. In many systems, integrins were shown to be the mediator of these effects. Other studies have demonstrated the involvement of other ECM proteins such as collagen IV, fibronectin, and tenascin, which are elevated in small cell lung cancer. These ECM proteins were found to protect small cell lung cancer cells from chemotherapy-induced apoptosis using various chemotherapeutic drugs, and there was evidence that the signaling was through the integrin β1 receptor. Interestingly, some evidence suggest that integrin-mediated changes in polarity and tissue architecture may be key in drug resistance as opposed to simple activation of the integrin receptors.

Weaver and collaborators recently showed that in breast cancer cells, reconstituted basement membrane, a specialized ECM for epithelial cells, can lead to the formation of polarized structures, which are essential in protecting these cells from chemotherapy-induced apoptosis. As a clue to possible downstream signals that may result from these interactions, chemoresistance was accompanied by nuclear factor (NF)-κB activation, a transcription factor that has been involved in mediating survival signals. Similarly, myeloma cells strongly induce NF-κB when adhered to fibronectin. The effects of ECM on myeloma cell survival has been studied extensively. It has been demonstrated that fibronectin interactions with the β-1 integrins may be important in inhibiting drug-induced apoptosis and therefore promoting cell survival. Again, similar to what was observed with MR, p27 appeared to play an important role, suggesting that these different mechanisms of drug resistance may converge at the molecular level.

In addition, up-regulation of Bcl-2 and, as mentioned above, activation of NF-κB may also play a role in fibronectin-mediated cell survival. Fibronectin has also been shown to promote survival in many solid tumors such as prostate cancer, breast cancer, and colon cancer. In addition to integrins, other receptors are known to interact with the ECM. For example, CD44 has been shown to bind hyaluronate, collagen, and fibronectin. In a fashion similar to the integrins, the CD44 receptor is involved in a wide range of physiological processes, including invasion and survival. In addition, many tumors have been shown to overexpress abnormal CD44 variant, especially colon cancer cells, where it has been suggested to contribute to drug resistance.

Again, the increased drug resistance was attributable to a decrease in sensitivity to drug-induced apoptosis. Similar findings have been reported in other cancers, including lymphoid malignancies. In the past several years, the tumor vasculature has emerged as a prime target for cancer therapy using specific antiangiogenic compounds. The tumor vasculature is also a target of conventional chemotherapy and the efficacy of cancer drugs may, at least in part, be because of their ability to destroy the tumor vasculature. It has been reported that fibronectin can inhibit drug-mediated apoptosis in tumor-derived endothelial cells, making this interaction an attractive new target for tumor therapy, and providing evidence that ECM interactions can also protect tumors through their ability to provide survival signals to endothelial cells during chemotherapy. Because tumor cells are known to produce ECM components, there is a possibility that they may produce their own favorable ECM, thereby increasing their resistance to chemotherapy.

Indeed, a study of ovarian cancer gene expression recently found that collagen VI is one of the genes most highly upregulated in a model of acquired cisplatin resistance in vitro. In addition, collagen VI is expressed in many ovarian tumors in vivo but absent from normal ovarian epithelial cells. It has been hypothesized that the tumor cells may produce collagen VI and other ECM components to reorganize the ECM and increase their resistance to apoptosis. Indeed, ovarian cells adhered to collagen VI exhibit an increased survival when exposed to a variety of anticancer drugs. In addition, it has been suggested that the tumor cells may secrete various cytokines and growth factors that may lead the neighboring stroma to produce protective ECM proteins. An important point that will need to be clarified is whether the ECM interactions necessary for tumor survival in the presence of drug are dependent on tumor type.

It appears likely that different cancers will have different optimal requirements for ECM-mediated protection. For example, collagen IV and laminin have been shown to increase resistance of lung cancer cells to various agents. Our own work shows that collagen VI, but not collagen I, may be crucial in establishing interactions leading to cisplatin resistance in ovarian cancer. A reconstituted basement membrane (Matrigel) is necessary for decreasing drug-induced apoptosis in breast cancer cells. Whereas these interactions are likely to be extremely complex and tumor specific, it is possible that they will lead to the activation of similar intracellular pathways that may in turn represent an attractive target for therapy aimed at reversing chemoresistance.

Growth Factors

The tumor microenvironment is also rich in soluble factors, cytokines, growth factors, secreted by the stroma or by the tumor cells themselves. Interleukin (IL)-6, a cytokine secreted by bone marrow stromal cells, has been observed to be a survival factor in various normal, leukemic, and myeloma cells. IL-6 can protect against chemotherapy-induced apoptosis and therefore induce drug resistance in myeloma cells. Interestingly, IL-6 has also been shown to promote drug resistance in prostate cancer cells, and autocrine production of IL-6 has been shown to cause drug resistance in breast cancer. Activation of the IL-6 receptor is believed to lead to downstream signaling through the Janus kinase/signal transducer and activator of transcription pathway, the mitogen-activated protein kinase cascade, the phosphoinositide 3 kinase/Akt pathway, and X-linked mammalian inhibitor of apoptosis protein, all of which are believed to be important in controlling survival and apoptosis. Growth factors also influence drug response in solid tumors. For example, pathways involving insulin-like growth factor I, epidermal growth factor, hepatocyte growth factor, and basic fibroblast growth factor have all been shown to influence the survival of various cancers in the presence of chemotherapeutic drugs.

Environment Affecting Drug Delivery

In addition to its effects on tumor cell signaling, the environment may also affect the ability of therapeutic agents to effectively reach the tumor cells. For example, the ECM can influence physical properties such as mechanical stiffness in the tumor, affecting diffusion of drugs. Overall, the limited ability of drugs to reach tumor cells, especially cells that are distant from blood vessels, may be an important component of the effect of the environment on drug resistance.

DRUG STRATEGIES

Drug resistance is a major problem that has attracted significant attention from researchers in the field of cancer research. However, from a clinical standpoint, most efforts have focused on the identification of alternative therapies against resistant tumors as opposed to the identification of targets that may reverse the resistance to conventional, proven therapy. Whereas, in principle, this represents a sound strategy, in practice it has been observed that tumor cells often exhibit crossresistance to many antitumor agents. This may be due, at least in part, to the fact that these cells tend to be generally more resistant to apoptosis, and that this phenomenon may not be compound dependent. Importantly, the identification of specific pathways responsible for drug resistance may provide targets for combination therapy aimed at circumventing or decreasing resistance. Therefore, intervention aimed at reducing the interactions between the environment and the tumor cells may increase the efficiency of cytotoxic therapy. For example, inhibition of the interactions between the ECM and the integrins or blocking of downstream signaling pathways represent promising avenues.

In addition, inhibition of various growth factors receptors such as epidermal growth factor receptor may also sensitize tumor cells to conventional chemotherapy. This area of research is crucial, as it is becoming increasingly evident that even treatment with new generation of mechanism-based targeted drugs such as Gleevec, a highly specific Bcr-Abl kinase inhibitor, can lead to the development of resistance in the treated cells. Whereas it is clear that conventional mechanisms such as altered expression or mutations of genes involved in DNA repair, growth, and apoptosis are crucial in the development of drug resistance, there is a broad range of evidence for a role of the microenvironment.

The resistant tumors arising from strong selection during chemotherapy will be a sum of the genetic changes that favor tumor growth in the presence of the drug, including those affecting the interactions of the cells with their environment. Although much energy has been devoted to identifying the cell-autonomous mechanisms leading to drug resistance, increasing efforts are being devoted to understanding the effects of the microenvironment on resistance. These efforts will likely lead to a more global understanding of the mechanisms involved in drug resistance and may provide novel specific targets for drugs aimed at reversing drug resistance.

IN VIVO RESISTANCE

The clinical problem of drug resistance was recognized when in the 1940s, Dr. Alfred Gilman's lymphoma patient failed to respond to a third course of nitrogen mustard after responding to the drug twice. Since then, elucidation of the mechanisms by which malignant cells develop a tolerance toward exposure to cytotoxic anticancer drugs has been a major area of investigation. Much of the research into drug resistance has been carried out in cells culture, most often using sublines cloned after repeated and/or chronic exposure of malignant cells to a specific agent. Although changes in cells developed in this manner were clear and often very well characterized, it has been more difficult to confirm that these changes correspond to the clinical problem. The second most commonly used model is the transplantable tumor. This system has added knowledge beyond that which can be learned from cell culture, especially in the area of solid tumor physiology.

Recognition of the physiological differences between solid tumors and normal tissues has permitted researchers to understand that a high degree of heterogeneity exists in the environmental conditions in which tumor cells survive in vivo. However, much of the work done with transplantable tumors has also focused on characteristics of singular tumor cells rather than on the tumor as a tissue interacting with the normal host tissues. The largest exception to this may be the study of interactions between the immune system and the tumor. One school of thought for overcoming drug resistance in the clinic is that drug resistance simply represents "*under treatment*" of the disease. From this notion, treatment regimens consisting of high-dose combination chemotherapy with hematopoietic stem cell transplantation were developed and have been under clinical investigation for more than 10 yr. Clinical protocols involving sequential high-dose regimens have also been reported. The efficacy of this treatment approach, however, remains an open question.

Preclinical in vivo modeling allows rigorous examination of many aspects of treatment, treatment combinations, sequences, and so on, that would be impossible to approach by clinical trial, because trial (because the number of variables involved) would require a prohibitively large number of patients. The scientific study of cancer therapy relevant to the high-dose setting has required the development of preclinical models that go beyond the conventional dose end points of increase in life span and tumor regression/ growth delay. High-dose therapy can be modeled using the tumor cell survival assay that allows tumor-bearing animals to be treated with "*supralethal*" doses of anticancer treatments with a quantitative measure of tumor cell killing and using a tumor growth delay assay with transplant of hematopoietic stem cells from syngeneic donors.

General Study

The high-dose setting is the most informative situation in which to examine the effect of drug sequence and drug combination, because it is in the high-dose setting where the greatest potential cell killing effects can be obtained. In the design of sequential high-dose chemotherapy regimens, the selection of antitumor alkylating agents to be included in each intensification, and the interval between the intensifications are critical to the design of the therapy. The tumor cell survival assay and tumor growth delay assay using the murine EMT-6 mammary carcinoma were used as a solid tumor model in which to address these issues. Tumor-bearing mice were treated with high-dose melphalan or cyclophosphamide (CTX) followed 7 or 12 d later by melphalan, CTX, thiotepa (THIO), or carboplatin (CARBO). After treatment with melphalan, both 7 and 12 d later, the tumor was resistant to each of the four drugs studied.

Treatment of the tumor-bearing animals with melphalan (30 mg/kg) on day 5 after tumor cell implantation followed by various doses of melphalan (20, 30, and 40 mg/kg) on day 12 or on day 17 resulted in marked resistance of the tumor to the second melphalan treatment, so that the first exposure to melphalan killed 2.5 logs of cells, but the second dose given 7 d later killed <1 log of EMT-6 tumor cells, and a dose given 12 d later killed about 1.5 logs of EMT-6 tumor cells. A similar pattern was seen if a dose of CTX was followed by a dose of melphalan. The findings for the killing of bone marrow granulocyte-macrophage colony-forming units (GM-CFU) paralleled the findings in the tumor cells. After treatment with CTX, both 7 and 12 d later, the tumor was resistant to melphalan and THIO but was not resistant to CTX or CARBO. When administered to the animals, a dose of CTX (400 mg/kg) killed about 2.5 logs of EMT-6 tumor cells.

Treatment of the tumor-bearing animals with melphalan (30 mg/kg) either 7 or 12 d before treatment with CTX resulted in <0.5 log and 1 log of tumor cell killing by CTX (400 mg/kg). On the other hand, treatment of the tumor-bearing animals with CTX (400 mg/kg) either 7 or 12 d before treatment with a second dose of CTX resulted in 3 logs and 2.5 logs of EMT-6 tumor cell killing by CTX (400 mg/kg). Again, parallel effects were seen with the response of bone marrow GM-CFU to these treatments. Administration of melphalan (30 mg/kg) to the tumor-bearing animals either 7 or 12 d before treatment with THIO or CARBO resulted in markedly decreased killing of EMT-6 tumor cells by THIO or CARBO. Administration of CTX (400 mg/kg) either 7 or 12 d before THIO resulted in 1.5-2 logs of EMT-6 tumor cell killing by THIO (30 mg/kg), respectively. Treatment with either melphalan or CTX decreased the killing of bone marrow GM-CFU by a subsequent dose of THIO.

Administration of CTX (400 mg/kg) either 7 or 12 d before CARBO decreased EMT-6 by the drug to about 1.5 logs. Neither prior treatment with melphalan nor prior treatment with CTX altered toxicity of CARBO toward bone marrow GM-CFU. To extend the interval between high-dose treatments to 14 and 21 d, after the first intensification, the tumor was transferred to second hosts that were either drug-treated or not drug-treated. When high-dose melphalan-treated tumors were treated with a second high dose of melphalan, the tumors were very resistant with the 14-d interval and less resistant with the 21-d interval. This small effect was evident in the bone marrow GM-CFU, except in the hosts pretreated with melphalan.

When high-dose CTX-treated tumors were treated with a second high dose of CTX, drug resistance was observed both with the 14- and 21-d interval if the host was nonpretreated or was pretreated with melphalan, but not if the host was pretreated with CTX. The same was true in the bone marrow GM-CFU. Tumor growth delay studies that allow the assessment of tumor response while the tumor remains in the host supported these findings the with high-dose CTX, melphalan, THIO, or CARBO resulted in additivity to greater-than-additive tumor growth delay. High-dose combination regimens required dose reduction of the drugs, which resulted in decreased tumor growth delays. Mice bearing the EMT-6 tumor received high-dose melphalan (30 mg/kg) or high-dose CTX (400 mg/kg) on day 5 and a second high-dose treatment on day 12. These animals received hematopoietic support consisting of peripheral blood cells from mobilized syngeneic donors on days 6 and 13 and recombinant GM-CFU on days 6 through 22. Tumor response to therapy was determined by tumor volume measurements.

When animals bearing EMT-6 tumor were treated with melphalan on day 5, a tumor growth delay of 5.1 d was produced. Followed this dose of melphalan 7 d later with a second dose of melphalan or with high-dose cyclophsophamide, THIO, or CARBO resulted in tumor growth delays between 6.9 and 11.3 d that were greater than for melphalan but less than expected for independent additive effects of the drugs. A single high-dose treatment with CTX (400 mg/kg) on day 5 produced a tumor growth delay of 19.6 d. Treatment of these animals 7 d later with high-dose melphalan, CTX, THIO, or CARBO resulted in tumor resulted in tumor growth delays of 42.8, 46.4, 29.8, and 30.8 d, respectively, which were additive to greater than additive for independent effects of the drugs.

Table 15.1 Growth delay of the EMT-6 murine mammary carcinoma after two high-dose chemotherapy treatments with stem cell support

Treatment group	*Tumor growth delay (days)*
Melphalan (30 mg/kg), day (d)5	5.1 ± 0.4
Melphalan (30 mg/kg), d5 → melphalan (30 mg/kg), d12	7.2 ± 0.7
Melphalan (30 mg/kg), d5 → cyclophosphamide(400 mg/kg), d12	11.3 ± 1.4
Melphalan (30 mg/kg), d5 → thiotepa (30 mg/kg), d12	7.6 ± 0.5
Melphalan (30 mg/kg), d5 → carboplatin (250 mg/kg), d12	6.9 ± 0.8
Cyclophosphamide (400 mg/kg), d5	19.6 ± 1.4
Cyclophosphamide (400 mg/kg), d5 → melphalan (30 mg/kg), d12	42.8 ± 2.3
Cyclophosphamide (400 mg/kg), d5 → cyclophosphamide (400 mg/kg), d12	46.4 ± 2.0
Cyclophosphamide (400 mg/kg), d5 → thiotepa (30 mg/kg), d12	29.8 ± 1.5
Cyclophosphamide (400 mg/kg), d5 → carboplatin (250 mg/kg), d12	30.8 ± 1.3

The effect of two drug combinations in the high-dose setting was assessed by tumor growth delay. The combination of high-dose CTX and high-dose THIO was toxic, in that more than 50% of the animals died within 2 wk after treatment. The tumor growth delay for the remaining animals was 6.3 d, much less than expected for additivity of the drug treatments. Treatment with high-dose CTX and high-dose CARBO was toxic. When the doses of the drugs were deduced to allow survival of the treated animals, the resulting tumor growth delays were less than those obtained with a full dose of the most effective single agent in each combination regimen.

The reason for this finding at in part may be the severely depleted condition of these animals. It can be concluded from the studies described above that the first administration of melphalan induced a metabolic condition in the host and in the tumor, which resulted in drug resistance that was slowly dissipating over a period of weeks. This finding is supported by a body of data from previous studies in cell culture and in vivo that show that exposure to melphalan resulted in decreased sensitivity to subsequent drug treatment, at least with antitumor alkylating agents. The sequential high-dose therapy tumor growth delay studies clearly show the diminished response of the EMT-6 tumor to treatment with high-dose melphalan, CTX, THIO, or CARBO 1 wk after a prior dose of melphalan.

Table 15.2 Growth delay of the EMT-6 mammary carcinoma and dose intensity after high-dose chemotherapy with stem cell support

Treatment group	*Dose intensity*	*Tumor growth delay*
Cyclophosphamide (400 mg/kg), (d)7	1	19.2 ± 1.2
Thiotepa (30 mg/kg), d7	1	4.2 ± 0.5
Carboplatin (250 mg/kg), d7	1	9.0 ± 1.1
Cyclophosphamide (400 mg/kg), d7 + thiotepa (30 mg/kg), d7	2	6.3 ± 0.6/toxic
Cyclophosphamide (400 mg/kg), d7 + carboplatin (250 mg/kg), d7	2	Toxic
Cyclophosphamide (225 mg/kg), d7 + thiotepa (20 mg/kg), d7	1.2	5.8 ± 0.4
Cyclophosphamide (225 mg/kg), d7 + carboplatin (125 mg/kg), d7	1.1	8.8 ± 0.7
Thiotepa (30 mg/kg), d7 + carboplatin (250 mg/kg), d7	2	5.0 ± 0.4
Thiotepa (20 mg/kg), d7 + carboplatin (125 mg/kg), d7	1.2	4.4 ± 0.4

The exception to this was the bone marrow GM-CFU of animals treated with melphalan, in which the sensitivity of this tissue to melphalan was increased when exposed to the drug again on day 21 but not on days 12, 14, or 17. Second, prior exposure to CTX results in much less or no drug resistance to administration of a second drug. In fact, administration of a second dose of CTX to animals that had already been treated with the drug can result in increased tumor cell killing compared with the initial treatment. This effect appears to be primarily an effect on the host, because when an untreated host or a host treated with melphalan bearing the CTX tumor was exposed to CTX, diminished sensitivity of the tumor to the drug was observed. Only when the host was treated with CTX was enhanced sensitivity to a second dose of CTX observed.

Although a much smaller effect, the melphalan-treated tumors in the hosts treated with CTX were the most sensitive to the second dose of melphalan. Exposure to cytotoxic therapy and especially to high-dose cytotoxic chemotherapy produces major metabolic responses in the host and in the tumor. Some of these responses are acute and short-lived, but other may last weeks or longer. These metabolic changes, although not permanent genetic alterations, clearly affect the response of the host and the tumor to subsequent exposures to cytotoxic anticancer agents. On the design of sequential and combination normal dose and high-dose treatment regimens, the response of the host and the tumor to prior therapy should be taken into account.

ROLE OF GROWTH FACTOR-β IN *IN VIVO* RESISTANCE

The EMT-6 mammary carcinoma sublines resistant to antitumor alkylating agents, CTX, THIO, cisplatin (CDDP), and CARBO were produced by repeated exposure of fresh tumor-bearing hosts to each drug. After 10 treatments, metastable, resistant tumors were produced. Although the tumors were resistant to drug treatment, the tumor cells in monolayer culture were not. As determined by the tumor cell survival assay from tumors treated in vivo at a level of 1 log (90%) of cell killing, the EMT-6/ CDDP tumor is fourfold resistant to CDDP, and the EMT-6/CTX tumor is threefold resistant to CTX as compared with the EMT-6/Parent tumor. When the survival of bone marrow GM-CFU–an alkylating agent sensitive normal tissue–was assessed in mice bearing the EMT-6 parental tumor or the in vivo resistant EMT-6/CDDP, EMT-6/ CTX, EMT-6/THIO, and EMT-6/CARBO tumors, the survival pattern of the bone marrow GM-CFU recapitulated the survival of the tumor cells, mimicking the development of resistance and reversion to sensitivity on removal of the selection pressure for each of the four alkylating agents.

When the EMT-6 parental tumor was implanted in the opposite hind limb of the animals bearing the EMT-6/CDDP or EMT-6/CTX tumor, the survival of the parental tumor cells after treatment of the animals with the appropriate antitumor alkylating agent was enhanced. The EMT-6/CDDP tumor was cross-resistant to CTX and melphalan, whereas the EMT-6/CTX tumor was somewhat resistant to CDDP and markedly sensitive to etopo side. In each case, the survival pattern of the bone marrow GM-CFU reflected the survival of the tumor cells. Thus, the presence of an alkylating agent-resistant tumor in an animal altered the drug response of the tissues distal to the resistant tumor. When the expression of several early response genes and genes associated with malignant disease was assessed in the EMT-6/Parent tumor and the EMT-6/CTX and EMT-6/ CDDP in vivo resistant lines growing as tumors, it was found that, in the absence of treatment, the levels of mRNA for the genes c-*jun*, c-*fos*, c-*myc*, Ha-*ras* and *p53* were increased in the EMT-6/ CTX and EMT-6/CDDP tumors as compared with the EMT-6/ Parent tumor, whereas the expression of *erb-2* was similar in all three tumors.

There was increased expression of both c-*jun* and *erb-2* in the livers of tumor-bearing animals. The highest expression of both c-*jun* and *erb-2* occurred in the livers of animals bearing the EMT-6/CDDP tumor. Treatment of the animals with CDDP or CTX, in general, resulted in increased expression of both genes at 6 h posttreatment. The increased expression of these genes may impart metabolic changes in the tumors and/or hosts that contribute to the resistance of these tumors to specific antitumor alkylating agents. Several observations, including the fibrous nature of the resistant tumors, the increased metastatic potential of the resistant tumors, and the altered pharmacokinetics of the drugs in the resistant tumor-bearing hosts, led to the hypothesis that transforming growth factor-β (TGF-β) might be integrally involved in in vivo antitumor alkylating agent resistant in the EMT-6 tumor lines.

Because it is difficult to maintain increased systemic levels of TGF-β by administering the protein to mimic the resistance phenotype, administration of TGF-β-neutralizing antibodies to animals bearing the resistant tumors in an attempt to reverse the resistance was chosen as the experimental design to address the hypothesis. The potential role of transforming growth factor-β in in vivo resistance was examined by administration of transforming growth factor-β-neutralizing antibodies to animals bearing the EMT-6/Parent tumor or the antitumor alkylating resistant tumors, EMT-6/ CTX or EMT-6/CDDP. Treatment of tumor-bearing animals with anti-TGF-β antibodies by intraperitoneal injection daily on days 0-8 post-tumor cell implantation increased the sensitivity of the EMT-6/Parent tumor to CTX and CDDP and markedly increased the sensitivity of the EMT-6/CTX tumor to CTX and the EMT-6/CDDP tumor to CDDP, as determined by tumor cell survival assay.

When animals bearing the EMT-6/Parent tumor or the EMT-6/CTX tumor were treated with a single dose of CTX, the tumor cell killing was obtained. CTX (300 mg/kg) killed 2 logs of EMT-6/Parent tumor cells and less than 1 log of EMT-6/CTX tumor cells. CTX (500 mg/kg) killed 3.5 logs of EMT-6/Parent tumor cells but only 1.5 logs of EMT-6/CTX tumor cells. To assess the possibility the TGF-β might have a role in in vivo alkylating agent-resistance, animals bearing the EMT-6/Parent or EMT-6/CTX tumor were treated with anti-TGF-β 2G7, anti-TGF-β 4A11, or anti-TGF-β 2G7, and anti-TGF-β 4A11 daily by intraperitoneal injection on days 4-8 or daily on days 0-8 post-tumor cell implantation. No significant difference was observed in tumor growth rate by administration of these antibody treatments.

There was no significant effect of the administration of anti-TGF-β 2G7 or anti-TGF-β 4A11 on days 4-8 on tumor cell killing by cyclophosphamide (300 mg/kg) in animals bearing either the EMT-6/Parent or EMT-/CTX tumor; however, treatment with the combination of anti-TGF-β2G7 and anti-TGF-β4A11 on days 4-8 post-tumor cell implantation resulted in increased tumor cell killing of both EMT-6/Parent tumor and EMT-6/CTX tumors by CTX (300 mg/kg). When treatment with antibodies to TGF-β was extended to the full period of tumor growth, days 0-8, a significant increase in the tumor cell killing of the EMT-6/Parent tumor by cyclophosphamide (300 mg/kg) was observed with each of the three antibody regimens ($p < 0.01$). A much greater effect on tumor cell killing by cyclophosphamide tumor was treated with any of the anti-TGF-β regimens on days 0-8.

The effect of anti-TGF-β treatment on EMT-6/CTX tumor cell killing was not only highly significant ($p < 0.001$), but also resulted in the restoration of drug sensitivity to the level of the parent tumor in the EMT-6/CTX tumor cells. A similar study was carried out with animals bearing the EMT-6/Parent tumor and animals bearing the EMT-6/CDDP tumor treated with CDDP. Treatment of animals bearing the EMT-6/Parent tumor or the EMT-6/CDDP tumor with a dose of CDDP (20 mg/kg) resulted in 1 log less killing of the EMT-6/CDDP tumor cells than of the EMT-6/Parent tumor cells. Treatment of animals bearing the EMT-6/Parent tumor or the EMT-6/CDDP tumor with 50 mg/kg of CDDP produced 4 logs less killing of the EMT-6/CDDP tumor cells than of the EMT-6/Parent tumor cells. The anti-TGF-β treatment regimens were the same as those described above.

Administration of the antibodies to TGF-β on days 4-8 did not alter the response of the EMT-6/Parent tumor to treatment with CDDP (20 mg/kg), but significantly increased the killing of the EMT-6/CDDP tumor cells by CDDP (20 mg/kg) ($p < 0.01$). Administration of the antibodies to TGF-β on days 0-8 increased the tumor cell killing of the EMT-6/Parent tumor cells by CDDP, but increased the tumor cell killing of the EMT-6/CDDP tumor cells by CDDP to a much greater degree ($p < 0.001$). The increase in the tumor cell killing in the EMT-6/CDDP tumor by treatment with anti-TGF-β on days 0-8 in addition to CDDP (20 mg/kg) was sufficient to produce cell killing of the EMT-6/CDDP tumor cells that was equivalent to GM-CFU survival was determined from these same animals.

The increase in the sensitivity in the tumors on treatment with the anti-TFG-β antibodies was also observed in increased sensitivity of the bone marrow GM-CFU to CTX and CDDP. Treatment of nontumor-bearing animals with the anti-TGF-β regimen did not alter blood ATP or serum glucose level but did decrease serum lactate levels. This treatment also decreased hepatic glutathione, glutathione *S*-transferase, glutathione reductase, and glutathione peroxidase in nontumor-bearing animals by 40-60%, but increased hepatic cytochrome P450 reductase in these normal animals. Animals bearing the EMT-6/CTX and EMT-6/CDDP tumors had higher serum lactate levels than normal or EMT-6/Parent tumor-bearing animals; these were decreased by the anti-TGF-β regimen. Treatment of animals bearing any of the three tumors with the anti-TGF-β regimen decreased by 30-50% the activity of hepatic glutathione *S*-transferase and glutathione peroxidase, and increased by 35-80% the activity of hepatic cytochrome P450 reductase.

In conclusion, treatment with TGF-β-neutralizing antibodies restored drug sensitivity in the alkylating agent-resistant tumors, altering both the tumor and host metabolic status. Affecting in a therapeutically meaningful way on the resistance of EMT-6/CTX and EMT-6/CDDP tumors has been very difficult. Traditional approaches to antitumor alkylating agent resistance such as administration of thiol-depleting agents or administration of "*chemosenstizer*" such as etanidazole were not effective in restoring drug sensitivity to these tumors. The fibrous nature of the resistant tumors and the increased metastatic potential of the resistant tumors compared with the EMT-6/ Parent tumor led to the hypothesis that TGF-β might be important in the genesis and maintenance of the in vivo resistant phenotype. The key to understanding mechanisms of therapeutic resistance in vivo lies in understanding the response of the tumor and host to exposure to cytotoxic therapy and how specific factors of that response after repeated fractions or courses of therapy result in a tumor that is no loner responsive to cytotoxic therapy.

The initiation of a cytokine cascade or storm by repeated exposure to cytotoxic therapy has been recognized for several years in radiation therapy and other disease areas. The connection between this early burst of cytokine production and a perpetual production of TGF-β leading to postirradiation pulmonary fibrosis has also been established. In the EMT-6 in vivo alkylating agent-resistant tumor lines, a connection has been established between increased levels of TGF-β and drug resistance. The repeated induction of a cytokine cascade by sequential fractions of radiation or courses of chemotherapy may result in a metastable increase in TGF-β levels and to therapeutic resistance. Animals bearing EMT-6/CDDP tumors have higher circulating levels of TGF-β than do animals bearing the EMT-6/Parent tumor; however, when exposed to a cytotoxic insult (CTX or CDDP), a rapid and sustained induction of TGF-β occurred in animals bearing the EMT-6/ Parent tumor, whereas in animals bearing the resistant tumors, a lesser and more short-lived induction of TGF-β occurred. In the tumor tissue, transcription of TGF-β was induced by the cytotoxic therapies. A marked increase in TGF-β protein occurred in the EMT-6/ Parent tumors paralleling the changes seen in the circulating blood.

PHYSIOLOGICAL APPROACH

The most clear-cut, direct-acting, most frequently found angiogenic factor in cancer patients is vascular endothelial growth factor (VEGF). The signal transduction pathways of the KDR/Flk-1 and Flt-1 receptors include tyrosine phosphorylation, downstream activation of protein kinase C and activation of the mitogen-activated protein kinase pathway. Protein kinase C isoforms are centrally involved in signaling transduction pathways related to regulation of the cell cycle, apoptosis, angiogenesis, differentiation, invasiveness, senescence, and drug efflux. When protein kinase C pathways were activated in human glioblastoma U973 cells by phorbol 12-myristate 13-acetate, VEGF mRNA expression was upregulated via a posttranscriptional mRNA stabilization mechanism. Recent results provide evidence for the involvement of protein kinase C in the invasiveness of breast cancer cells through regulation of urokinase plasminogen activator.

Several studies have associated specific isoforms of protein kinase C with metabolic pathways in prostate cancer cells. Protein kinase C has also been identified as an interesting therapeutic target for the treatment of malignant gliomas. To assess the contribution of protein kinase C activation to VEGF signal transduction, the effects of a protein kinase Cβ selective inhibitor, LY333531, which blocks the kinase activity of conventional and novel protein kinase C isoforms was studied. LY333531 demonstrated antitumor activity alone and in combination with standard cancer therapies in the murine Lewis lung carcinoma and in several human tumor xenografts.

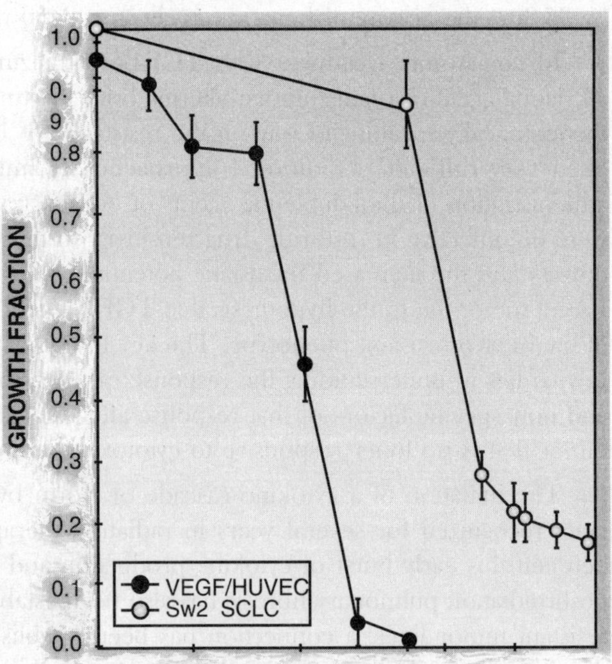

Fig. 15.3. Concentration-dependent growth inhibition of human umbilical vein endothelial cells and human SW2 small cell lung carcinoma cells after 72 h of exposure to various concentrations of LY317615 as determined by WST-1 assay.

The National Cancer Institute 60-cell line identified the protein kinase C inhibitor UCN-01, 7-hydroxy-staurosporine. UCN-01 has undergone phase I clinical trial. UCN-01 has been shown to the in vitro and in vivo growth of many types of tumor cells including breast, lung, and colon cancer. The compound LY3 17615 (enzastaurin) is a potent and selective inhibitor of protein kinase Cβ. When various concentrations of LY3 17615 were added to the cultures for 72 h, the proliferation of the VEGF (20 ng/mL)-stimulated human umbilical vein endothelial cells was profoundly inhibited by 600 n*M* of the compound. In a similar experiment, when human SW2 small cell lung carcinoma cells were exposed to various concentrations of LY3 17615 for 72 h, a potency differential in the effect of the compound on the malignant cells versus the human umbilical vein endothelial cells was apparent. Administration of LY317615 orally twice per day on days 1-10 postsurgical implant of VEGF impregnated filters resulted in markedly decreased vascular growth in the cornea of Fisher 344 female rats. A dose of 10 mg/kg of LY317615 or decreased vascular growth to about one half of the VEGF stimulated controls; whereas a dose of 30 mg/kg of LY317615 decreased vascular growth to the level of the unstimulated surgical control.

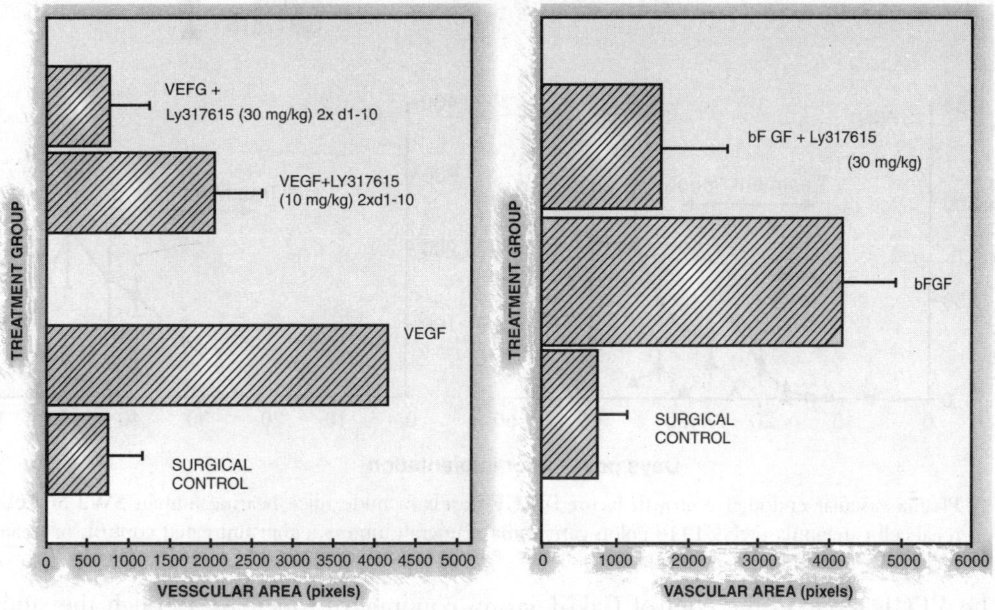

Fig. 15.4. Vascular area determined by image analysis and described in pixel number for Fisher 344 female rats implanted with a small filter disc impregnated with vascular endothelial growth factor (VEGF) or basic fibroblast growth factor (bFGF).

Administration of LY317615 (30 mg/kg) orally twice per day on days 1-10 postsurgical implantation of bFGF resulted in decreased vascular growth to a level of 26% of that of the bFGF control. Nude mice bearing human tumor sc xenografts were treated with LY317615 orally twice daily on days 4-14 or 14-30 post-tumor cell implantation. Tumors were collected and immunohistochemically stained for expression of endothelial specific markers, either CD105 or CD31. The number of intratumoral vessels in the samples was quantified by counting stained regions in 10 high power microscope fields (X200). The number of intratumoral vessels was decreased from one half to one quarter of the controls in animals treated with LY317615 (30 mg/kg). Although some of the tumors responded to LY317615 as an antiangiogenic agent, in no case was angiogenesis completely blocked as in the corneal micropocket neoangiogenesis model.

The tumor growth delay in the tested tumors did not correlate with intratumoral vessel decrease. The plasma levels of VEGF in mice bearing the human SW2 SCLC and Caki-1 renal cell carcinomas treated or untreated with LY317615 were measured by the Luminex assay. Plasma samples were obtained every 3 d starting on day 7 postimplantation and carried through treatment, as well as after the termination of treatment. Plasma VEGF levels were undetectable until tumor volumes were 500-600 mm^3. Plasma VEGF levels were similar between the treated and untreated groups through day 20, when plasma VEGF levels reached 75 pg/mL. Plasma VEGF levels in the SW2 control group continued to increase throughout the study reaching values of 400 pg/mL on day 40 postimplantation. On termination of treatment, plasma VEGF levels slightly increased to 100 ng/mL, which were still significantly decreased compared to the untreated control group.

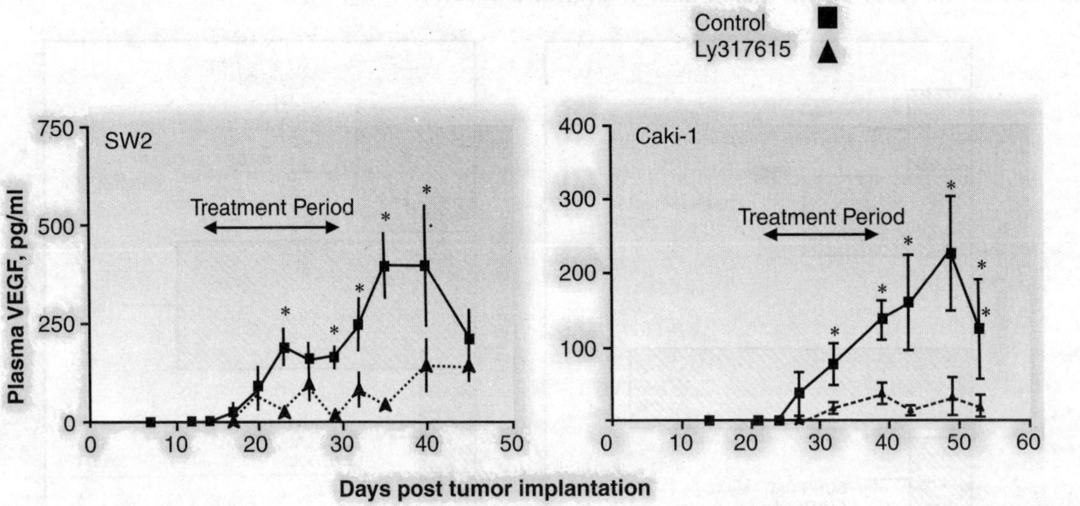

Fig. 15.5. Plasma vascular endothelial growth factor (VEGF) levels in nude mice bearing human SW2 SCLC, Caki-1 renal cell carcinoma or HCT116 colon carcinoma xenograft tumors, either untreated controls or treated with LY317615 orally twice daily on days 14-30.

The VEGF levels in the control Caki-1 group continued to increase through the study and peaked at 225 pg/mL on day 49 post-tumor implantation. In the treatment group, the plasma levels remained suppressed compared to controls throughout the treatment period (days 21-39). The plasma VEGF levels, reaching a maximum of 37 pg/mL, remained suppressed out to day 53, which was 14 d after terminating treatment. A sequential treatment regimen was used to examine the efficacy of LY317615 in the SW2 small cell lung cancer xenograft. Administration of LY317615 alone on days 14-30 post-tumor implantation over a dosage range from 3 to 30 mg/kg produced tumor growth delays between 7.4 and 9.7 d in the SW2 small cell lung cancer.

The SW2 tumor is responsive to paclitaxel, and treatment with that drug alone produced a 25-d tumor growth delay. Treatment with paclitaxel followed by LY317615 (30 mg/kg) resulted in over 60 d of tumor growth delay, a 2.5-fold increased in the duration of tumor response. The SW2 small cell lung cancer was less responsive to carboplatin, which produced a tumor growth delay of 4.5 d in that tumor. Sequential treatment with carboplatin followed by LY317615 resulted in 13.1 d of tumor growth delay. The antitumor activity of LY317615 alone and in combination with cytotoxic antitumor agents has been explored in several human tumor xenografts growing subcutaneously in nude mice. Whereas in most of the tumor models the tumor growth delay produced by treatment with LY317615 as a single agent was not sufficient to predict single agent activity in the clinic, in combination regimens LY317615 was a useful addition to the therapeutic regimen. LY317615 is currently in phase I clinical trial.

CONCLUDING REMARKS

The reason why most patients with cancer are not cured by cytotoxic anticancer therapies is that their disease becomes less responsive to the therapy and/or their normal tissues reach a limit of tolerance to the therapy. It appears that malignant cells can become tolerant to the cytotoxic therapies. Many mechanisms for the development of this "*drug tolerance*" or "*drug resistance*" have been elucidated in cell culture; however, the translation of these observations in isolated malignant cells to drug resistance in tumors has been problematic. In vivo survival advantage to repeated cytotoxic insults may be achieved by the induction of factors that are operative in complex tissues and require normal cells. It is becoming clear, as our knowledge of growth factors and cytokines increases, that the proliferation and metabolism of tumor cells, like those of normal cells, are influenced by these naturally occurring growth regulators. Our findings and those of others support the notion that the metabolism of tumor cells can be altered to enhance their survival via mechanisms that involve the autocrine and paracrine functions of growth factors and cytokines.

Therapeutic resistance of a tumor in a host organism can evolve by a phenotypic change in the tumor cells that does not confer drug resistance on the isolated tumor cells but which, through alterations in the handling of the drug by host tissues, leads to therapeutic resistance. The implications of this type of resistance to the treatment of clinical disease are vast. The signal transduction pathways that control cellular responses to these proteins and peptides provide new targets for the medicinal chemist. Therapeutic strategies that focus on enhanced drug activation (such as cyclophosphamide in the liver) or decreased drug catabolism may provide new ways of improving the efficacy of chemotherapeutic agents.

16

Antiestrogenic Resistance

In 1962, Jensen and colleagues discovered estrogen receptor (ER)-α. Jensen demonstrated that 17β-estradiol (E_2), the circulating female hormone that promoted breast cancer growth, binds to diverse tissue sites around a woman's body, but is retained in estrogen-target tissues, for example, the uterus and vagina. The identification of the ER-α as the target of E_2 action in the breast, and the discovery that antiestrogens blocked the binding of E_2 to the ER-α, provided a therapeutic target and an approach for the treatment of breast cancer. Nonsteroidal antiestrogens were initially developed as contraceptives in the 1960s. Walpole and colleagues synthesized tamoxifen (termed ICI 46, 474), a potent antiestrogen with antifertility properties in rats. However, in humans, tamoxifen induced ovulation in subfertile women, and consequently, its development as a contraceptive was discontinued. Fortunately, Walpole also patented the application of tamoxifen as a drug treatment for hormone-dependent cancers. Although many antiestrogens were discovered and tested during the 1960s and 1970s, only tamoxifen was considered safe enough for extensive clinical evaluation. Clinical trials were started to evaluate tamoxifen compared to the standard endocrine treatment at the time, diethylstilbestrol (DES), for the treatment of advanced breast cancer in postmenopausal women.

Tamoxifen was as effective as DES for the treatment of advanced breast cancer, but with fewer side effects. This advantage in side effect profile of tamoxifen compared to DES was crucial for its subsequent development as a treatment for all stages of breast cancer. Further clinical trials demonstrated that tamoxifen was truly a targeted therapy; it was highly effective in patients with tumors expressing hormone receptors, but ineffective when neither ER nor *progesterone receptor* (PR) were expressed. Following its approval as a treatment for advanced breast cancer, tamoxifen was evaluated as an adjuvant therapy in early stage breast cancer, and was demonstrated to reduce recurrence rates by 50%. Five years of adjuvant tamoxifen treatment was found to be more effective than less than 5 yr in improving time to tumor recurrence and overall survival. In contrast to the beneficial effects of tamoxifen as a treatment for breast cancer, both laboratory and clinical results demonstrated that tamoxifen acted as an estrogen agonist on the uterus, and increased the risk of endometrial cancer by 0.1%. These results strongly suggested that tamoxifen was not a pure antiestrogen, but had selective functions depending on the target tissue. Tamoxifen was demonstrated to reduce the incidence of contralateral breast cancer, in patients with ER-positive breast tumors. Based on this observation, and its acceptable side effect profile, tamoxifen was evaluated as a breast cancer preventive in high-risk women, and was approved for this indication in 1998.

PHYSIOLOGICAL ACTIVITIES

The concept of selective estrogen receptor modulators (SERMs) was first recognized in the 1 980s from research that investigated the pharmacological properties of tamoxifen at distinct tissue locations around a woman's body. For example, tamoxifen has estrogen-like properties on bone in ovariectomized rats, is both a partial estrogen and an antiestrogen on the uterus, and an antiestrogen/antitumor agent in the rat mammary gland. In athymic mice bitransplanted with human breast cancers and endometrial cancers, tamoxifen inhibits growth of the breast cancers, while simultaneously stimulating growth of the endometrial cancers. This concept of the selective actions by nonsteroidal antiestrogens first recognized in the laboratory was then applied to clinical drug development.

Tamoxifen maintains bone density and lowers circulating cholesterol levels in postmenopausal women. More importantly, tamoxifen lowers the recurrence of ER-positive breast cancer and decreases the incidence of contralateral breast by 50%, whereas it causes a modest increase in the risk of endometrial cancer by 0.1% in postmenopausal women. The beneficial effects of tamoxifen around a woman's body were used to justify the evaluation of tamoxifen as an adjuvant treatment for ductal carcinoma *in situ,* and as a preventive in high-risk pre- and postmenopausal women. To date, there is no definitive explanation as to why tamoxifen has differential agonist and antagonist effects at different target tissues. As this chapter outlines, possible explanations include differential expression of ER subtypes, and/or coregulatory proteins, in different target tissues.

Role of Endoplasmic Reticulum

An understanding of the selective nature of antiestrogens in patients first requires an understanding of the function of ER at the molecular level. Human ER is a member of the nuclear receptor family of ligand-inducible transcription factors. The classical model of ER activation involves the binding of the ligand E_2 to the ligand-binding domain (LBD) of the nuclear ER in target tissues that induces a conformational change in the three-dimensional structure of the complex. The E_2-ER complex dimerizes and subsequently binds to DNA sequences referred to as estrogen response elements (EREs). These EREs are *cis*-acting enhancer elements located within regulatory regions of target genes. The DNA-bound ER interacts with the preinitiation transcriptional machinery either directly or indirectly via coregulatory proteins. In contrast, the nonclassical or tethered pathway for activation of the ER involves the indirect interaction of the nuclear E_2-ER complex with transcription factors including Jun/Fos (activating protein [AP]- 1) or specificity protein (Sp)1 and subsequent activation of AP-1 or Sp1-driven genes, respectively. Thus, the ER can regulate the transcription of a wide variety of genes by both classical and nonclassical pathways. It is also now apparent that ER-α can exist both in the nucleus and in the cytoplasm, and can have differential effects on gene transcription, for example at AP-1 sites, depending on its cellular location.

Different Types of ER Isoforms

The identification of different ER isoforms is a recognized method in pharmacology for developing tissue-selective drugs. The discovery of two genetically distinct ERs, ER-α and ER-β, provides a possible explanation for the tissue selective nature of antiestrogens. The two receptors have some degree of homology in that both receptors have a LBD and DNA-binding domain. However, there are notable differences in the two *activating functions* (AFs). First, ER-β does not have an AF1 region but retains an AF2 region. These differences in the AF regions alter the activity of the antiestrogen–ER complex, resulting in increased or decreased estrogenic activity. For example, tamoxifen is more antiestrogenic when it is complexed with ER-β. However, to date, little is known about the role of ER-β in antiestrogen action. What is known from mRNA expression profiles is that ER-β is expressed in a wide variety of tissues, including the breast and uterus.

However, the existence of significant levels of ER-β protein in all mRNA-expressing tissues has been more difficult to interpret and requires further investigation. Studies in vitro indicate that both ER-α and ER-β activate transcription in response to E_2 at estrogen-response elements (ERE)-containing promoters, with ER-α being a more efficient transcriptional activator than ER-β. However, when both receptors are expressed in cells, a splice variant of ER-β decreases the activity of ER-α and decreases the overall sensitivity of agonists. The mechanism is not known, and little in vivo data exist showing competition between the two receptors for the same DNA-sequence and/or heterodimerization of ER-β with ER-α. In cells where tamoxifen activates ER-α-mediated, ERE-driven transcriptional activity, coexpression of ER-β completely suppresses this activity. Thus, expression levels of ER-β might regulate the magnitude of the tamoxifen–ER-α complex on ERE-driven genes, but only if the ratio of receptors was appropriate.

In AP-1-containing elements/ER-α systems, where ER-α binds indirectly to AP-1 promoters by interaction with AP-1, estrogens activate transcription and antiestrogens display a range of activities from partial to full agonists. In contrast, estrogens antagonize and all antiestrogens-tested activate transcription in cells expressing AP- 1–ER-β systems. However, demonstration of the physiological significance of the AP-1–ER interaction in vivo has yet to be shown. Studies in vitro have demonstrated that tamoxifen appears to inhibit E2-activated transcription by both ER-α and ER-β. However, in some cell and promoter contexts, the tamoxifen–ER-α complex manifests partial agonist activity. Therefore, it remains to be explained how the same ER-α–ligand complex can be recognized differently in different cells. There is accumulating evidence to suggest that both ERs exist, not only in the nucleus, but in the cytoplasm, under specific conditions, and that their functions in response to estrogen may differ depending on their location. Nongenomic actions of ER-α have been demonstrated to activate mitogen-activated protein kinase (MAPK), in response to estrogen, resulting in cell proliferation. Additionally, binding of estrogen to cytoplasmic ER-α has been demonstrated to activate AP-1 sites, whereas AP-1 activation is inhibited by binding of estrogen to nuclear ER-α. It is possible that location of ER-α in the cell may contribute to differential effects of tamoxifen on different target tissues.

Coregulators of ER

Another explanation for the selective activities of antiestrogens in different cell types is the expression pattern of coregulatory proteins. ERs directly or indirectly activate or repress target genes by binding to hormone response elements in promoter or enhancer regions. Modulation of transcription by ERs requires the recruitment of coregulators. Coregulators are transcription modifiers that can either activate or repress the activity of the ER complex. A basic mechanism for the switching of target genes from off to on requires a ligand-dependent exchange of corepressors for coactivators. The family of steroid receptor coactivators ([SRC]-1, -2, and -3) compliment the activity of the ER.

Ligand-dependent recruitment of coactivators is dependent on the AF-2 region within the C terminus of the LBD of both ER-α and ER-β. The precise region of the coactivators that interacts with the AF-2 region of the ER is an LXXLL domain, where "L" represents a leucine, and "X" represents any amino acid. The ER–coactivator complex functions to stabilize the preinitiation transcriptional machinery and promotes chromatin remodeling by recruiting histone acetyltransferases at the promoter region of the target gene. Thereafter, transcription of the target gene is either activated or repressed, depending on the promoter and cellular context. More than 30 additional putative coactivators have been identified to date. Thus, expression levels of coactivators might explain the selective actions of antiestrogens in different cells. The estrogen-like activity of the tamoxifen–ER complex has been shown to be because of higher levels of SRC-1 in endometrial cancer cells vs MCF-7 breast cancer cells, where the tamoxifen–ER complex is mostly antiestrogenic. ER-α and ER-β also play critical roles by repressing gene transcription. They function as ligand-independent repressors on some target genes or ligand-dependent repressors on others. Corepressors are coregulators that repress the activity of both ER-α and ER-β. The known corepressors are N-CoR, SMRT, and REA.

These coregulators contain multi-independent repressor domains that interact with histone deacetylases to mediate deacetylation of histones and promote condensation of chromatin repressing transcription of the target gene. The ER–corepressor complex prevents the interaction of the ligandless ER to the transcriptional machinery resulting in transcriptional repression. Transcriptional repression of target genes by antiestrogen–ER complexes might depend on the structure of the selective antiestrogen, the three-dimensional structure of the antiestrogen–ER complex, and the availability of corepressors vs coactivators. In summary, the mechanism resulting in differential effects of SERMs on different target tissues is not definitely known. Differential expression of coregulators in different target tissues probably plays a role in this process.

EFFECT OF TAMOXIFEN

Tamoxifen is widely used for all stages of breast cancer. Approximately 50% of patients with hormone receptor-positive breast cancer respond to tamoxifen treatment. Unfortunately, the other 50% of patients do not respond, and many who initially respond will eventually experience disease relapse. Therefore, a major problem with the use of tamoxifen therapy is that many breast tumors will either be resistant (intrinsic or *de novo*) before endocrine treatment or become resistant (acquired) during therapy. Considerable research has examined possible mechanisms responsible for the tamoxifen-resistant phenotype. Examples of this research include loss of the *ER-α* gene and/or protein in tumors during short-term and long-term treatment, identification of mutations within the *ER-α* gene, changes in the pharmacologic response of resistant breast cancer cells to tamoxifen, activation of ER-α-mediated gene transcription in the absence of ligand, modifications in gene transcription by changes in interactions between ER-α and coregulators, and the development of enhanced surface-to-intracellular signaling crosstalk between growth factor receptors and ER-α. Current evidence suggests that tamoxifen resistance (both intrinsic and acquired) is likely multifactorial.

Resistance Against Breast Cancer

Breast cancers initially lacking expression of ER and/or PR have intrinsic resistance to endocrine therapy. In addition, loss of ER-α expression on recurrence of breast cancer during tamoxifen therapy has been reported in 25% of tumors. However, most of acquired resistance occurring during tamoxifen treatment is not because of changes in ER-α expression, and most tamoxifen-resistant breast tumors retain ER-α, and, therefore, may respond to secondary endocrine treatments. In summary, absence of ER-α does predict for *de novo* tamoxifen resistance, but does not appear to be a major mechanism for acquired resistance.

Phylogenic Effects of ER-α and ER-β

Several mutant forms of both ER-α and ER-β have been identified and previously reviewed. However, the functional significance of these mutants as a mechanism of antiestrogen resistance is yet to be elucidated. Most of the data demonstrate existence of splice variants for ER-α and ER-β mRNA, without evidence of translation into functional proteins. In most cases, breast tumors express both mutant and wild-type ER, with the wild-type being the predominant species. A single point mutation in the LBD of ER-α has been reported (D351Y) that converts tamoxifen and raloxifene from antiestrogen to estrogens in in vivo and in vitro models of antiestrogen-stimulated MCF-7 tumor cells. However, the D351Y ER-α mutant has yet to be detected in either *de novo* or acquired resistant breast tumors from patients.

In addition, mutations in the F-region of the ER have been shown to affect the activities of both E_2 and 4-hydroxytamoxifen (4OHT), the active metabolite of tamoxifen. In summary, the clinical relevance of ER mutants is unclear to date as a mechanism of resistance to antiestrogens. However, an understanding of ER mutations in response to antiestrogens may lead to the development of better antiestrogens or other drugs for breast cancer treatment.

Influence of Nongenomic ER

There is increasing evidence to support the existence of ERs outside the nucleus. Nongenomic actions of estradiol have only been recognized recently, and encompass activation of MAPK, Ras, Raf-1, phosphokinase C and phosphokinase A. Santen et al. have suggested that estradiol binds to an ER near or in the cell membrane, initiating binding of Shc, and resultant association with Grb-2 and Sos. Ultimately, this series of events results in activation of Ras and MAPK, resulting in cell proliferation. Therefore, it seems possible that two estradiol-regulated pathways exist, a genomic pathway where binding to nuclear ER results in gene transcription, and a nongenomic pathway, that results in cell proliferation through MAPK.

Santen et al. have also demonstrated that estrogen deprivation of breast cancer cells in vitro results in hypersensitivity to estradiol. Treatment of these estrogen-deprived cells with estradiol resulted in an increase in MAPK phosphorylation in minutes because of activation of nongenomic

ER. Hypersensitivity to estradiol may occur in tamoxifen-exposed breast cancers, and may explain the activity of aromatase inhibitors, which reduce circulating estrogen levels in postmenopausal women, in tamoxifen-refractory breast tumors. Therefore, increasing evidence suggests that activation of nongenomic ER pathways, perhaps associated with an increase in growth factor expression, may play a role in tamoxifen-resistance.

Physiological Function of PR

PR has been demonstrated to play an important role in determining sensitivity of breast cancer cells to tamoxifen. Horwitz demonstrated that patients with metastatic breast cancers expressing both ER and PR were more likely to respond to tamoxifen, than breast tumors expressing only ER. Additionally, they showed that patients with metastatic breast tumors that were ER-negative, but PR-positive, were more likely to respond to tamoxifen compared to patients with ER-positive, PR-negative tumors. There are less data available regarding the predictive value of PR to tamoxifen in the adjuvant setting. The Oxford Overview analysis does not correlate patient outcome specifically with PR-status, because many of the trials do not report PR-status at all, or report it independent of ER-status. However, over the past 12 mo, there has been renewed interest in PR as a predictor for outcome from tamoxifen. The Arimidex, Tamoxifen, Alone or in Combination trial randomized 9000 post-menopausal patients with early-stage breast cancer to the aromatase inhibitor, anastrozole, alone, tamoxifen alone, or to a combination of the two drugs, each given for 5 yr. Overall, at a follow-up of 4 yr, patients treated with anastrozole were significantly less likely to have a disease recurrence compared to patients treated with tamoxifen.

A retrospective analysis of this trial evaluated the role of PR in predicting outcome from the individual therapies, in a large subset of patients enrolled in the trial. This analysis revealed no statistically significant difference in disease-free survival between anastrozole and tamoxifen, in patients with ER-positive, PR-positive breast cancers. In contrast, patients with ER-positive, PR-negative tumors treated with anastrozole had a highly statistically significantly improved disease-free survival, with a relative improvement of 50%, compared to patients with the same tumor phenotype treated with tamoxifen. In the small amount of patients with ER-negative, PR-positive breast cancers, there was no significant difference in outcome between treatment groups. Therefore, it appears the absence of PR is associated with a decrease in sensitivity to tamoxifen, and/or an increase in sensitivity to aromatase inhibitors. PR exists in two isoforms, PRA and PRB, resulting from posttranslationally modifications. Commercially available PR antibodies, used in immunohistochemistry, detect either isoform, but are unable to differentiate between them.

Using Western blot analysis, however, the two distinct isoforms can be detected based on size. Fuqua et al. recently correlated expression of PRA and PRB with 5-yr disease-free survival in patients with early-stage breast cancer treated with tamoxifen. They demonstrated that the majority of breast cancers express both isoforms in varying ratios. Tumors with higher expression of PRB, compared to PRA, were associated with a significantly improved 5-yr disease-free survival compared to tumors with higher expression of PRA, to PRB. To date, it remains unclear as to why the absence of PR, or higher expression of PRA to PRB, results in likely *de novo* resistance to tamoxifen. It is possible that the absence of PR allows a nongenomic ER pathway to be more dominant, which may in turn result in increased activation of the MAPK, and/or Akt pathways, leading to increased cell proliferation, and/or inhibition ofapoptosis. These hypotheses are being actively researched, and may lead to new therapeutic possibilities in patients with tumors expressing ER, but lacking PR.

Coregulative Effect in Resistance

The recruitment of coactivators or corepressors (coregulators) to the ER determines the switch between ER activation and repression. In addition, the coordinated action of ligand, ER, and coregulators determines that genes are transcribed or repressed depending on the cellular context, and thus, which cells will or will not proliferate. Available data suggest that intricate modulation of the ER-to-coregulator ratio in breast cancer cells could determine whether a breast cancer cell is sensitive or resistant to antiestrogens. For example, the coactivator SRC-1 activates ER in a ligand-independent manner while increasing 4OHT's agonist activity. On the other hand, the corepressor SMRT blocks the agonist activity of 4OHT-induced by SRC-1. The other member of the p160 coactivator family is amplified in breast cancer (AIB 1, SRC-3, RAC3). AIB1 mRNA was found to be amplified in 60% of breast cancers, and its protein product was found to be overexpressed in about 10% of tumors.

Recently, an association was discovered between high AIB1 and HER-2/*neu*, a member of the epidermal growth factor receptor (EGFR) family of receptor tyrosine kinases, and tamoxifen-resistant breast cancers. The potential mechanism for this resistance requires activation of the MAPK signaling cascade by HER-2/*neu*, which in turn potentially phosphorylates and activates AIB1. Activated AIB1 in turn can convert 4OHT from an antiestrogen to an estrogen with regard to ER-α activity. In contrast, low levels of N-CoR mRNA, a corepressor, have been associated with tamoxifen resistance. In summary, the specific expression of coregulators in breast cancer cells could well determine, at least in part, whether a patient is likely to benefit from tamoxifen, or other antiestrogens. Based on available data, it would seem reasonable to consider measuring levels of coregulators, particularly AIB1, in breast tumors, before initiating tamoxifen therapy. Additionally, the identification of abnormally expressed coregulators will probably assist in the design of future therapies that improve the efficacy of endocrine treatment without the development of resistance.

Clinical Studies

Several laboratory and clinical studies suggest that overexpression and/or aberrant activity of the HER-2/*neu* (ErbB2) signaling pathway is associated with antiestrogen resistance in breast cancer. The HER-2/*neu* receptor is a member of the EGFR family of receptor tyrosine kinases, which include HER-3 (ErbB3) and HER-4 (ErbB4). Ligand (i.e., EGF, transforming growth factor-α, heregulin, or HB-EGF) binding to EGFR, HER-3, or HER-4 results in homodimerization of individual receptors or heterodimerization with the preferred partner, HER-2/*neu*. On dimerization, the tyrosine kinase domains-located within the COOH-terminal regions of receptors are activated by an autophosphorylation cascade on specific tyrosine residues. The phosphotyrosine residues recruit SH_2 domain-containing intracellular adaptor molecules, including Shc, Grb-2, and the p85 subunit of phosphatidylinositol 3 kinase (PI3-K), that in turn activate downstream effectors such as MAPK and Akt that promote cellular proliferation, survival, transformation, and inhibit apoptosis.

HER-2/*neu* has been suggested to be a proto-oncogene, which when overexpressed, transforms normal mammary epithelial cells. HER-2/*neu* is overexpressed and/or amplified in 25-30% of breast tumors and is associated with a more aggressive phenotype and poor prognosis. HER-2/*neu* was first noted to be associated with tamoxifen resistance, when transfection of HER-2/*neu* into MCF7 cells rendered the cells resistant to the inhibitory effects of tamoxifen. Subsequently, retrospective analyses of trials in which patients with metastatic breast cancer yielded conflicting results on whether HER-2/*neu* predicted for a worse outcome with tamoxifen.

In the neoadjuvant setting, clinical response rates to tamoxifen have been noted to be lower in patients with HER-2/*neu*-overexpressing breast cancers. Based on these data, one possible mechanism for *de novo* resistance to tamoxifen is overexpression of HER-2/*neu* in ER-α-positive breast cancers. However, as outlined here, a recent study suggests that overexpression of HER-2/*neu* alone is not sufficient to produce the tamoxifen resistant phenotype. This retrospective study of patients treated with adjuvant tamoxifen demonstrated that high levels of HER-2/*neu* alone was associated with a similar 5-yr disease-free survival compared to patients with breast cancers with low levels of HER-2/*neu* expression. However, patients with tumors expressing high levels of HER-2/*neu* along with high levels of the coactivator AIB1, had a significantly shorter disease-free survival. Expression of HER-2/*neu* also appears to play a role in acquired resistance to tamoxifen. Several groups have demonstrated an increase in expression of HER-2/*neu* in SERM-resistant tumors compared to wild-type tumors.

We have observed an increase in both HER-2/*neu* and EGFR in tamoxifen and raloxifene-resistant breast cancers compared to tamoxifen-sensitive breast cancers. Iressa (ZD 1839, gefitinib), an EGFR-specific tyrosine kinase inhibitor has been shown to inhibit growth of breast cancer cell lines in vitro that are resistant to tamoxifen. Osborne et al. have noted an increase in EGFR in tamoxifen-resistant tumors in vivo, and have demonstrated that gefitinib, a tyrosine kinase inhibitor that targets EGFR, can inhibit the growth of these tamoxifen-resistant breast cancers, and can delay the onset of tamoxifen resistance. This finding is being evaluated in clinical trials. Therefore, it appears that increased expression of EGFR and HER-2/*neu* may play a role in acquired SERM resistance. Based on these findings, clinical trials are being developed using agents in combination that target the ER and the EGFR pathways. Strategies used to target EGFR and HER-2/*neu* include the use of humanized monoclonal antibodies to the receptors, tyrosine kinase inhibitors that block reduction of ATP to ADP + P_i, and receptor antisense molecules.

Trastuzumab (Herceptin) is a humanized monoclonal antibody directed against the ectodomain of the HER-2/*neu* receptor. It has been shown to restore breast cancer cell sensitivity to tamoxifen in HER-2/*neu*-overexpressing cells. One small clinical trial evaluated the use of trastuzumab with the aromatase inhibitor, letrozole, in patients with tamoxifen-refractory, HER-2/*neu*-positive metastatic breast cancer. Response rates were not notably different from historical trials utilizing the agents separately. EGFR and/or HER-2/*neu* inhibitors may prove to be beneficial in preventing and/or treating tamoxifen-resistant breast cancer.

Influence of Serine/Threonine Kinase

MAPK is a serine/threonine kinase that is activated by phosphorylation cascades originating from GTP-bound Ras, a downstream effector of EGFR and HER-2/*neu*. MAPK has been shown to be hyperactivated in MCF-7 breast cancer cells overexpressing either EGFR or HER-2/*neu*. This increased activity of MAPK promotes an enhanced association of ER-α with coactivators and decreased interaction with corepressors, thereby enhancing hormone-dependent gene transcription and possibly leading to tamoxifen resistant breast cancer. The role of the MAPK signaling pathway in tamoxifen resistance has been demonstrated primarily in vitro using specific inhibitors, such as U0126 (MAPK kinase, MAPK kinase/extracellular signal- related kinase 1/2 inhibitor), AG1478 (EGFR and HER-2/*neu* inhibitors) and PD98059 (MAPK, extracellular signal-related kinase 1/2 inhibitor).

The exact mechanism by which hyperactivated MAPK converts the antiestrogenic activity of the tamoxifen–ER-α complex to more estrogenic in resistant breast cancer cells is yet unclear. However, it has been demonstrated that activation of the Ras/MAPK pathway by EGFR/HER-2/*neu* receptor activation could phosphorylate Ser-118 in the AF-1 domain of ER-α, thus promoting ligand-independent transcription of the ER-α and possibly loss of tamoxifen-induced inhibition of ER-α-mediated gene transcription. Conversely, there is increasing evidence to suggest that nongenomic ER plays a role in MAPK phosphorylation. There is evidence in vitro that estrogen can bind to nongenomic ER, and result in MAPK phosphorylation. Based on these preclinical data, several clinical trials are underway to determine whether treating patients with EGFR inhibitors (for example, gefitinib) and/or MAPK-specific inhibitors (for example, U0126) can effectively treat and/or prevent antiestrogen resistance in breast tumors.

Effect of Heterodimer Complex

PI3-K is a heterodimer complex consisting of a regulatory subunit, p85, and a catalytic subunit, p110. The regulatory subunit p85 is phosphorylated by either receptor tyrosine kinases such as EGFR/HER-2/*neu* or intracellular adapter kinases such as Shc or Src, and thereafter interacts and activates the catalytic subunit p110. Activated PI3-K subsequently activates downstream effector kinases such as Akt (phosphokinase B), a serine/threonine kinase. Akt has been shown to activate, either directly or indirectly, proteins responsible for survival and inhibition of apoptosis in cancer cells. Several reports indicate that the PI3-K/Akt pathway interacts with the ER-α pathway resulting in bidirectional crosstalk. It has been shown that PI3-K/Akt can mediate E_2-induced transcription of cyclin D1 and entry of MCF-7 breast cancer cells into S phase.

In addition, the PI3-K/Akt pathway can induce ER-α phosphorylation on Ser-167 to promote ER-α-mediated transcription, thereby, protecting cells against tamoxifen-induced apoptosis. Accumulating data suggest that ER-α-positive breast tumors with alterations of PI3-K and Akt signaling, whether dependent or independent of EGFR/ HER-2/*neu*, might be insensitive to antiestrogens and thus lead to resistance. Future studies are needed to confirm whether blocking PI3-K and/or Akt signaling in ER-α-positive breast cancer will be beneficial in subverting resistance to antiestrogens.

PHARMACOLOGICAL FINDINGS

New drug discovery for SERMs is currently driven by the known side effects of tamoxifen, which include the increase in the incidence of endometrial cancer, and the development of resistance. Additionally, recent reports of the lack of efficacy of hormone-replacement therapy on coronary heart disease, and its association with increased breast cancer risk, have led to a search for "the ideal SERM". Several approaches are being pursued including altering the antiestrogenic side chain or improving the pharmacokinetics of existing SERMs. Several novel antiestrogenic compounds have been developed as treatments for ER-α-positive breast cancer, without the agonist effects associated with tamoxifen. These new compounds have the potential to be at least as effective as tamoxifen, possibly without the development of acquired resistance during therapy, and may offer a reduction in incidence of endometrial cancer.

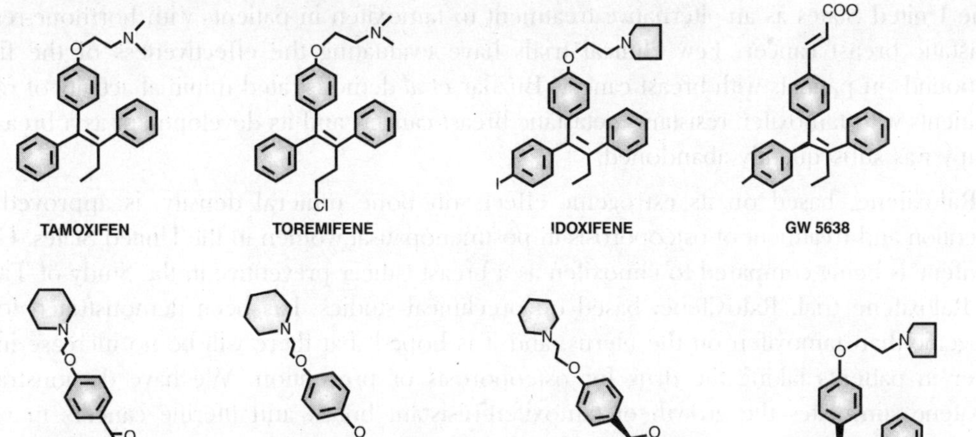

Fig. 16.1. Structures of available selective estrogen receptor modulators.

These newer agents can be broadly divided into two groups, newer SERMs, which include tamoxifen-like, triphenylethylene, agents such as toremifene (Farneston), idoxifene, and GW 5638, and the fixed-ring compounds (benzothiophenes) such as raloxifene (Evista), arzoxifene, EM 652, and CP 336,156; and selective estrogen receptor down-regulators (SERDs) such as ICI 182,780 (fulvestrant, Faslodex), SR 16234, and ZK 191703. Both of these classes of newer agents act directly on nuclear ER. The new SERMs, like tamoxifen, bind to ER-α, but result in incomplete receptor dimerization. SERDs also bind to the ER-α, destabilize the protein structure, and induce ubiquitin-mediated degradation of the protein, resulting in marked down-regulation of the receptor.

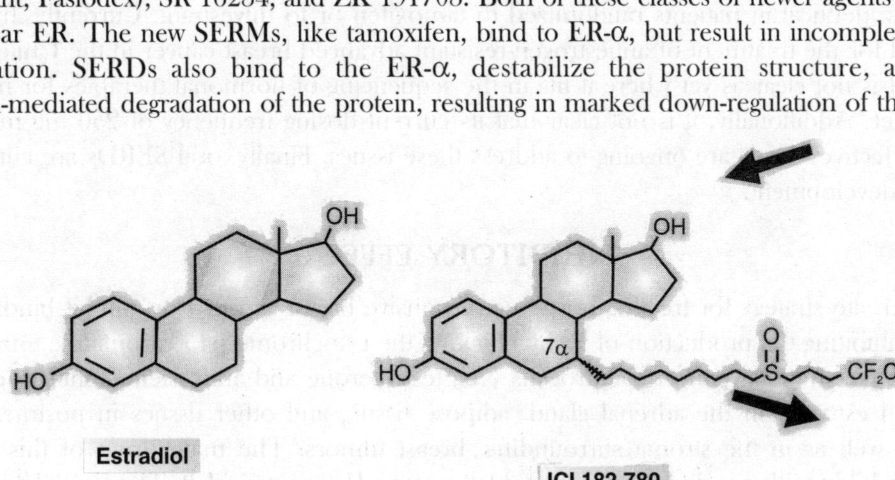

Fig. 16.2. Structure of fulvestrant compared to estradiol.

In phase II trials that enrolled patients with tamoxifen-resistant metastatic breast cancer, the new SERMs all showed low response rates (0-15%), suggesting cross-resistance to tamoxifen. Additionally, phase III trials that randomized over 1500 patients with metastatic breast cancer to either toremifene and idoxifene first line, or to tamoxifen, no significant difference in outcome was observed in patients treated with the new SERMs compared to tamoxifen. Based on these results, toremifene is approved in the United States as an alternative treatment to tamoxifen in patients with hormone-responsive metastatic breast cancer. Few clinical trials have evaluating the effectiveness of the fixed-ring compounds in patients with breast cancer. Buzdar *et al* demonstrated minimal activity of raloxifene in patients with tamoxifen-resistant metastatic breast cancer, and its development as a breast cancer therapy was subsequently abandoned.

Raloxifene, based on its estrogenic effects on bone mineral density, is approved for the prevention and treatment of osteoporosis in postmenopausal women in the United States. Currently, raloxifene is being compared to tamoxifen as a breast cancer preventive in the Study of Tamoxifen and Raloxifene trial. Raloxifene, based on preclinical studies, has been demonstrated to be less estrogenic than tamoxifen on the uterus, and it is hoped that there will be no increase in uterine cancer in patients taking the drug for osteoporosis or prevention. We have demonstrated that raloxifene stimulates the growth of tamoxifen-resistant breast and uterine cancers in vivo, and therefore, it is not recommended that raloxifene be used after adjuvant tamoxifen. Fulvestrant and other SERDs have been shown to have high affinity for the ER compared to tamoxifen, with none of the agonist activities.

Unlike the currently available hormonal agents, fulvestrant has poor oral bioavailability and must be given by intramuscular inj ection. In clinical trials, fulvestrant showed promise for the treatment of advanced tamoxifen-refractory breast cancer. However, in a phase III clinical trial where patients were randomized to fulvestrant or tamoxifen as first-line therapy of advanced breast cancer, fulvestrant did not demonstrate superiority to tamoxifen in terms of disease progression. When only patients with hormone receptor-positive breast cancer were evaluated, progression-free survival was identical in patients randomized to tamoxifen or to fulvestrant. Currently, fulvestrant is approved for the treatment of antiestrogen-resistant advanced breast cancer in the United States. However, it is not clear as yet where it fits in the sequencing of hormonal therapies for metastatic breast cancer. Additionally, it is not clear that its current dosing frequency of 250 mg monthly is the most effective. Trials are ongoing to address these issues. Finally, oral SERDs are currently in preclinical development.

INHIBITORY EFFECT

An alternate strategy for treating hormone-responsive breast cancer is to inhibit binding of E_2 to ER by inhibiting the production of E_2 by blocking the cytochrome p450 aromatase enzyme, the rate-limiting enzyme that converts androgens (i.e., testosterone and androstenedione) to estrogens (i.e., E_2 and estrone) in the adrenal gland, adipose tissue, and other tissues in postmenopausal women, as well as in the stroma surrounding, breast tumors. The main drugs of this type are aromatase inhibitors that include type I (steroidal) or type II (nonsteroidal). The steroidal inhibitors are competitive-substrate mimics of androstenedione. These include formestane and exemestane, which are irreversible inhibitors that bind with high affinity to the binding site of aromatase and become converted to a covalently bound intermediate.

Nonsteroidal inhibitors include the first-generation aromatase inhibitor aminoglutethimide and the third-generation compounds anastrozole and letrozole. The nonsteroidal aromatase inhibitors act by binding reversibly to the enzyme and competitively inhibiting binding of the substrate, androstenedione. The benefits of using aromatase inhibitors over tamoxifen are believed to be the complete deprivation of E_2 and perhaps better efficacy for ER-α- positive breast cancer. Recent clinical data have demonstrated that anastrozole, letrozole, and exemestane are more effective than tamoxifen as first-line treatments for metastatic breast cancer. Based on these clinical results currently both anastrozole and letrozole are approved as first-line treatment for post-menopausal ER-α-positive advanced breast cancer. These encouraging results from the metastatic disease setting led to a number of trial examining aromatase inhibitors as adjuvant therapies for early stage breast cancer. These adjuvant trials have two different designs, head-to-head comparisons of the aromatase inhibitors and tamoxifen; aromatase inhibitors after some duration of tamoxifen. The first head-to-head trial to report efficacy data to date is the Arimidex, Tamoxifen, Alone or in Combination trial. This trial randomized over 9000 post-menopausal patients with early stage breast cancer to anastrozole, tamoxifen, or the combination for 5 yr.

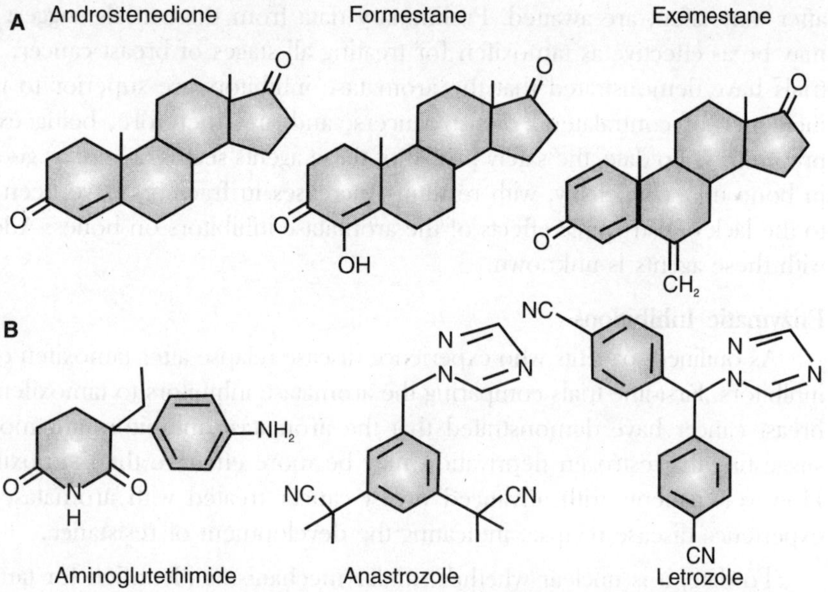

Fig. 16.3. Structures of steroidal and nonsteroidal aromatase inhibitors.

As outlined above, after 4 yr of follow-up, the absolute difference in recurrence rate in favor of patients treated with anastrozole, compared to tamoxifen, is 2.7% in patients with known hormone receptor-positive tumors. Although anastrozole was approved as an adjuvant therapy for hormone receptor-positive breast cancer in postmenopausal patients, no survival data are available as yet from this trial. Data from several other head-to-head trials are awaited. Continuing tamoxifen for more than 5 yr has not been demonstrated to improve outcome in patients with node-negative early-stage breast cancer. The MA-17 trial evaluated the use of letrozole after 5 yr of tamoxifen. At a follow-up of 30 mo, patients treated with letrozole had a significantly improved disease-free survival compared to patients treated with placebo. Additionally, patients treated with letrozole had a significantly improved survival compared to those treated with placebo.

Letrozole was recently approved for extended adjuvant therapy after 5 yr of tamoxifen. The Intergroup Exemestane trial evaluated the use of exemestane after 2 to 3 yr of adjuvant taimoxifen. Disease-free survival was significantly better in patients who switched to exemestane after 2 to 3 yr of tamoxifen, compared to patients who remained on tamoxifen for the remainder of the 5 yr. The results of several other trials, evaluating the optimal way of sequencing aromatase inhibitors after tamoxifen, are awaited. Preliminary data from these trials suggest that aromatase inhibitors may be as effective as tamoxifen for treating all stages of breast cancer. Additionally, the adjuvant trials have demonstrated that the aromatase inhibitors are superior to tamoxifen in reducing the incidence of contralateral breast cancers, and are, therefore, being examined as breast cancer preventives. To date, the safety profile of these agents seems at least as good as tamoxifen. Decreases in bone mineral density, with resulting increases in fractures have been demonstrated, secondary to the lack of estrogenic effects of the aromatase inhibitors on bones. Additionally, long-term safety with these agents is unknown.

Enzymatic Inhibitions

As outlined, patients who experience disease relapse after tamoxifen can benefit from aromatase inhibitors. First-line trials comparing the aromatase inhibitors to tamoxifen in patients with advanced breast cancer have demonstrated that the aromatase inhibitors are more effective to tamoxifen, suggesting that estrogen deprivation may be more effective than tamoxifen in delaying resistance. However, patients with advanced breast cancer treated with aromatase inhibitors will eventually experience disease relapse, indicating the development of resistance.

To date, it is unclear whether similar mechanisms identified for tamoxifen resistance are also involved in resistance to aromatase inhibitors. Patients treated with aromatase inhibitors clearly benefit from second-line endocrine therapy, suggesting that ER-α continues to be expressed and functional. Although the exact mechanisms contributing to aromatase inhibitor resistance have yet to be elucidated fully, in vitro studies have identified mutations within the aromatase gene that confer resistance. However, these mutations have not yet been identified in human breast carcinomas. Other studies have demonstrated that estrogen deprivation supersensitizes breast cancer cells to low levels of estrogen, thus creating a hypersensitive environment to overcome estrogen deprivation resulting in resistance.

In addition, results suggest that there is increased crosstalk between growth factor receptor signaling pathways and ER-α. ER-α has been shown to become activated and supersensitized by several different intracellular kinases, including MAPKs, insulin-like growth factors, and the PI3-K/Akt pathway. Therefore, the data suggest that the ER-α continues to be an integral part of the breast cancer cell-signaling pathway, even after resistance to aromatase inhibitors has developed.

CONCLUDING REMARKS

It is clear that the central regulator of growth for ER-α-positive breast tumors is ER-α. ER signaling does not appear to be independent of other cellular signaling pathways. Research studies have demonstrated that there is bidirectional crosstalk between EGFR/ HER-2/*neu*, MAPK, PI3-K/Akt, and ER-α that ultimately affects the cellular response of a cell to estrogens and mitogens. ER-α integrates numerous signals from hormones, growth factors, and intracellular kinases along with the array of coregulators to modulate cellular physiology and tumor pathology. The first generation antiestrogen/SERM, tamoxifen, has been effective at increasing overall survival of patients with ER-α-positive breast cancer. However, the eventual development of resistance to tamoxifen is common because of the multiple signaling networks affecting the function of the ER-α. As a result of this limitation, intense research over the last 25 yr has revealed the need for alternate treatment strategies to tamoxifen such as newer and better SERMs with less agonist activity in the uterus while being full antagonists in the breast. In addition to SERMs, other therapeutic approaches have emerged, such as the use of aromatase inhibitors to prevent the synthesis of E_2, and thus, deprive breast cancer cells and other cells of E_2.

Estrogen deprivation, however, may result in resistance, bone loss, and possibly decreases in cognitive brain function. The long-term effects of estrogen deprivation have not been realized, but with longer follow-up in the adjuvant aromatase inhibitor clinical trials, the consequences of using aromatase inhibitors will hopefully be clearer. Therapeutic opportunities also exist from research studies investigating the role of growth factor receptors, EGFR and HER-2/*neu*, and their intracellular signaling communicators, MAPK and PI3-K/Akt, inthe development oftamoxifen resistance. Trastuzumab, the HER-2/*neu*-specific antibody, has been shown to be highly effective at treating breast tumors-overexpressing HER-2/*neu*. Moreover, research data indicate a role for HER-2/*neu* in antiestrogen resistance demonstrating a rationale for using trastuzumab to treat or prevent antiestrogen-resistant breast cancer.

Gefitinib and other receptor tyrosine kinase inhibitors may also prove to be beneficial in preventing resistance when given in combination with tamoxifen or other SERMs or SERDs, such as fulvestrant. In addition to EGFR and HER-2/*neu* as important modulators of tamoxifen action, dissecting the specific interrelationship between ER-α and its coregulators may open the door for novel therapeutics. An understanding of how ER-α-mediated gene transcription is dys-regulated in breast cancer and how this changes in resistance during endocrine therapy will hopefully lead to improved strategies for the use of current and future antiestrogen therapies.

17

Chemotherapic Resistance

RESISTANCE TO RADIATION

For many years, the impact of ionizing radiation on cell biology and survival was not fully understood, and thought to be solely dependent on the levels of DNA damage caused following radiation exposure. Similarly, the mechanisms by which growth factors and cytokines modulate cell behavior were largely unknown. In the mid-1980s, with the discovery of the first *mitogen-activated protein kinase* (MAPK) pathway and with subsequent discoveries of other MAPK family pathways in the early 1990s, our understanding of the hormonal control of cell biology was provided with a greater degree of molecular underpinning. In light of these findings, the ability of ionizing radiation to control the activity of MAPK family (and other) signaling pathways was first investigated in the mid-1990s. It was discovered that ionizing radiation in a cell type-dependent manner simultaneously activates multiple intracellular signal transduction pathways: the activation of some pathways has been reported to be DNA-damage dependent, that of others by generation of lipids such as ceramide, whereas others have been noted to be dependent on mitochondria-derived reactive oxygen/nitrogen species and the activation of growth factor receptor tyrosine kinases. The precise roles of growth factor receptors and signal transduction pathways in cellular responses to radiation exposure are presently under intense investigation.

Generally, in a cell type and dose-dependent manner; inhibition of the *extracellular-regulated kinase* (ERK) 1/2, and to a greater extent, phosphatidylinositol 3 kinase (PI3K)/Akt, pathways can enhance radiosensitivity. The modulation of radiosensitivity by the ERK1/2 and Akt pathways has been correlated, in part, to the expression of both mutant active Ras isoforms and to growth factor receptors of the ErbB and insulin/insulin-like growth factor (IGF) families. The activation of the c-Jun NH_2-terminal kinase (JNK) 1/2, ERK1/2, and PI3K/Akt pathways in tumor cells has also been linked to the expression of paracrine ligands such as *transforming growth factor* (TGF)-α, ligands that can promote cell growth and survival after irradiation and that are generally only expressed at high levels in transformed cells. This chapter discusses the signal transduction pathways activated by ionizing radiation, the roles each pathway can potentially play in cellular responses after irradiation, and the molecular approaches being taken to radiosensitize tumor cells in vitro and in vivo.

History

Ionizing radiation has been shown to simultaneously activate multiple signal transduction pathways; the specific cassettes of pathways that are activated is cell type dependent. In the following subheadings, the roles of growth factor receptors, MAPK pathways, PI3 kinase/Akt, and nuclear factor (NF)-κB in cell survival after irradiation are described.

Enzymatic pathway

"MAP-2 kinase" was first reported by the laboratory of Dr. Thomas Sturgill in 1986. This protein kinase was originally described as a 42-kDa insulin-stimulated protein kinase activity whose tyrosine phosphorylation increased after insulin exposure, and which phosphorylated the cytoskeletal protein MAP-2. Contemporaneous studies from the laboratory of Dr. Melanie Cobb identified an additional 44-kDa isoform of this enzyme, termed ERK1. Because many growth factors and mitogens could activate these enzymes, the acronym for this enzyme was subsequently changed to denote mitogen-activated protein, or MAP kinase. Additional studies demonstrated that the p42 (ERK2) and p44 (ERK1) MAP kinases regulated another protein kinase activity ($p90^{rsk}$), and that they were themselves regulated by protein kinase activities designated MAPK kinase (MKK)1/2 (MAP2K), also termed MAPK extracellular-regulated kinase (MEK) 1/2. MKK1 and MKK2 were also regulated by reversible phosphorylation. The protein kinase responsible for catalyzing MKK1/2 activation was initially described as the proto-oncogene Raf-1. This was soon followed by another MEK1/2 activating kinase, termed MEKK1, which was a mammalian homologue with similarity to the yeast *Ste11* and *Byr2* genes. However, further studies have shown that the primary function of MEKK1 is to regulate the JNK1/2, rather than the ERK1/2, pathway.

More-recent studies have suggested that other enzymes at the level of MKK1/2 can also phosphorylate and activate ERK1/2, e.g., RIP2, which plays a role in tumor necrosis factor (TNF)-α-induced, but not EGF-induced, ERK1/2 activation. Plasma membrane receptors transduce signals through the membrane to its inner leaflet, leading to the recruitment and activation of guanine nucleotide exchange factors, which increases the amount of GTP bound to membrane-associated GTP-binding proteins, in particular Ras family proteins. There are four widely recognized isoforms of Ras: Harvey (H), Kirsten (K4A, K4B), and neuroblastoma (N). Receptor-stimulated guanine nucleotide exchange of Ras to the GTP-bound form permits Raf proteins and P110 PI3K to associate with Ras, resulting in kinase translocation to the plasma membrane environment, where activation of these kinases takes place. Ras contains a GTPase activity that converts bound GTP to GDP, resulting in inactivation of the Ras molecule.

Mutation of Ras results in a loss of GTPase activity, generating a constitutively active Ras molecule that can lead to elevated activity within downstream signaling pathways. Raf-1 is a member of a family of serine-threonine protein kinases comprising also B-Raf and A-Raf. All Raf family members can phosphorylate MKK1/2 and activate the ERK1/2 pathway. Thus, the Raf kinases act at the level of a MAPK kinase kinase (MAP3K). The NH_2 domain of Raf-1 can reversibly interact with Ras family members in the plasma membrane, and the ability of Raf-1 to associate with Ras is dependent on the Ras molecule being in the GTP-bound state. Additional protein serine/threonine and tyrosine phosphorylation(s) are also known to play a role increasing Raf-1 activity when in the plasma membrane environment. Protein kinase C (PKC) isoforms, which can be activated by radiation, have been proposed to be Raf-1-activating kinases. The initial biochemical analyses of purified Raf-1 demonstrated constitutive Y340-Y341 phosphorylation when the protein was coexpressed with Src. PI3 kinase-dependent phosphorylation of S338 may facilitate Raf-1 tyrosine phosphorylation by Src family members, leading to full activation of Raf-1. In this regard, ionizing radiation, compared to epidermal growth factor (EGF), potently enhances tyrosine phosphorylation of Raf-1.

Radiation activates Raf-1 but not B-Raf; B-Raf is not tyrosine phosphorylated. Of note, activating B-Raf mutations have been shown to play a role in tumorigenesis in melanoma, thyroid, colorectal and cholangiocarcinoma. Phosphorylation of Raf-1 at S259 by either Akt or the cAMP-dependent protein kinase (PKA) can inhibit Raf-1 activity and its activation by upstream stimuli. Phosphorylation of Raf-1 at S43 by PKA inhibits the interaction of Raf-1 with Ras molecules, thereby blocking Raf-1 translocation to the plasma membrane and its Ras-dependent activation. In contrast to Raf-1, the B-Raf isoform does not contain an equivalent to S43, but contains multiple sites of Akt-mediated (and potentially PKA-mediated) phosphorylation in addition to the B-Raf equivalent of S259. B-Raf can be activated by both Ras and by cAMP via the RAP1 GTPase. Thus, the regulation of the MEK1/2-ERK1/2 pathway is very complex, and in some cell types, may be both inhibited by cAMP/PKA, through Raf-1, as well as being stimulated by cAMP/RAP1 ,through B-Raf. Gene deletion studies suggest additional complexity that may have significant implications for therapeutic interventions: loss of Raf-1 function was embryonically lethal because of weak placental angiogenesis and hepatoblast apoptosis.

In Raf-1-null hepatocytes, ERK1/2 was activated by B-Raf rather than Raf-1, which suggests (i) fetal hepatocytes may utilize different Raf molecules to activate the ERK1/2 pathway in comparison to adult hepatocytes, and (ii) deletion of Raf-1 caused the hepatocytes to survive by recruiting in compensation B-Raf as the MAP3K activator for ERK1/2, which will tend to promote growth arrest over proliferation. With reference to (i), the relative ability of Raf-1 and B-Raf to activate the ERK1/2 pathway in primary hepatocytes and established HepG2 and HuH7 hepatoma cells is also different: in primary hepatocytes and HepG2 cells, ERK1/2 pathway activation by growth factors is dependent on Raf-1 and inhibited by cAMP, whereas in HuH7 cells B-Raf, and ERK1/2 activation can be enhanced by cAMP. This implies primary and fetal hepatocytes, and established cell lines can have very different signaling behavioral characteristics.

Thus, inhibition of one Raf family member, e.g., by a specific antisense approach, will be unlikely to have a profound prolonged growth inhibitory or radio sensitizing effect on tumor cells because of a compensatory utilization of another Raf family member. In light of this, whereas antisense oligonucleotides to Raf-1 have been shown in vitro and in vivo to enhance the radiosensitivity of tumors cells, such studies have been largely abandoned in the clinic because of poor response rates. In addition to playing a role in the activation of the ERK1/2 pathway, it is important to note that Raf-1 may act on substrates other than MEK1/2, such as the myosin phosphatase-binding protein. In this instance, radiation-stimulated Raf-1 was shown to regulate this process. Raf-1 has been proposed to act as an inhibitor of apoptosis signaling kinase 1 (ASK1) by binding to ASK1: the inhibitory actions of Raf-1 were reported to be independent of Raf-1 protein kinase activity. Several groups have shown that the epidermal growth factor receptor ([EGFR], also called ErbB1 and Her-1) is activated in response to irradiation of various carcinoma cell types.

Radiation exposure in the range of 1 to 2 Gy, via activation of ErbB1, can activate the ERK1/2 pathway to a level similar to that observed by physiologic, growth stimulatory, EGF concentrations (~0.1 n*M*). Recent publications argue that radiation- induced reactive oxygen and nitrogen species (ROS/RNS) appear to play an important role in the activation of ErbB family receptors and the ERK1/2 pathway. The primary target of ROS/RNS is likely to be protein tyrosine phosphatases, each of which contain an ROS/RNS sensitive cysteine residue within their active sites, and which is a residue that is essential for phosphatase activity. The actions of ErbB receptor autocrine ligands have also been shown to play important roles in the activation of the ERK1/2 pathway after radiation exposure. TGF-α has been shown to mediate secondary activation of ErbB1 and the downstream ERK1/2 and JNK1/2 pathways after irradiation of several carcinoma cell lines. Radiation caused an ERK1/2-dependent cleavage of pro-TGF-α in the plasma membrane within 2 h that led to its release into the growth media. Increasing the radiation dose from 2 up to 10 Gy enhanced both the secondary activation of ErbB1 and the secondary activation of the ERK1/2 and JNK1/2 pathways, suggesting that radiation can promote a dose-dependent increase in the cleavage of pro-TGF-α that reaches a plateau at ~10 Gy.

It should be noted that in contrast to the secondary activation, primary activation of the receptor and signaling pathways appeared to have come to a plateau at 3-5 Gy. In addition, signaling by Ras proteins ERK1/2 and p53, the activities of which can be increased following radiation exposure, has been shown over many hours/days in a variety of cell systems to increase the expression of autocrine factors such as heparin-binding EGF and epiregulin. More-recent findings from our group have shown that loss of mutant active K-Ras expression in HCT116 cells not only causes a reduction in epiregulin expression, but also causes a compensatory increase in heregulin expression, resulting in a switch from radiation-induced ERK1/2 signaling to radiation-induced PI3K/Akt signaling in this cell type. These findings argue that the activation of ErbB family receptors and the ERK1/2 pathway by radiation will be influenced by both the Ras and p53 status (mutant or wild type) of a given tumor cell.

Stress-activated pathway

JNK1 and JNK2 were initially described biochemically to be stress-induced protein kinase activities that phosphorylated the NH_2-terminus of the transcription factor c-Jun; hence, the pathway is often called the stress-activated protein kinase pathway. Multiple stresses increase JNK1/2 (and the subsequently discovered JNK3) activity including UV- and γ-irradiation, cytotoxic drugs, bile acids, and ROS (e.g., H_2O_2). Phosphorylation of the NH_2-terminal sites Ser-63 and Ser-73 in c-Jun increases its ability to transactivate activating protein 1 enhancer elements in the promoters of many genes. It has been suggested that JNK1/2 can phosphorylate the NH_2-terminus of c-Myc, potentially playing a role in both proliferative and apoptotic signaling. In a similar manner to the previously described ERK1/2 MAPK pathway, JNK1/2 activities were regulated by dual threonine and tyrosine phosphorylation, which were found to be catalyzed by a protein kinase analogous to MKK1/2, termed stress-activated extracellular-regulated kinase 1, also called MKK4. An additional isoform of MKK4, termed MKK7, was subsequently discovered.

As in the case of MKK1 and MKK2, MKK4 and MKK7 were regulated by dual-serine phosphorylation. Recent studies have also indicated that Akt can phosphorylate and inhibit the activity of MKK4, demonstrating crosstalk between the PI3 kinase and JNK pathways. In contrast to the ERK1/2 pathway, however, which appears to primarily utilize the 3 protein kinases of the Raf family to activate MKK1/2, at least 10 protein kinases are known to phosphorylate and activate MKK4/7, including the Ste11/Byr2-homologues MKKK1-4, as well as proteins such as TGF-β-activated protein kinase 1 and tumor progression locus 2. Cleavage of MEKK1 by caspase molecules into a constitutively active molecule may play an amplifying role in the execution of apoptotic processes. Upstream of the MAP3K enzymes is another layer of JNK1/2 pathway protein kinases, e.g., Ste20-homologues and low-molecular-weight GTP-binding proteins of the Rho family, in particular CDC42 and RAC1. It is not clear how growth factor receptors, e.g. ErbB1, activate the Rho family low-molecular-weight GTP-binding proteins; one mechanism may be via the Ras proto-oncogene, whereas others have suggested via PI3K and/or PKC isoforms.

In addition, others have shown that agonists acting through the TNF-α and fatty acid synthase receptors, via sphingomyelinase enzymes generating the messenger ceramide, can activate the JNK1/2 pathway by mechanisms that may act through Rho family GTPases. Thus, there appear to be at least three distinct mechanisms by which ionizing radiation activates the JNK1/2 pathway. Initial reports demonstrated that radiation-induced ceramide generation, and that the clustering of death receptors on the plasma membrane of cells played an important role in JNK1/2 activation. This was linked to a proapoptotic role for JNK1/2 signaling following irradiation of cells. Other studies have argued that radiation-induced JNK1/2 activation was dependent on the *ataxia telagectasia mutated* (ATM) and c-Abl proteins. Studies by our group of laboratories have shown that low dose radiation activates JNK1/2 in two waves in carcinoma cells.

The first wave of JNK1/2 activation was dependent on activation of the TNF-α receptor, whereas the second wave of JNK1/2 activity was dependent on ErbB1 and autocrine TGF-α. In HCT116 cells, radiation also activates JNK1/2 in two waves; however, in these cells the second wave of activation is most likely dependent on autocrine epiregulin. Finally, it is also possible that radiation-induced JNK1/2 activation could be a secondary event to the activation of effector procaspases: cleavage of the upstream activator MEKK1 can lead to constitutive activation of this enzyme and the downstream JNK1/2 pathway, which, in some cell types, plays a key role in full commitment to apoptotic cell death.

p38 MAPK pathway

The p38 MAPK pathway was originally described as a mammalian homologue to a yeast osmolarity-sensing pathway. It was soon discovered that many cellular stresses activated the p38 MAPK pathway, in a manner similar to that described for the JNK pathway. Rho family GTPases appear to play an important role as upstream activators of the p38 MAPK pathway, and via several MAP3K enzymes, e.g., the PAK family, regulate the MAP2K enzymes MKK3 and MKK6. At least four isoforms of p38 MAPK exist, termed p38-α, -β, -γ, and -δ. There are several protein kinases downstream of p38 MAPK enzymes that are activated following phosphorylation by p38 isoforms including $p90^{MAPKAPK2}$ and mitogen- and stress-activated protein kinase 1/2. $P90^{MAPKAPK2}$ phosphorylates and activates HSP27, whereas mitogen- and stress-activated protein kinase 1/2 can phosphorylate and activate transcription factors that regulate survival such as cyclic-AMP response element binding protein. The role of p38 MAPK signaling in cellular responses is diverse, depending on the cell type and stimulus. For example, p38 MAPK signaling has been shown to both promote cell death as well as enhance cell growth and survival.

The ability of ionizing radiation to regulate p38 MAPK activity appears to be highly variable, with different groups reporting no activation, weak activation, or strong activation. This is in contrast to the ERK1/2 and JNK1/2 pathways where radiation-induced activation has been observed by many groups, in diverse cell types, in response to low and high radiation doses. In studies where p38 MAPK activation has been observed following exposure to ionizing radiation, the p38-γ isoform has been proposed to play an essential role in causing radiation-induced G_2/M arrest. In these studies, p38-γ signaling was dependent on expression of a functional ATM protein. In support of this finding, overexpression of constitutively active MKK6 also enhanced cell numbers in G_2/M phase. Other groups have argued that p38-α also plays a role in UV radiation-induced G_2/M arrest. Collectively, these findings suggest that specific inhibitors of p38-γ may have therapeutic benefit.

ERK5 pathway

The "*big MAP kinase*" pathway was first described in 1995. The term "big" derives from the fact that whereas the molecular masses of ERK1/2 and JNK1/2 are 42/44 and 46/54 kDa, respectively, ERK5 has a mass of ~90 kDa. The upstream activators of ERK5, the MEK5 isoforms, have a similar molecular mass to other MAP2K molecules and display different subcellular locations. The response of the MEK5-ERK5 pathway to growth factors such as EGF is similar to that of the MEK1/2-ERK1/2 pathway, including in many, but not all cell types, a dependency on Ras signaling. Furthermore, MEK5 is also partially inhibited by the MEK1/2 inhibitors PD98059 and U0126. PD98059 acts as inhibitor of Raf-mediated phosphorylation of MEK1/2/5 molecules, but it is neither a kinase domain inhibitor of Raf proteins nor of activated MEK1: in cells PD98059 is a relatively good inhibitor of MEK1 and MEK5 (IC_{50} ~5 μ*M*) but a much poorer inhibitor of MEK2 (IC_{50} ~40 μ*M*). U0126 is reported as a more equipotent inhibitor of MEK1/2/5 that blocks both activating phosphorylations on the MEK1/2/5 proteins as well as their kinase domain activity, with IC_{50}s in the ~0.5 μ*M* range.

The clinically used MEK1/2 inhibitor PD184352 (CI1040) acts in a similar manner to U0126, and at concentrations below 10 μ*M*, has shown specificity for MEK1/2 over MEK5; its use may be a useful tool to demarcate between MEK1/2-ERK1/2 and MEK5-ERK5 signaling dependencies. The MAP3K enzymes recently shown to phosphorylate MEK5, MEKK2, and MEKK3 have been previously linked to signaling through the JNK1/2 pathway. ERK5 has been proposed to phosphorylate and activate the transcription factors MEF2C, SAP1a, and c-Myc. In a similar manner to the ERK1/2 pathway, the ERK5 cascade has been proposed to play a key role in growth factor-stimulated cell growth and in cell survival processes. In growth factor-deprived PC12 cells, the ERK1/2 and ERK5 pathways appeared to each contribute ~50% of a PD98059/U0126-inhibitable survival signal. The ability of ionizing radiation to activate MEK5-ERK5 is unknown.

Based on the fact that ErbB1/Ras signaling can promote MEK5-ERK5 activation, and radiation also activates these molecules, it seems likely this pathway will be stimulated following exposure of cells to ionizing radiation. One report has shown in drug-resistant MCF-7 cells that MEK5 is overexpressed and plays a protective role against chemotherapeutic agents and death receptor activation. Thus, it is possible that inhibition of the MEK5-ERK5 signaling module may promote radiation-induced death.

PI3 Kinase/Akt Signaling

PI3K enzymes consist of two subunits, a catalytic p110 subunit and a regulatory subunit, p85; several different classes of PI3K enzymes exist. The p85 subunit of PI3K enzymes contains a phosphotyro sine-binding domain. The major catalytic function of the phosphatidyl inositol 3 kinase enzymes is in the p110 subunit that acts to phosphorylate inositol phospholipids (phosphatidyl inositol 4,5 bis-phosphate) at the 3 position within the inositol sugar ring. Mitogens such as TGF-α and heregulin stimulate tyrosine phosphorylation of ErbB family receptors, providing acceptor sites for the phosphotyrosine domain of p85. Binding of p85 to active ErbB receptors (predominantly ErbB3) results in p110 PI3K activation. Other studies have suggested that mutant oncogenic H-Ras, or serpentine receptors that are stimulated by mitogens in which the p110 subunit of PI3K can directly bind to Ras-GTP, lead to catalytic activation of the kinase. The molecule inositol 3,4,5 trisphosphate is an acceptor site in the plasma membrane for molecules that contain a plecstrin-binding domain, in particular, the protein kinases phosphoinositide-dependent kinase (PDK) 1 and Akt (also called protein kinase B). PDK1 is proposed to phosphorylate and activate Akt.

Signaling by PDK1 to Akt and by PDK1 and Akt downstream to other protein kinases such as PKC isoforms, GSK3, mTOR, $p70^{S6K}$, and $p90^{rsk}$, has been shown to play a key role in mitogenic and metabolic responses of cells as well as protection of cells from noxious stresses. The antiapoptotic role of the PI3K/Akt pathway has been well documented by many investigators in response to numerous noxious stimuli, and in some cell types, the antiapoptotic effects of ErbB signaling have been attributed to activation of the PI3K/Akt pathway. ErbB signaling to PI3K/Akt has been proposed to enhance the expression of the mitochondrial antiapoptosis proteins BCL-XL, MCL-1, and caspase inhibitor proteins such as c-FLIP isoforms. Enhanced expression of BCL-XL and MCL-1 will protect cells from apoptosis via the intrinsic/mitochondrial pathway, whereas expression of c-FLIP isoforms will block killing from the extrinsic pathway via death receptors. In addition, Akt has been shown to phosphorylate Bcl2-antagonist of cell death and human procaspase 9, thereby rendering these proteins inactive in apoptotic processes.

Inhibitors of ErbB signaling have been shown to decrease the activity of the PI3K/Akt pathway in a variety of cell types and to increase the sensitivity of cells to a wide range of toxic stresses including cytotoxic drugs and radiation. Activation of Akt was shown to protect cells from death in the presence of ErbB receptor inhibition. These findings strongly argue that PI3K/Akt signaling is a key cytoprotective response in many cell types downstream of ErbB family receptors. Inhibition of PI3 kinase signaling in cells has been achieved by use of several drugs, notably wortmannin and LY294002. Wortmannin inhibits PI3K isoforms with IC_{50} values in the low-anomolar range and LY294002 with IC_{50} values in the low-micromolar range. Concentrations of 10 μM or less of either drug is sufficient to block PI3K signaling. The therapeutic usefulness of PI3K inhibitors is limited because of systemic toxicities, e.g., wortmannin was noted to cause hemorrhages, and the development of effective PI3K inhibitors without dose-limiting side effects is being undertaken by many pharmaceutical companies.

Signaling Pathway

Inhibition of growth factor receptors

Signaling by ErbB family of receptors is, in general, believed to be proproliferative and cytoprotective. In some cell types, however, signaling by ErbB1 is known to promote growth arrest and apoptosis. Because both receptor expression as well as autocrine growth factor levels are often increased in carcinoma cells compared to normal tissue, many laboratories have studied signaling by the ErbB family in tumor cell growth and survival control. Thus, it has been discovered that when signaling from ErbB family receptors is blocked, either by use of inhibitory antibodies (e.g., C225, 4D5 Herceptin, monoclonal antibody 806), small-molecular-weight inhibitors of receptor tyrosine kinases (e.g., PD183805 [also called CI1033], PKI166, AG1478, PD153035, ZD1839, PD169414, OSI774, AG825, AG879), dominant-negative truncated receptors (e.g., dominant-negative EGFR-CD533, dominant-negative ErbB2) or antisense approaches (antisense EGFR), that tumor cell growth can be reduced, and the sensitivity of these cells to being killed by noxious stresses increased. The antibodies C225 and 4D5 Herceptin bind to the extracellular portions of ErbB1 and ErbB2, respectively.

In the instance of ErbB1, C225 appears to bind to the portion of the molecule that associates with growth factor ligands such as EGF and TGF-α. Thus the ability of growth factors, in the presence of receptor bound C225, to stimulate ErbB1 receptor function is abolished. C225 does not block the primary activation of the receptor or ERK1/2 following irradiation, in general agreement with the ligand-independent nature of this process. The antiproliferative and antisurvival mechanisms of action of Herceptin appear to be more complex, in as much as whereas Herceptin binds to ErbB2, this receptor has no known ligand. Instead, it appears that Herceptin acts by causing the internalization and degradation of ErbB2, as well as by blocking ErbB2 heterodimerization with other ErbB family members. Both C225 and Herceptin have been shown to individually kill cells, and to interact in a synergistic fashion in combination with standard therapeutic regimens such as ionizing radiation, cisplatin and taxol to reduce tumor cell survival both in vitro and in vivo. Both C225 and Herceptin are entering the clinic and it is likely, despite setbacks for C225 in Food and Drug Administration approval, that both agents will become standard tools in the treatment of epithelial cell cancers. More-recent studies have used monoclonal antibodies to target truncated forms of ErbB1, e.g., EGFR VIII. In these studies, a novel monoclonal antibody, 806, was found to potently inhibit truncated forms of ErbB1 and more weakly inhibit full-length receptors.

The inhibition of receptor function correlated with reduced tumor cell growth in vitro and in vivo. Of note, however, it is presently unclear whether all of the antitumor effects of anti-ErbB receptor antibodies are mediated solely via receptor inhibition or by a combination of receptor inhibition and enhanced immunological reactivity in vivo because of the Fc portion of the antibody.

Small-molecule inhibitors of the tyrosine kinase domains of the ErbB family of receptors have been used with some success in blocking tumor cell growth and survival both in vitro and in vivo. The inhibitors AG1478, ZD1839, PD 153035 (also called AG1517), PKI1 66, OSI774, CI1033 (PD183805), and PD169414 (an irreversible inhibitor), all bind to the catalytic kinase domain of ErbB1 and inhibit tyrosine kinase activity. Some studies have suggested that CI1033 binds to, and inhibits, all ErbB kinase domains. Inhibition of ErbB1 kinase activity not only blocks phosphorylation of ErbB1 itself in response to the growth factors that it binds, but also inhibits the transphosphorylation of other ErbB family members by ErbB1.

In addition to inhibiting ErbB1, the tyrphostin AG1478 has been shown to inhibit ErbB4. The tyrphostin inhibitors AG825 and AG879 are ErbB2 inhibitors with apparently weaker kinase specificity than AG1478 for ErbB1/4, and they can inhibit Trk receptors. Thus, AG825/AG879, together with AG1478, have the potential to not only impact on EGF/TGFα signaling through ErbB1, but also neuregulin/heregulin signaling through ErbB4 and ErbB3. Small-molecular-weight ErbB inhibitors are currently in clinical trials, both as stand-alone agents and in combination with ionizing radiation and other standard chemotherapeutic agents. In addition to use of antibodies and small-molecular-weight inhibitors, the ErbB family of receptors has been inhibited by the use of dominant-negative and antisense approaches. In particular, expression of truncated forms of ErbB1 (EGFR-CD533), ErbB2 and ErbB3 in a variety of cell types has been shown to reduce proliferation and survival of both normal and tumor cells in vitro and in vivo.

The dominant-negative approaches are believed to act by blocking homo- and hetero-dimerization of ErbB family members, reducing receptor transphosphorylation and thus, downstream signaling by the receptors. Initial studies demonstrated that radiation could activate ErbB1 and subsequent investigations using dominant-negative EGFR-CD533 demonstrated that it could block this effect. Whether such adenoviral approaches, which lack a true "*bystander effect*" on uninfected tumor cells, can be translated into the clinic, is unknown. In addition to the ErbB family, other growth factor and cytokine receptors are believed to play an important role in cellular radiation responses, such as IGF1R. Prolonged exposure of MCF-7 cells to tamoxifen results in the development of a cell line that expresses high levels of ErbB1 and its ligand TGF-α.

Exposure of these cells to ZD1839, results in the development of a cell line resistant to the anti-proliferative and cytotoxic effects of tamoxifen and ErbB receptor inhibitors, but that has become sensitive to inhibitors of the IGF1R such as AG1024. Radiation can enhance tyrosine phosphorylation of the IGF1R, and can radiosensitize tumor cells. Cytokines such as TNF-α, interleukin-6, urokinase-type plasminogen activator, and TGF-β have also been proposed to control cell survival responses following irradiation. Radiation has been shown to cause rapid activation of the TNF-α receptor and in addition, radiation stimulated signaling modules such as the classical MAPK and p38 pathways are known to enhance the synthesis of TNF-α ligand. TNF-α signaling following irradiation may lead to the activation of both procaspase enzymes as well as the cytoprotective transcription factor NF-κB. Thus, the cellular outcome of radiation- induced TNF-α receptor signaling will be a complex summation of opposing cellular signals.

TGF-β in nontransformed cells can cause growth arrest and differentiation. In tumor cells, TGF-β has been shown to cause in a cell-type-dependent manner either cytoprotection or apoptosis.

In some cells, TGF-β appears to confer a protective effect via ERK1/2 signaling and potentially the expression of molecules such as heparin-binding EGF and BCL-X_L. In contrast, TGF-β in other cell types appears to protect cells in a Ras- and ERK1/2-independent manner that is dependent on PI3K signaling. Thus, multiple cytokines, in addition to those that bind to ErbB family receptor, play a role in the radiation responses of both nontransformed as well as tumor cells.

Inhibition of Ras function

Mutated active *H-Ras* was one of the first oncogenes discovered, in bladder cancer cells, and approximately one third of human cancers have *Ras* mutations, primarily the *K-Ras* isoforms. Early studies examining Ras transformation noted that Ras proteins were prenylated and proteolytically processed, causing their translocation into the plasma membranes of cells. Ras proteins that could not be prenylated were not located in membranes and could not cause transformation. Subsequently, pharmaceutical companies developed inhibitors of Ras prenylation (farnesyltransferase and geranylgeranyltransferase inhibitors) that block the prenylation of Ras proteins as well as other small GTP-binding proteins including RAC/Rho/CDC42. Antisense techniques have also been used to manipulate Ras expression. Inhibition of small GTP-binding protein prenylation has been shown to both growth arrest and radiosensitize tumor cells in a mutated active Ras-independent fashion, arguing that Ras may not be the primary target of this therapeutic approach. In the clinic, as single agents, farnesyltransferase inhibitors have not proven highly successful and the long-term future of such agents remains unclear.

AKT function

Data from several groups supports the argument that a key radioprotective pathway downstream of plasma membrane receptors is the PI3K pathway. Inhibition of p110 PI3K function by use of the inhibitors LY294002 and wortmannin radiosensitzes tumor cells expressing mutant Ras molecules or wild-type Ras molecules that are constitutively active. It is possible that these inhibitors may exert a portion of their radiosensitizing properties by inhibiting proteins with PI3K-like kinase domains such as ATM, ATM–Rad3-related, and DNA-PK. Expression of a constitutively active p110 PI3K molecule was able to recapitulate partially the expression of mutant H-Ras in protecting cells from radiation toxicity in these studies. In the same studies that PI3K p110 inhibitors were shown to radiosensitize cells, p38 inhibitors, e.g., SB203580, were used at concentrations that also have been shown to blunt PDK1 andAkt activity, did not radiosensitize cells. Recent data have also suggested that antisense PDK1 oligonucleotides do not radiosensitize cells.

In contrast, dominant-negative Akt radiosensitizes HCT116 cells expressing mutant active H-Ras. Thus, LY294002 and wortmannin may sensitize tumor cells to radiation through PI3K-dependent, PDK1- independent pathway(s). Of note, in these cell lines and culture conditions, inhibition of the ERK1/2 pathway did not appear to alter the radiosensitivity of cells. The cytotoxic effects of drugs, as well as radiation, can be magnified by inhibition of ErbB receptors that is paralleled by a reduced ability of cells to activate the ERK1/2 and PI3K pathways. For example, expression of dominant-negative EGFR-CD533 enhanced apoptosis and radiosensitized MDA-MB-231 mammary carcinoma cells that were dependent on, at least in part, inhibition of radiation-induced ERK1/2 signaling: neither basal activity nor activation of PI3K/Akt was blocked under these conditions.

Expression of this dominant-negative ErbB1 molecule could also radio sensitize glioblastoma cells that correlated with both reduced basal ERK1/2 activity and radiation- induced ERK1/2 activation. In many cell types, ERK1/2 signaling does not appear to play a role in controlling radiosensitivity. In those cells where effects have been observed, the abilities of MEK1/2 inhibitors to enhance cell killing by radiation was originally linked to a derangement of radiation-induced G_2/M growth arrest and enhanced apoptosis. Although others have not noted this finding, despite observing radiosensitization. In general agreement with this notion, other groups have argued that prolonged activation of the ERK1/2 pathway in fibroblasts following irradiation promotes radiosensitivity in some cell types by abrogating the G_2/M checkpoint. In agreement with this concept, and of particular note, we have recently shown that the G_2/M checkpoint abrogator UCN-01 potently activates the ERK1/2 pathway.

MEK1/2 inhibitors synergize with UCN-01 to promote cell death, but radiation does not further enhance the apoptotic response of these cells, indicative that MEK1/2 inhibitors require an intact G_2/M checkpoint to enhance radiation-induced apoptosis. The dual positive and negative nature of ERK1/2 signaling in the control of cell survival has also been observed for other DNA damaging agents such as adriamycin and UV radiation. An additional example of this dual nature of ERK1/2 signaling is displayed by DU145 human prostate cancer cells. These cells secrete the ErbB1 ligand TGF-α that confers autocrine growth through ERK1/2 signaling. Ionizing radiation markedly increases the release of TGF-α providing a growth stimulus that is at odds with cellular repair mechanisms. If ErbB1-ERK1/2 signaling is *transiently* blocked either by the tyrphostin AG1478 or by a MEK1/2 inhibitor before irradiation, then DU145 cell growth is retarded and radiation-induced cell killing is decreased. Moreover, if ErbB1 is strongly activated by EGF or TGF-α immediately after irradiation, then cell killing is increased, as would be expected, i.e., transient inhibition of radiation-induced ERK1/2 signaling or supra-stimulation of ERK1/2 activity at the time of irradiation radiosensitizes tumor cells.

Removal of MEK1/2 inhibitor from the growth media 24 and 48 h after irradiation results in a null effect on DU145 cell radiosensitivity. Similar data were also obtained in A431 squamous carcinoma cells. On the other hand, we have previously reported that following irradiation, *prolonged* inhibition of ERK1/2 (≥72 h) can significantly increase the apoptotic response of DU145 and A431 cells and reduce clonogenic survival. Therefore, the interruption of ERK1/2 signaling can either enhance or degrade carcinoma cell survival depending on its timing and duration. The transcription factor NF-κB has been shown to be downstream of ErbB and TNF-α receptors and was proposed to act in radioprotection. However, other studies have argued against NF-κB as a direct radioprotective factor. Several manuscripts have suggested that NF-κB signaling is regulated by the PI3K pathway, whereas others have suggested ERK1/2 signaling can regulate this transcription factor through autocrine mechanisms, e.g., enhanced TNF-α expression. Radiation-induced inhibitory κB kinase activation has been argued to be PI3K-independent. ERK1/2 signaling has the potential to inhibit expression of the protein PAR4 that is downstream of mutant Ras molecules.

Proteinase-activated receptor (PAR)4 is a protein inhibitor of PKC-ζ and NF-κB function. More recently, PAR4 has been shown to radiosensitize prostate tumor cells. This may be, in part, because of enhanced signaling from death receptors. Thus, PAR4 may be a link between ERK1/2 signaling,

NF-κB, function, and radiosensitivity. Hence in a cell-type-dependent manner, PI3K, NF-κB, or ERK1/2 signaling, downstream of receptors and Ras molecules, or a combination of these signals, may play a radioprotective role. Tumor samples isolated from patients by fine-needle aspiration are often subjected to a battery of tests to determine some of the molecular characteristics known to predispose tumor cells to invasion and chemotherapy/radiation resistance, e.g., ErbB2 expression, Ras protein levels, and estrogen/progesterone receptor expression. Based on knowledge gained in the last 10 yr, it should be straightforward to utilize such samples to also determine the expression of other growth factor receptors, e.g., IGF1R; paracrine ligands, e.g., TGF-α; and the activity status of pathways such as ERK1/2 and PI3K/Akt. Armed with such knowledge, it should then be possible to design rationally therapeutic strategies that are tailored to the malignancy in each individual.

RESISTANCE TO TAXANES

Resistance is a relatively new chapter in the long history of taxanes and their use in oncology. Originally mentioned as toxic substances isolated from the leaves of the yew tree, such as *Taxus baccata* L. in 1856, as a component of clarified butter for use in treating cancer in 1912, and as a cytotoxic isolated from the bark of *T. brevifolia* in the National Cancer Institute's screening program in 1962, the first isolated taxane that had anticancer activity in tissue culture systems, designated taxol, was reported in 1971 by Wall and coworkers. A breakthrough came in 1979 when Susan Horwitz's laboratory found that taxol stabilized microtubules, which was in contrast to the depolymerizing agents: vinca alkaloids and colchicine. It took an additional 13 years to develop a suitable formulation and source of taxol that ultimately led to its approval for use in cancer patients beginning in 1992. Taxol is also known as paclitaxel; the latter name will be used in this chapter. As early as 1978, and reaffirmed over the following decade, it was apparent that paclitaxel was not equally effective in human tumor xenograft model systems. This was reemphasized in the early phase II trials that took place in the mid 1980s, where paclitaxel had good, weak, and poor activity when tested in patients with ovarian carcinoma, melanoma, and renal carcinoma, respectively. Ultimately, paclitaxel was approved for use in patients with ovarian, lung, or breast cancers.

Paclitaxel given as monotherapy to patients with these types of advanced cancers induces a 15-25% response rate and a 10- to 13-mo median survival time. It is now apparent that patients with certain tumor types (i.e., colon and renal) have very low response rates to taxane-containing regimens and therefore, have inherent resistance. More insidious is the case where some patients initially respond to paclitaxel, but then fail to respond on repeated courses of therapy. These patients have acquired resistance. Therefore, resistance to taxanes is a major limiting factor in successful taxane therapy. A second taxane, docetaxel (also known as Taxotere), with good anticancer activity in preclinical models was reported in 1991 and has also been used to treat lung and breast cancer since 1996 and 1999, respectively. Paclitaxel and docetaxel are highly related molecules. Both contain a baccatin III core composed of four fused rings including a four-membered oxetane ring; the C-10 position is deacetylated in docetaxel.

Both baccatin cores are esterified at the C-13 position of the A ring; paclitaxel and docetaxel have an *N*-benzoyl-β-phenylisoserine and an *N-tert*-butoxycarbonyl-β-phenylisoserine side chain, respectively. It stands to reason that if the molecular basis for resistance to taxane were understood,

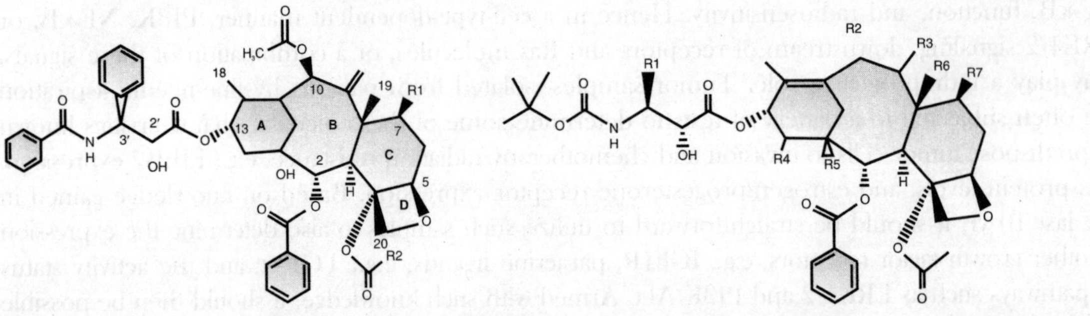

Taxane	R1	R2
Paclitaxel	OH	CH_3
BSM 184476	OCH_2SCH_3	CH_3
BSM 188797	OH	OCH_3

Taxane	R1	R2	R3	R4	R5	R6	R7
Docetaxel	phenyl	OH	O	H	OH	CH_3	OH
BAY 59-8862	ι-phenyl	$OCOCH_3$	O	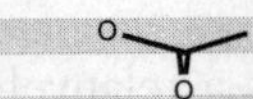		CH_3	OH
BSM-275183	*t*-butyl	$OCOCH_3$	O	H	OH	CH_3	OH
XRP9881	phenyl	$OCOCH_3$	O	H	OH	--- CH_2 ---	
XRP6258	phenyl	OCH_3	O	H	OH	CH_3	OCH_3
MAC-321	2-furanyl	OH	O	H	OH	CH_3	$OCOCH_2CH_3$
DJ-927	3-F-2-pyridyl			H	OH	CH_3	OH

Fig. 17.1. Chemical structure of taxanes.

then resistance to paclitaxel or docetaxel might be avoided. This chapter provides an overview of the mechanisms of resistance to taxanes. In the final analysis, one striking observation is clear. Resistance to taxanes is complex and can occur by multiple mechanisms including alteration in microtubules, transporters, proteins that regulate the cell cycle, apoptotic machinery, and signal transduction pathways. The importance of these mechanisms regarding the treatment of human disease will require careful evaluation of qualitative and quantitative changes for each mechanism. Reassembling all the data, or perhaps finding key markers, will be necessary to identify those patients with tumors that will respond well to taxanes or require therapeutic strategies to circumvent resistance.

ROLE OF MICROTUBULE

A working knowledge of how taxanes interact with its primary (but probably not only) target, the microtubule, and how it induces cell death is needed before discussing resistance mechanism. Microtubules are highly dynamic polymers whose essential element is the α/β-tubulin heterodimer. Each subunit binds GTP. One molecule is tightly bound to α-tubulin at a nonexchangeable site. The second molecule binds to a homologous site in β-tubulin at the exchangeable site; it undergoes hydrolysis and is required for normal tubulin polymerization. Under certain conditions, the dimer binds in a head to tail fashion to form linear protofilaments. About 13 protofilaments associate in parallel to a cylindrical axis to form the microtubule of 25 nm in diameter and up to micrometers in length. The polymer is initially formed from a short microtubule nucleus located at the so-called "–" end; in the case of the dividing cell this is found within the centrosome. Although the protofilament can grow from both ends, α-/β-tubulin dimers are added more quickly to the opposite or "+" end. The microtubule can elongate slowly or shorten rapidly by adding or removing tubulin dimers from the existing polymer, respectively.

The conversion from growing to shrinking is called catastrophe, whereas the conversion from shrinking to growing is called rescue. The transition between these two behaviors is known as dynamic instability. How do microtubules participate in cell division? After chromosome replication during cell division, microtubules emanating from a centro some undergo pronounced elongation and capture one of the sister chromatids at the kinetochore (central region of the chromosome). This can allow for rapid poleward movement of the bound chromosome. However, simultaneously or soon after this event, the sister kinetochore becomes attached to microtubules emanating from the opposite centrosome and the chromosomes align at the metaphase plate. Then, during anaphase, the sister chromatids separate and move to the opposite poles; this is coordinated with widening of the distance between the poles. These events require rapid elongation and contraction of microtubules and are distinct compared with the relatively quiescent microtubules in the cytoplasm present during interphase.

All antimicrotubule agents induce a blockage at the G_2/M phase of the cell cycle during the metaphase/anaphase transition by inhibiting the microtubule-dependent movement of sister chromatids to the opposite pole of the cell, as well as inhibiting the lengthening of the interpolar distance. At substoichiometric ratios compared with tubulin, all antimitotics including taxanes act in a similar manner. They inhibit dynamic instability, causing improper tensioning of the mitotic apparatus and/or insufficient microtubule contraction required to move the chromosomes to the pole. At equal stoichiometric ratios, taxanes induce the polymerization of microtubules (in a GTP-independent manner) from purified tubulin or in cells and stabilize the microtubule against depolymerization. They do so by binding to the β-subunit within polymerized tubulin. A great deal of information on the molecular interaction of taxanes with tubulin have been obtained by:

1. Mapping the sites of interaction of bovine brain tubulin with photoaffinity labeling analogs of paclitaxel,
2. Examining electron crystallographic interaction of taxanes with zinc-induced sheets of tubulin, and
3. Reevaluating the crystallographic data using fluorescence energy transfer spectroscopy and solid-state rotational echo double-resonance or nuclear magnetic resonance studies.

The paclitaxel-binding site resides near the so-called microtubule (M)-loop within β-tubulin. This region interacts with the H3 helix of the adjacent β-tubulin within the adjacent protofilament. It has been proposed that conformational changes in the M-loop that are induced by paclitaxel would stabilize the interaction between protofilaments. Five other classes of structurally diverse natural products, epothilones, discodermolide, laulimalide, eleutherobin, sarcodictyins, also enhance tubulin polymerization. The binding sites for epothilone A and paclitaxel overlap within tubulin but are distinct.

Taxane Resistance

Our understanding of the molecular basis of resistance to taxanes is founded largely on experimental work done using tumor cells grown in tissue culture or animals. Once established, the clinical correlation with the mechanism has been explored. This approach has some obvious limitations. In much of the work, when tumor cells in tissue culture are grown in the presence of continuous and multiple ascending concentrations of paclitaxel or docetaxel over a period of many weeks or months, they develop resistance to taxanes. It is clear that this type of exposure does not approximate the periodic exposure to taxanes in animals or humans (e.g., from every day for a short period up to once every 21 d), nor does it replicate the tumor environment. More-recent work has suggested relationships between taxane resistance and changes in the control of apoptosis, cell division, and signaling pathways. This work is also largely based on experiments in tissue culture where certain proteins have been manipulated or correlated with taxane resistance.

VARIATIONS IN MICROTUBULES

At least four factors immediately related to tubulin can influence the polymerization state of microtubules and have been associated with taxane resistance. First, mutations in tubulin have been associated with taxane resistance. Second, changes in the expression of tubulin isomers can occur. In humans, six β-tubulin and at least five α-tubulin isoforms have been described. The isomers are highly conserved with differences within a single species of 4-16% for β-tubulins and approximately 10% for α-tubulins. The most variable regions are located in the last 15-20 amino acids of the carboxy terminal.

The human (H)β-isoforms are grouped according to classes: class I, HM40; class II, Hβ9; class III, Hβ4, class IVa, Hβ5; class IVb, Hβ2; and class VI, HJβ1. The α-isoforms are classified as follows: group 1, bα-1 and kα-1; group 2, TUBA4; group 3, TUBA2; and group 4, TUBA8. Third, numerous other proteins that bind to tubulin, such as microtubule-associated proteins (MAPs) and stathmins stabilize and destabilize the microtubule, respectively. Fourth, posttranslational modifications of tubulin including glutamylation, glycylation, acetylation, tyrosination, and phosphorylation may alter polymerization of tubulin. All of these modifications, with the exception of acetylation, occur in the C-terminal 20 amino acids of tubulin. This region also is the same region that interacts with many MAPs. An excellent detailed review on the mechanisms of paclitaxel resistance related to microtubules is available. An overview is provided here.

Point Mutations

Numerous point mutations in tubulin have been found in a variety of tumor cell lines selected for resistance to paclitaxel or have crossresistance to paclitaxel after cells are selected for resistance to epothilones. Cells selected for survival in the presence of paclitaxel are often coselected with a so-called P-glycoprotein-inhibitor, because P-glycoprotein frequently mediates resistance to paclitaxel by itself. Most of the mutations are found in class I β-tubulin, which is the predominate (and therefore, most easily assessed) isoform found in cancer cells.

Mutations are often found in a single tubulin allele and therefore, may be coexpressed with wild-type tubulin or if solely expressed, require that the wild-type gene be silenced. The mutations in β-tubulin associated with paclitaxel selection are: human (*h*)$26^{Asp \rightarrow Glu}$, hamster (*ha*)$26^{Asp \rightarrow Glu}$, *ha* $60^{Val \rightarrow Ala}$, *ha*$215^{Leu \rightarrow His}$, *ha*$215^{Leu \rightarrow Phe}$, *ha*$215^{Leu \rightarrow Arg}$, *ha*$217^{Leu \rightarrow Arg}$, *ha*$228^{Leu \rightarrow Phe}$, *ha*$228^{Leu \rightarrow His}$, and *h*$270^{Phe \rightarrow Val}$. The mutations in β-tubulin associated with epothilone A- or B-selection are: *h*$173^{Pro \rightarrow Ala}$, *h*$231^{Ala \rightarrow Thr}$, *h*$274^{Thr \rightarrow Ile}$, *h*$282^{Arg \rightarrow Gln}$, *h*$292^{Gln \rightarrow Glu}$, *h*$364^{Ala \rightarrow Thr}$, and *h*$422^{Tyr \rightarrow Cys}$. However, mutations are not limited to β-tubulin, because an A549 lung carcinoma selected for resistance to paclitaxel contains a mutation in kα-1 tubulin ($h379^{Ser\rightarrow Arg}$). The differences between the mutations observed inpaclitaxel- and epothilone-selected cells compared with the preferential resistance to selecting agent has been used, in conjunction with crystallographic data, to argue that paclitaxel and epothilone have similar binding domains in β-tubulin, but the binding pockets are unique. Mutations may reduce the affinity of paclitaxel to tubulin or alter the stability of tubulin.

Consistent with the first possibility, residue 274^{Thr} makes contact with the oxetane ring of taxanes, residue 26^{Asp} makes contact with the C-3′ NHCO-phenyl group of paclitaxel, residue 270^{Phe} makes contact with paclitaxel's C3′ phenyl and C4-OAc groups, and residues 217^{Leu} and 219^{Leu} make hydrophobic contact with the two-phenyl ring of paclitaxel. Consistent with the second possibility, mutations within the M-loop ($h282^{Arg\rightarrow Gln}$) or residues that influence the orientation of the M-loop ($h292^{Gln\rightarrow Glu}$) may influence lateral interaction between adjacent protofilaments. In addition, mutations near the C terminus (α-$379^{Ser\rightarrow Arg}$, β-$364^{Ala\rightarrow Thr}$, and β-$422^{Tyr\rightarrow Cys}$) reside in regions that interact with MAPs, which by themselves alter microtubule stability. Finally, the Pro-to-Ala mutaton at β-173 occurs at a site that is involved with ribose binding at the exchangeable GTP site as well as longitudinal contacts between tubulin dimers. GTP binding and hydrolysis is an essential regulatory mechansm of microtubule stability.

The way that mutations would alter the stability is explained as follows. Normally, cells control the ratio of polymerized microtubules to depolymerized tubulin. Any shift from a set point ratio could induce resistance. Therefore, paclitaxel-resistant cells may shift the equilibrium towards the depolymerized form of tubulin. This may explain why some paclitaxel-resistant cells with tubulin mutations actually become dependent on paclitaxel for growth because the drug would be required to restore the set point ratio for proper cell growth. It would also explain why some paclitaxel-resistant cells display increase sensitivity to depolymerizing agents. If the latter occurred in patients, it may suggest novel therapeutic strategies using a sequence of polymerizing and depolymerizing agents. It is important to note that a causal link between tubulin mutation and resistance has only been substantiated in a few cases where enforced expression of the mutant tubulin mediates paclitaxel resistance, production of the putative mutated protein (predicted by cDNA analysis) is documented, or the mutant tubulin and/or microtubules that presumably contained the mutant tubulin has altered polymerization properties.

Similar to the way mutations in the epidermal growth factor receptor (EGFR) have been correlated with enhanced sensitivity to an EGFR inhibitor in patients, one might expect to find mutations in tubulin in resistant patients. One study reported mutations in class I β-tubulin in the serum DNA isolated from 33% of patients with non-small cell lung cancer that were associated with resistance to paclitaxel. However, the results have not been confirmed by subsequent studies where DNA or cDNA was obtained from tumor or serum samples. In addition, no mutations in β-tubulin that encode a different protein structure have been found in 62 human breast cancers. The discrepancy between the original report and subsequent studies is likely attributed to the use of nonselective primers used during polymerase chain reaction (PCR) amplification of β-tubulin that would allow hybridization of probes to tubulin pseudogenes present in genomic DNA. The lack of positive results does not exclude the possibility that clinical resistance to paclitaxel is correlated with mutations in other isomers of α- or β-tubulin.

Role of Tubulin Isoform

It is likely that each isomer has different functional properties, and therefore, differential expression could mediate resistance. This is based on numerous experimental findings. *First*, each β-tubulin isotype has a unique pattern of expression in tissue. For example, class βII predominates in brain tissue, βIII and IVa are found primarily in neurons, and class I and IVb are constitutively expressed. Based on PCR analysis, 17 human tumor cell lines of diverse origin predominately expressed class βI (range 76.0– 99.7%) and class βIII (range: 0.2–11.0%). In contrast, based on protein analysis using human breast tissue, class βII is the most highly expressed isoform (58% compared with total tubulin derived from classes I, II, and IV), no differences are observed between tumor and breast tissue, and protein expression does not correlate with tubulin mRNA levels using PCR analysis. A similar lack of correlation between tubulin isotype protein expression and mRNA levels has been noted in other paclitaxel-resistant cell lines. Based on protein expression analysis, human breast, lung, and cervical lung carcinoma cell lines predominately express Kβ-1, β6-tubulins (β6-tubulin has a distinct amino acid sequence compared with all the tubulin isoforms mentioned above). *Second*, the rate of tubulin polymerization and microtubule dynamics depends on the isotype composition.

In particular, microtubules assembled from purified αβIII isotype (from bovine brain) are more dynamic than microtubules made from αβII or αβIV isotypes, and even mixing αβII and αβIII isomers (1:1) behave differently compared with monoisomeric studies. *Third*, changes in isoform expression, particularly in βIII and βIV, are correlated with paclitaxel resistance. A list of alterations in isoform expression and its association with paclitaxel resistance across approximately 15 paclitaxel-resistant tumor cell lines has been previously summarized. In most paclitaxel-selected cell lines, an elevation in the expression of the βIII or βIV isoforms (range: two- to ninefold), which is based on reverse transcription-PCR methodology, is correlated with low-level of paclitaxel resistance (usually under 10-fold) compared with parental cells. In addition, the expression level of more than one isoform can be altered. Overexpression of the βIII and βIV isoforms are also found in non-drug-selected, paclitaxel-resistant cell lines. For example, in 17 nondrug-selected cancer cell lines, a significant correlation with paclitaxel resistance is found only with the level of βIII expression ($r^2 = 0.61$, $p < 0.007$) and is not associated with resistance to three depolymerizing agents.

More importantly, a clinical correlation has been reported by Kavallaris et al., who found a significant increase in classes βI (3.6-fold), III (4.4-fold), and IVa (7.6-fold) isotypes in tumors from paclitaxel resistant patients compared with untreated primary ovarian tumors. One group has reported that changes in α-tubulin may influence paclitaxel resistance because it is overexpressed (along with β-tubulin) in a multidrug resistance protein (MDR)1-expressing H460 lung carcinoma cell line that was 1000-fold resistant to paclitaxel. Consistent with this, downregulation of Kα-1 tubulin by antisense methodologies in the drug resistant cell line increases the sensitivity to paclitaxel by 50%, whereas overexpression of Kα-1 in drug parental cells confers approximately a two- to threefold increase in resistance to a few anti microtubule agents including paclitaxel. Does this association with altered β-tubulin isoform expression imply a causal relationship?

The evidence is controversial. In favor of this possibility, microtubules composed of βIII or βIV tubulin are less sensitive to the suppressive effects of paclitaxel on microtubule dynamics, and bovine brain depleted of βIII tubulin has enhanced sensitivity to paclitaxel. In addition, downregulation of βIII tubulin by antisense oligonucleotides in paclitaxel-resistant A549-T24 cells increases the sensitivity to paclitaxel, albeit a modest 1.6-fold, whereas moderate overexpression of human βIII tubulin in CHO cells increases resistance to paclitaxel 1.5- to 2-fold. Arguing against this relationship are the facts that stable overexpression of class I, II, III (that contained a point mutation), or IVb by transfection methods fails to confer resistance to paclitaxel.

The meaning of the negative data remains controversial, because among other problems, tubulin expression is tightly controlled in cells and compensatory changes in response to over- or underexpression of the class II or IV isoforms may confound the analysis. Beyond this, many MAPs bind to the variable C-terminal regions of tubulin isoforms, so differential MAP expression may also contribute to resistance. Ultimately, detecting quantitative differences in the expression of tubulin isoforms in patients who may not be responsive to paclitaxel will be a challenge. This is underscored by the work of Dozier et al., who used immunocytochemistry, enzyme-linked immunosorbent assay methodology (to measure protein levels), and PCR methodology (to measure mRNA levels) to assess tubulin isoform expression in normal and tumor breast tissue. He found that βIII expression, which is low compared to other isoforms, is predominately isolated to nerve fibers and not tumor cells. In addition, the level of expression of mRNA does not correlate with βIII protein expression.

Molecular Changes

MAPs can alter the state of polymerization and therefore, may influence sensitivity to taxanes. Two MAPs, τ and MAP-2, are primarily found in neurons and are not very relevant to the discussion here, although, Veitia et al. report higher τ expression in docetaxel-sensitive tumors. MAP-4 is present in all nonneuronal tissue. It stabilizes microtubules by increasing the rescue frequency without altering the catastrophe frequency. The protein is localized to mitotic spindle and interphase microtubules. Like other MAPs, phosphorylation of MAP-4 results in loss of binding and stabilization of microtubules and can undergo cell-cycle dependent phosphorylation. Therefore, a decrease in MAP-4 expression or an increase in MAP-4 phosphorylation may shift microtubules to a more destabilized state and therefore, mediate resistance to paclitaxel. Consistent with this hypothesis, an increase in MAP-4 phosphorylation is associated with paclitaxel resistance in paclitaxel-selected ovarian cell lines and an A549 paclitaxelselected cell line.

Furthermore, enhanced expression of MAP-4 by transfection methods or as occurring in p53 mutant cells stabilizes microtubules and slightly enhances paclitaxel sensitivity. However, contrary to this prediction, overexpression of MAP-4 is found in epothilone-selected leukemic cell lines that are also 15-fold resistant to paclitaxel. In contrast to MAP-4, stathmin destabilizes microtubules. It does so by either sequestering tubulin dimers, which would block the formation of microtubules, or by stimulating microtubule +-end catastrophe. Stathmin associates longitudinally along the lateral border of tubulin dimers causing a bend in the microtubule that does not allow polymerization. Consistent with this, stathmin inhibits paclitaxel-induced polymerization of microtubules from purified tubulin. The protein is deactivated by phosphorylation mediated by a variety of kinases, including those that participate in G_2/M phase of the cell cycle. Therefore, overexpression of stathmin or dephosphorylation of the protein would shift microtubules to a depolymerized state and may mediate resistance to paclitaxel. Experimental data fit this hypothesis. Stathmin levels increase in paclitaxel-selected A549 lung carcinoma cells (by twofold) or in A2780/1A9 ovarian cells (by two- to threefold) that are resistant to paclitaxel or epothilone.

Of course, other changes including mutations in tubulin and changes in MAP-4 levels have been observed in these same cells. Furthermore, downregulation of stathmin by antisense methodologies enhances sensitivity to paclitaxel. Other MAPs may influence paclitaxel sensitivity and are just beginning to be explored. These include the microtubule-based motor proteins, the kinesins and dyneins, an apoptotic regulatory protein that binds to microtubules, survivin, and many recently described microtubule plus-end tracking proteins.

Posttranslational Changes

A few associations between changes in posttranslational modifications of tubulin and paclitaxel resistance have been observed. Enhanced acetylation of α-tubulin has been found in a low-level paclitaxel-resistant small cell lung carcinoma cell line. Enhanced tyrosinyated α-tubulin coordinated with an increase in total α-tubulin has been observed in an MCF-7/paclitaxel-selected cell line. In contrast, tubulin isolated from A549 or HeLa cell lines selected for resistance to paclitaxel and epothilone, respectively, are not extensively posttranslationally modified and even so, little changes in posttranslational modifications of tubulin isoforms exist in resistant cells. It remains unclear if post- translational modifications of tubulin might effect microtubule stability associated with taxane resistance attributable to a direct effect on the microtubule or its interaction with regulatory proteins that might bind in the same region as the posttranslational modification.

RESISTANCE TO TRANSPORTERS

Unlike water-soluble drugs, hydrophobic drugs such as paclitaxel and docetaxel probably diffuse into the cell membrane without the need for any specific carrier protein. However, entry and exit of hydrophobic agents from the cell membrane can be regulated by a variety of ATP-binding cassette (ABC)-transporters. Of the 48 ABC transporters in the human genome, four have been associated with resistance to paclitaxel and docetaxel. They are P-glycoprotein encoded by MDR1 (according to the ABC classification system, designated ABCB1); sister of P-gylcoprotein (also known as the bile salt export protein, encoded by ABCB11), multidrug resistance protein MRP7 (encoded by ABCC 10), and MDR3 (sometimes called MDR2 and encoded by ABCB4). Transfection of cells with ABC11 confers low level of resistance to paclitaxel in one of two reports. Cells transfected with MDR3, which primarily transports phosphatidylcholine, transports paclitaxel inefficiently and does not display resistance to paclitaxel.

Recently, Kruh's laboratory reported that transfection of HEK-293 cells with MRP7 mediates approximately 10-fold resistance to docetaxel, threefold resistance to paclitaxel, and vinca alkaloids, but no resistance to doxorubuicin or cisplatin. The latter three transporters will not be discussed further. P-glycoprotein, like many ABC transporters, is composed of two halves, each containing six trans-membrane domains and an ATP-binding site, connected by a linker region. On binding paclitaxel, docetaxel, or a variety of other cancer chemotherapeutic agents, such as vinca alkaloids, anthracyclines, mitoxantrone, and bisantrene to the trans-membrane region of P-glycoprotein, ATP hydrolysis is used to expel these agents from cells. The evidence that P-glycoprotein mediates resistance to paclitaxel and docetaxel in experimental models is strong for the following reasons. First, expression of P-glycoprotein by transfection methods or as it occurs endogenously in tumor cell lines (e.g., certain colon carcinomas) is associated with resistance to paclitaxel or docetaxel. Second, P-glycoprotein expression increases remarkably in numerous tumor cell lines selected for resistance to paclitaxel, docetaxel, or other substrates for P-glycoprotein.

The level of P-glycoprotein increases with the amount of resistance to paclitaxel, although there may not be linear relationship. In extreme cases, where resistance to paclitaxel or docetaxel can exceed 1000-fold, P-glycoprotein overexpression is extraordinarily pronounced and is usually mediated by gene amplification. This is aphysiological. However, KB-8-5 HeLa cells (which were selected for resistance to continuous low-level exposure to colchicine), or MES-SA sarcoma cells (which were selected for only 7 d in the presence of 10 n*M* paclitaxel) have low to moderate P-glycoprotein expression and are 20- to 40-fold resistant to paclitaxel. This is likely to be physiological, because the level of P-glycoprotein expression in KB-8-5 cells is at the upper end of the range found in certain types of tumors. Remarkably, the growth of tumors derived from KB-8-5 cells in nude mice are completely unresponsive to paclitaxel given at the maximum tolerated dose (and optimal schedule), whereas the growth of tumors derived from KB, parental cells that are treated the same way are completely inhibited by paclitaxel. Third, agents that inhibit the function of P-glycoprotein, the so-called reversal agents such as verapamil and cyclosporine A, resensitize P-glycoprotein-expressing cells to paclitaxel, whereas they have little or no effect on cells that do not overexpress P-glycoprotein. Fourth, paclitaxel can be used to enrich for cells that express viral-directed exogenous P-glycoprotein in vivo, even many months after insertion of the gene and apparent loss of protein expression.

Finally, further evidence is obtained in mice where P-glycoprotein has been deleted by genetic methods. Normally, P-glycoprotein is expressed at the blood–brain barrier as well as the luminal (apical) surface of cells that line the gastrointestinal tract, the fetal–maternal interface, and hepatobiliary interface (in hepatocytes). However, mice that have a deletion of one or more P-glycoprotein isoforms (known as mdr1a and mdr1b) have altered pharmacokinetics of paclitaxel. In particular, the oral bioavailability of paclitaxel increases from 11% in wild-type mice to 35% in the mdr1 a knockout mice. Furthermore, when paclitaxel is given orally, fecal excretion (the normal route of elimination of paclitaxel) is reduced from 40 (after intravenous administration) and 87% (after oral administration) in wild-type animals compared with <3% observed in knockout animals. Consistent with this, oral absorption of paclitaxel markedly improves if it is coadministered with a P-glycoprotein reversal agent such as cyclosporine A or its derivative, PSZ-833, in mice and in humans. The interaction of P-glycoprotein with docetaxel is not exactly the same as paclitaxel. Similar to paclitaxel, docetaxel can be transported by P-glycoprotein, cells that overexpress P-glycoprotein can be highly resistant to docetaxel, and cyclosporine A enhances the bioavailability of docetaxel in patients ninefold.

However, the bioavailability of docetaxel is increased to 23% in mdr1a/b$^{-/-}$ animals, which is lower compared with paclitaxel. In addition, total plasma exposure is unchanged for docetaxel, but increases twofold for paclitaxel in the knockout animals compared with wild-type animals. Furthermore, an inhibitor of CYP3A4, ritonavir, which has only minor inhibiting properties of P-glycoprotein, increases the exposure of docetaxel by 50-fold in mice. Cyclosporine A also inhibits cytochrome P450 3A4, and therefore, it may inhibit both P-glycoprotein-mediated absorption and one of the two cytochrome P450 enzymes that inactivate paclitaxel or docetaxel (the other enzyme is cytochrome P450 2C8). Therefore, alterations in taxane metabolism may influence drug resistance. If P-glycoprotein plays an important role in resistance to paclitaxel in patients, then the protein should be overexpressed or have increased functionality in resistant patients. Is this true?

In support of this, the highest P-glycoprotein levels are found in many (but not all) colon and renal carcinomas that are resistant to paclitaxel or docetaxel. However, expression of P-glycoprotein is much more subtle and variable in breast, lung, and ovarian carcinomas that are 16-55%, 15-30%, and 16-47% positive for P-expression by immunocytochemical methods, respectively. In a meta-analysis using 31 reports, Trock et al. found that breast cancer patients who are positive for P-glycoprotein expression are three times more likely to fail to respond to chemotherapy (which contained taxanes in some regimens) than patients whose tumors are P-glycoprotein-negative. However, the association was not confirmed in a recent study where the expression of P-glycoprotein was compared in greater than 60 breast carcinoma patients before and after chemotherapy. Exactly how much P-glycoprotein or what increase in P-glycoprotein functionality is needed to cause patients to failure to respond to paclitaxel is not known. If a small change in expression is relevant, can it be reliably detected with the current histological methods? Despite efforts to standardize the methods, they may not be able to discern subtle changes with certainty.

Glycoprotein-Mediated Resistance

Assuming P-glycoprotein is relevant in the clinic, then there might be two ways to circumvent such resistance. The first method is to inhibit the expression or function of P-glycoprotein. The second method would be to use chemotherapeutic agents that do not interact with P-glycoprotein. In the first case, numerous P-glycoprotein reversal agents have been coadministered with substrates for P-glycoprotein, including taxanes. First-generation reversal agents described in the 1980s, such as verapamil, cyclosporine A, and quinidine, have been followed by second- and third-generation agents that include PSC-833 (valspodar), dexniguldipine, VX-710 (biricodar), GF-120918 (elacridar), XR9576 (tariquidar), LY-335979 (zosuquidar), R-101933 (laniquidar), and OC144-093 (ONT-093). The latter agents are devoid of unwanted side effects (i.e., calcium channel blocking or immunosuppressive activity) and are more potent than original reversal agents. The agents work by binding to P-glycoprotein and blocking its ability to transport substrates. Trials with reversal agents in patients with acute myelogenous leukemia have been logical starting point, because an inverse relationship has been established between P-glycoprotein expression and function with response to chemotherapy.

Whereas addition of cyclosporine A to daunorubicin and cytarabine therapy for acute myelogenous leukemia reduces resistance, improves duration of remission, and enhances overall survival, studies with the a nonimmunosuppressive analog of cyclosporine, PSC-833, in combination with the same agents, daunorubicin and etoposide, or mitoxantrone, etoposide, and cytarabine do not produce marked benefit. In addition, reversal agents combined with cytotoxic agents do not add significant benefit for the treatment of drug-resistant solid tumors, including a regimen that used paclitaxel monotherapy combined with PSC-833 in patients with refractory ovarian carcinoma.

The interpretation of these clinical studies is obscured because in many cases a dose reduction of paclitaxel (as well as other chemotherapeutic agents) is required to achieve acceptable toxicity. This is because of an alteration in the clearance of such agents mediated by P-glycoprotein, other transporters, or metabolizing enzymes as described above.

Beyond this, many of the clinical trials have not determined P-glycoprotein expression levels, nor correlated P-glycoprotein expression levels with response to reversal agent therapy. In theory, P-glycoprotein positive patients would be the only ones to benefit from such therapy. It may be possible to identify these patients and the activity of reversal agents in the tumor using imaging methods. Some of the newer P-glycoprotein reversal agents, such as XR-9576, LY-335979, R101933, and OC144-093, which do not influence the plasma pharmocokinetics of clinically used taxanes and/or have reduced interaction with metabolic enzymes involved in taxane metabolism, may prove to be more promising agents. Clinical evaluation of some of these agents is in progress. However, based on the experiments using knockout animals, all reversal agents would be expected to enhance the toxicity of paclitaxel or docetaxel in any cell or compartment expressing P-glycoprotein.

It is also important to note that GF120918 also inhibits the ABCG2 transporter; this transporter is not involved with taxane transport but its use may confound certain combination studies. Other strategies designed to inhibit P-glycoprotein expression at the mRNA or DNA levels are summarized elsewhere. An alternative approach to overcome P-glycoprotein-mediated resistance to taxanes is to identify taxanes or taxane-like molecules that do not interact with P-glycoprotein. Several such taxane analogs exist including SBT-1213, BAY 59-8862 (IDN5109, SB-T-101131, ortataxel), DJ-927, MAC-321 (TL-139), MST-997 (TL-909), XRP9881 (RPR 109881A), XRP6258 (RPR 116258A), BMS-184476 and BMS-188797, and BMS-275183. Because the structures of these compounds are highly related to paclitaxel or docetaxel, subtle changes in the structure of taxanes markedly influences their interaction with P-glycoprotein. For example, resistance to MAC-321 and MST-997 is 1- and 40-fold in cells that express low and high-levels of P-glycoprotein (compared with parental cells), whereas resistance to paclitaxel is 20- and 1000-fold in the same cells, respectively. All of these novel taxanes when given intravenously retain antitumor activity in animal models using tumors that overexpress P-glycoprotein and are insensitive to paclitaxel.

Oral activity with BMS-275813, XRP6258, BAY 59-8862, MAC-321, MST-997, and DJ-927 (but not paclitaxel or docetaxel) has been demonstrated in experimental tumor models. This is presumably because of their weak interaction with P-glycoprotein in the gastrointestinal tract or hepatocytes located at the biliary interface. One of the most advanced agents, XRP9881, demonstrates a 29% partial response rate in metastatic breast cancer patients who have failed previous taxane therapy. A phase III trial in metastatic breast cancer with XRP9881 is underway. BMS-275183, BMS-184476, BMS-188797, MAC-321, MST-997, BAY 59-8862, and DJ927 are in phase I or II clinical trials.

A variation of this approach is to use novel microtubule polymerizing agents besides taxane analogs that do not interact with P-glycoprotein. Three types of such agents, epothilones, discodermolide, and laulimalide, induce microtubule polymerization, are poor substrates for P-glycoprotein, and retain some activity in paclitaxel-selected cells that have mutations in tubulin. Epothilone and discodermolide bind to a similar, but distinct, site compared with paclitaxel, whereas laulimalide probably does not because it cannot inhibit the binding of paclitaxel to tubulin. Several epothilone analogs are in clinical trials.

REGULATORY PROCESS

Because cell division is a highly regulated event, resistance to taxanes could also be mediated by changes in the regulation of cell division beyond microtubules. When the mitotic apparatus is impaired cells can arrest at prometaphase or pass through mitosis without cell division and become aneuploid. This signals a cascade of phosphorylation- and caspase-dependent events leading to apoptosis or programmed cell death. Apoptosis, in turn, is also a highly ordered event orchestrated by multiple pathways that utilize positive and negative regulatory proteins. Therefore, it is conceivable that changes in the activity or expression of proteins that control cell cycle progression and apoptosis can influence sensitivity to taxanes, as well as a variety of other chemotherapeutic drugs. Mammalian cells use two main pathways to undergo apoptosis: the extrinsic and intrinsic pathways. The extrinsic signaling pathway is mediated by death receptors (DR) such as subsets of the prototypical tumor necrosis factor (TNF) receptor, which binds TNF-related apoptosis-inducing ligand (TRAIL), also known as TRAIL/Apo-2L, and Fas/CD95, which binds Fas ligand (FasL) or agonistic anti-Fas antibodies.

Ligand-induced oligomerization of DRs mediates the recruitment and assembly of the adaptor protein, Fas-associated protein with death domain, and cysteine aspartyl-specific proteases (caspase) 8 into a death-inducing signaling complex. Assembly of the death-inducing signaling complex juxtaposes multiple procaspase 8 zymogens, resulting in autocatalytic cleavage and activation of caspase 8, leading to further down-stream cleavage of the effector procaspases 3, 6, and 7. Interestingly, there is tantalizing evidence that suggests paclitaxel resistance is associated with TNF-α or TRAIL expression. First, relatively high concentrations of paclitaxel induces TNF-α expression in murine macrophages associated with apoptosis that is blocked by antibodies against TNF-α. Second, nanomolar concentrations of paclitaxel promotes TRAIL-induced apoptosis through Akt inactivation in a renal carcinoma cell line, but not a normal renal epithelial cell line. Similarly, TRAIL is ineffective as a single agent but enhances sensitivity to paclitaxel in two ovarian carcinoma cell lines.

Finally, paclitaxel as well as a variety of chemotherapeutic agents induce DR family members and sensitivity to TRAIL-induced apoptosis in breast cancer cell line grown in vitro or in vivo. However, the utility of TRAIL in the clinic (and its association with paclitaxel resistance in patients) needs further exploration. The mitochondrial or intrinsic pathway is primarily initiated by release of cytochrome *c* and protein factors from the mitochondrial inner membrane space. These factors include second mitochondria-derived activator of caspases and Omi/HtrA2 (a serine protease) and mediate caspase disinhibtion. Accumulation of cytochrome *c* in the cytoplasm results in binding to the scaffolding protein, apoptotic protease activation factor-1 (Apaf-1), causing a dATP/ATP conformational change that allows binding and autocatalytic cleavage of procaspase 9. The formation of this macromolecular complex, termed the apoptosome, leads to effector procaspases 3, 6, and 7 cleavage and downstream activation of sentinel cell death events. Within the cytosol, both the extrinsic and intrinsic pathways are regulated by the *Bcl-2* super family of genes, which are defined by the conserved bcl homology domains. Based on their ability to either induce or inhibit the apoptosis cascade, the Bcl-2 proteins can be subdivided into three subfamily members.

The first subset inhibits apoptosis and includes Bcl-2, Bcl-x_L, and Mcl-1. The second family members, Bax and Bak, translocate into the outer mitochondrial membrane space and oligomerize into voltage- gated pore complexes resulting in the release of cytochrome *c* and other polypeptides in the intrinsic pathway. Lastly, Bid, Bim, and Bad are proapoptotic proteins that function primarily as transdominant inhibitors that bind and neutralize Bcl-2 cell survival activity. In addition, activated caspase 8 can cleave Bid at the C terminus (tBid) and facilitate oligomerization of Bax and Bak at the mitochondrial membrane thereby linking the extrinsic and intrinsic death pathways. Conversely, antiapoptotic Bcl-2 and Bcl-x_L, can bind and antagonize proapoptotic Bax and Bak, thus preventing dimerization and subsequent pore formation; a salient feature in tumor cell survival. Bcl-2 members are typically regulated posttranslationally by a number of different growth factor and cytokine signaling pathways. Changes in the levels or posttranslational modifications of Bcl-2 family members have been associated with taxane resistance. For example, the primary mechanism of paclitaxel- and docetaxel-induced cell death is hyperphosphorylation of Bcl-2 and Bcl-x_L as well as Bax upregulation. Moreover, biotinylated paclitaxel can bind and sequester Bcl-2, and therefore suggests that taxane-induced cell death may be mediated by microtubule-independent mechanisms.

The net effect of hyperphosphorylation of, or direct binding to, these proteins result in the inability of Bcl-2 or Bcl-x_L to antagonize Bak or Bax, thereby leading to apoptosis. Alternatively, upregulation of Bcl-2 or Bcl-x_L by ectopic expression or continuous exposure to paclitaxel can lead to resistance. The primary mechanism of resistance involves the inhibition of the mitochondrial pathway as increased levels of Bcl-2 sequestered Bax and Bak and prevented cytochrome *c* release. Growth factors or hormone driven induction of Bcl-2 can also lead to paclitaxel resistance. For example, exposure to estrogen in estrogen receptor-positive MCF-7 breast carcinoma cell lines leads to an increase in Bcl-2 levels and resistance to paclitaxel-induced apoptosis, whereas cotreatments with antiestrogens restored sensitivity. However, whereas the overall levels of Bcl-2 increased, Bax and Bak remained unchanged.

At the level of the apoptosome, loss of Apaf-1 expression in Apaf-$1^{-/-}$ cells has also been shown to confer resistance to apoptosis mediated by paclitaxel, whereas ectopic overexpression of Apaf-1 sensitized leukemia cells to paclitaxel-induced apoptosis. Thus altering the relative ratio of Bcl-2 family members either by overexpression, growth factor induction, or inhibition by paclitaxel itself can lead to a loss of drug sensitivity. Given that the rate-limiting step in apoptosis is caspase activation, a class of proteins termed inhibitors of apoptosis proteins (IAPs) that directly bind and inhibit caspases, provide another level of regulation. This class includes XIAP, cIAP-1, cIAP-2, and survivin, which are linked by conserved baculoviral inhibitor repeats domains that are responsible for caspase inhibition. Survivin is particularly relevant to taxane resistance, because it is localized to the mitotic apparatus and plays a dual role monitoring spindle integrity and regulating cell death.

More specifically, survivin is upregulated during the G_2/M phase of the cell cycle and is subsequently localized to the centrosome, polymerized microtubules of the metaphase and anaphase spindle, and midbodies. Inhibition of phosphorylation of Thr-34 within survivin by purvanalol A (an inhibitor of cyclin dependent kinase 1) or overexpression of dominant negative survivin enhances paclitaxel-induced cell death in vitro and in vivo in MCF7 breast tumor lines. In addition, ectopic expression of survivin is able to counteract apoptosis induced by paclitaxel in NIH3T3 fibroblasts. Moreover, high levels of survivin are associated with advanced ovarian carcinomas that are resistant to paclitaxel/platinum therapy but not nonpaclitaxel-based regimens.

In the same study, when ovarian carcinoma cells are selected for paclitaxel resistance by continuous exposure, the levels of phosphorylated survivin are elevated implying increased activation. Clinically, survivin expression is of prognostic importance, given that it is upregulated in a wide variety of solid and liquid tumor types, it is associated with more-aggressive disease, and in colorectal cancer, decreased 5-yr survival rates are observed in tumors that have elevated levels of survivin. Another interesting mediator of taxane resistance is XIAP, which is overexpressed in a variety of tumors including those derived from lung, breast, and colon. Moreover, small molecule inhibitors of XIAP that potentiate the toxicity of paclitaxel and other chemotherapeutic agents have been identified. Thus, modulating the rate-limiting enzymatic activity of caspases by altering the levels and activity of inhibitors, such as the IAPs, can lead to paclitaxel resistance.

TRANSCRIPTIONAL CHANGES

Drug resistance is typically associated with alterations in gene expression by activation or inhibition of transcription factors. For example, pathways that regulate the transcription factor, nuclear factor of κB (NF-κB), plays an important role in governing multiple apoptotic and survival mechanisms. NF-κB is often constitutively activated in many types of tumor types. Currently, there are five known members of the NF-κB family members which are distinguished by their reticuloendotheliosis homology domain, the portion of the protein that controls DNA binding and dimerization. The most common form observed in many human tumor cells is the p50/p65 heterodimer. In its inactive form, NF-κB is sequestered and bound to inhibitor (I)κBα, an endogenous repressor, that is phosphorylated by the IκBα kinase (IKK) complex on stimulation with TNF-α or interleukin (IL)-1β. Phosphorylation of IκBα leads to degradation by the 26S proteo some complex, thereby liberating NF-κB for nuclear translocation and transcriptional activation of pro- and antiapoptotic genes.

The antiapoptotic proteins induced include XIAP, cIAP1 and 2, Bcl-x_L and caspase 8/Fas-associated death domain- like IL-1B-converting enzyme inhibitory protein. Thus, aberrant regulation of the NF-κB system would be expected to mediate resistance to a variety of agents that induce apoptosis. Modulation of NF-κB alters the sensitivity of the cell to paclitaxel. For example, constitutive activation of NF-κB in MDA-MB-231 breast tumor cells renders them resistant to paclitaxel as result of induction of antiapoptotic genes such as *c-IAP2*, manganese superoxide dismutase, TNF receptor-associated factor, and defender against cell death-1. Conversely, downregulation of NF-κB activity either by antisense or ectopic overexpression of an IκBα super repressor or nondegradable mutant protein sensitizes breast cancer cells to paclitaxel. Furthermore, paclitaxel resistance in human pancreatic tumor cell lines AsPC-1, Capan-1, and Panc-1 is because of paclitaxel-induced upregulation of IKK, downstream activation NF-κB, and subsequent upregulation of Bcl-x_L.

In addition, ectopic expression of mitogen-activated protein kinase kinase kinase 3, which is upstream of IKK, mediates more than a 25-fold increase in resistance to paclitaxel (albeit from 1 to 25 μ*M*, an aphysiological concentration), as well as other chemotherapeutic agents. It is interesting to note that mitogen-activated protein kinase kinase kinase 3 is overexpressed in 39% of breast tumor biopsies compared with matched tissue controls. However, it remains to be determined whether its overexpression in tumors in humans has any effect on taxane responsiveness in the clinic. Recently, a novel non-cytokine-dependent NF-κB pathway was reported in MCF-7 cells that overexpress cysteine-rich extracellular matrix protein (Cyr)61. Cyr61 primarily transmits signals via integrin receptors such as $\alpha_v\beta_1$, $\alpha_{II}\beta_3$, and $\alpha_v\beta_5$. These cells are resistant to physiological concentrations of paclitaxel (10-100 n*M*) as a result of NF-κB activation by integrin-dependent PI3/Akt signaling and subsequent induction of XIAP.

The later discovery is intriguing given that this is the first example of an extracellular matrix/integrin signaling pathway driving NF-κB-mediated paclitaxel resistance. Moreover, elevated Cyr61 protein and mRNA levels have been detected in breast carcinomas and positively correlated with more advanced features of breast cancer. However, correlations between Cyr61 expression and taxane resistance in patients have not been reported. Collectively, these data suggest that inhibition of the NFκB signaling pathway may selectively enhance the efficacy of paclitaxel in tumor cells. Other transcription factors such as interferon regulatory factor (IRF)9 can also lead to paclitaxel resistance. For example, Luker and colleagues demonstrated that after single- step selection of MCF-7 breast adenocarcinoma cells in 72-n*M* paclitaxel, transcriptional activation of IRF9 and other interferon-responsive genes occurs, is independent of interferon treatment, and enforced expression of IRF9 mediates 13-fold resistance to paclitaxel. Interestingly, approximately 50% of donor-matched breast and uterine tumors overexpress IRF9.

However, tumors were not subclassified into paclitaxel-sensitive or -resistant, so it is unclear whether overexpression is associated with resistance. Moreover, the subset of IRF9 target genes that may lead to resistance in vitro remains to be determined. A novel class of nuclear receptors termed xenobiotic receptors represents another family of transcription factors that may be involved in paclitaxel resistance. Paclitaxel (but not docetaxel) can bind to xenobiotic receptors, which results in ligand-dependent nuclear translocation and induction of P-glycoprotein and the detoxifying enzyme family of cytochrome p450. This model suggests that resistance to paclitaxel is mediated by a novel feedback mechanism in which drug export or increased taxane metabolism may lead to decreased sensitivity. Thus, it is clear that in various tumor models that alterations in transcription of genes involved in the control apoptosis, cell-cycle, drug efflux pumps, and detoxifying enzyme can lead to taxane resistance.

CYTOKINES AND CHEMOKINES

The cytokine IL-6 and chemokine IL-8 are classically recognized as immune regulators during inflammatory responses. However, they may also play a potential role in tumorigeneis, angiogenesis, and resistance. Indeed, treatment with paclitaxel is associated with an increase in expression of IL-6, IL-8, and the chemokine monocyte chemoattractant protein 1 in patients with ovarian cancer. Moreover, in two independent studies, increases in IL-6, IL-8, and monocyte chemoattractant protein 1 levels are detected in a 600-gene cDNA array in SKOV-3 ovarian carcinoma cells that acquire paclitaxel resistance as a result of continuous exposure and selection. In addition, transfection of IL-6 cDNAs into paclitaxel-sensitive U2-OS human osteosarcoma cell lines mediates fivefold resistance to paclitaxel. The lack of MDR-1, MRP-1, and lung resistance-related protein overexpression and activity in these IL-6-transfected lines suggest that the multidrug-resistance phenotype is not a result of increased drug efflux.

A possible consequence of increase in IL-6 expression may be a direct consequence of activation of signal transducer and activator of transcription 3, the effector transcription factor in IL-6 signaling and an important mediator in tumor cell survival in a number of tumor cell lines. Despite the observation that paclitaxel induces IL-8 in ovarian carcinoma cell lines and elevated serum IL-8 levels in ovarian cancer patients correlates with more aggressive disease, there is no evidence to date that IL-8 contributes directly to paclitaxel resistance.

KINASES IN DRUG RESISTANCE

It is well established that networks of constitutively activated kinases governs growth factor independence in tumor cell proliferation. Therefore, one mechanism in which tumor cells can become drug resistant is through positive selection for kinases that, as a result of gain of function mutations or amplification, can subvert the apoptotic response. For example, increased levels of mitotic kinases that play a role in monitoring spindle checkpoints can confer drug resistance. One such example is the serine/threonine kinase, Aurora A or breast-tumor-amplified kinase, which is localized to chromosome 20q13, a region commonly amplified in human breast and colon tumors and correlates with poor prognosis. In addition, Aurora A is overexpressed in ovarian, prostate, neuroblastoma, and cervical cancer lines. The primary functions of Aurora A are centromsome separation, bipolar spindle formation, and chromosomal kinetochore attachment to the mitotic spindle.

Accordingly, overexpression of Aurora A leads to centrosome amplification, chromosomal instability, and transformation in epithelial tumor lines. Moreover, it has been demonstrated recently that enforced overexpresssion of Aurora A in a human epithelial tumor cell line confers resistance to paclitaxel-induced apoptosis. Inhibition of the expression of two other proteins, budding uninhibited benzimidazole receptor (a kinase) and mitotic arrest deficient 2, which regulate the spindle assembly checkpoint, have also been associated with paclitaxel resistance. Given that ligand-independent activation of growth factor receptor tyrosine kinases are critical for oncogenic transformation and tumor cell proliferation, it is conceivable that gene amplification or overexpression or receptor tyrosine kinases may also generate drug resistance. The erb/EGFR family members, HER-2/*neu*, and EGFRvIII (an EGFR variant with a truncated extracellular domain), are amplified and overexpressed in a primary breast and ovarian carcinomas, tumors in which taxane treatment is frequently used.

Ectopic expression of EGFRvIII or HER-2/*neu* in NIH3T3 murine fibroblasts induce transformation and resistance to paclitaxel-mediated cytotoxicty. The mechanism of resistance is not fully understood, although both EGFRvIII- and HER-2-expressing cells do not overexpress P-glycoprotein, and in one case, have increases in type IVa β-tubulin isoforms. Moreover, exogenous expression of HER-2/*neu* in MDA-MB-435 human breast cancer cells is associated with resistance to paclitaxel-induced apoptosis and inhibition of the cell cycle-regulatory protein p34^{Cdc2}, which controls entry into mitosis and subsequent apoptosis. Consistent with this, inhibition of erbB-2 by Herceptin reverses both p34^{cdc2}-mediated inhibition and paclitaxel resistance.

It will be interesting to determine whether a causal link exists between EGFR overexpression and paclitaxel resistance by ascertaining the effects of the emerging class of EGFR inhibitors, such as gefitinib (Iressa), in breast, lung, and ovarian cancers patients that have failed previous taxane therapy. Decreased paclitaxel sensitivity can also be because of increased signal transduction through a number of different cascades as opposed to direct amplification of kinases. For example, numerous extracellular and intracellular oncogenic signals intersect with the phosphatidylinositol 3-kinase (PI3K)/Akt pathway, which is often constitutively activated as a result of decreased expression of the endogenous Akt inhibitor, phosphatase and tensin homolog on chromosome ten, a tumor suppressor gene. Indeed, an increase in resistance to paclitaxel in MDA-MB-231 and MDA-MB-435 human breast tumor lines is because of extracellular matrix ligand-dependent activation of β_1 integrin signaling via the PI3K/Akt pathway. In addition, the activation status of Akt may be critical for taxane sensitivity because paclitaxel-mediated cytotoxicty is reversed on treatment with the PI3K inhibitor, LY294002, and that HER-2/*neu* and EGFR signal cascades can trigger Akt-mediated control of NF-κB.

CONCLUDING REMARKS

One of the greatest challenges facing both researchers and clinician alike is the ability to predict drug resistance in a given population of patients. Which of the mechanisms mentioned above account for resistance in patients, and can the results of this analysis be used to tailor future therapies? Two basic problems limit the answer to these questions. First, methods need to be developed to measure accurately the expression or functionality of individual mechanisms that contribute to resistance. Second, given the cellular heterogeneity in human tumors, it is highly likely that more than one mechanism may contribute to resistance, possibly within the same tumor. Beyond this, it must be recognized that paclitaxel resistance found in patients may not replicate the mechanisms defined, for the most part, by in vitro studies described here. Additional contributing factors that make the analysis complex include (i) greater genetic heterogeneity in tumors compared with cell lines; (ii) local environmental factors such as hypoxia, interstitial hypertension, drug delivery; and (iii) bioavailability of taxanes related to metabolism, compartmentalization, and formulation.

In fact, it was recently reported that a reformulated version of Taxol in a non-cremophor-based vehicle was effective in paclitaxel-resistant patients. As with much of the successful oncological therapies to date, empirical investigation in patients with selective agents that inhibit certain transporters, tubulin isoforms, apoptotic regulators, or signal transduction when used alone or in combination with standard taxane therapies will be necessary to bridge the gap between possible mechanisms of taxane resistance in the laboratory and prolonged survival in patients.

Index